Advances in Human–Machine Systems, Human–Machine Interfaces and Human Wearable Device Performance

Advances in Human–Machine Systems, Human–Machine Interfaces and Human Wearable Device Performance

Guest Editors

Kaiway Li

Lu Peng

MDPI

Basel • Beijing • Wuhan • Barcelona • Belgrade • Novi Sad • Cluj • Manchester

Guest Editors
Kaiway Li
Department of Industrial Management
Chung Hua University
Hsin-Chu
Taiwan

Lu Peng
Department of Engineering and Management
Nanjing Agricultural University
Nanjing
China

Editorial Office
MDPI AG
Grosspeteranlage 5
4052 Basel, Switzerland

This is a reprint of the Special Issue, published open access by the journal *Applied Sciences* (ISSN 2076-3417), freely accessible at: https://www.mdpi.com/journal/applsci/special_issues/GRTLU92NP0.

For citation purposes, cite each article independently as indicated on the article page online and as indicated below:

Lastname, A.A.; Lastname, B.B. Article Title. *Journal Name* **Year**, *Volume Number*, Page Range.

ISBN 978-3-7258-5153-9 (Hbk)
ISBN 978-3-7258-5154-6 (PDF)
https://doi.org/10.3390/books978-3-7258-5154-6

Contents

About the Editors

Kaiway Li

Kaiway Li is a professor emeritus with the Department of Industrial Management at Chung Hua University, Taiwan, and a visiting professor with the Master of Engineering Management Program of Tsinghua University in Beijing. He holds a B.S. in Civil Engineering from National Chiao Tung University (Taiwan), an M.S. in industrial engineering from the University of Texas at Arlington, and a Ph.D. in industrial engineering from Texas Tech University. His distinguished career includes research positions at the Liberty Mutual Research Institute for Safety in the U.S.A., as well as multiple leadership roles at Chung Hua University, such as Department Chair, Director of Environmental Safety and Health, and Deputy Director of Continuing Education. As an active scholar, he has served as a visiting scientist at Harvard University and Tsinghua University, held distinguished professorships at several Chinese universities, and contributed extensively to the ergonomics field through editorial roles—including Editor-in-Chief of Taiwan's *Journal of Ergonomics* and Guest Editor for *Symmetry* and *Applied Sciences*. With over 200 published journal and conference papers and three books, his expertise bridges academia and practical applications in human–machine systems and occupational safety and health.

Lu Peng

Lu Peng obtained her PhD in Systems Engineering and Engineering Management from the City University of Hong Kong in 2020. She is currently an associate professor with the Department of Engineering Management at Nanjing Agricultural University. She has a wide range of research interests, such as, human–computer interactions, human factors, work safety, and unmanned aerial vehicle operation behaviors

Preface

The rapid evolution of virtual reality (VR), augmented reality (AR), and wearable technologies has transformed how humans interact with machines, unlocking unprecedented capabilities across industries. This Special Issue compiles ten cutting-edge research articles that explore the performance, usability, and challenges of human–machine interfaces and wearable devices. Covering topics from cognitive ergonomics and adaptive control to real-time biosignal monitoring, these studies provide critical insights into optimizing human–system integration. By investigating user adaptability, interface design, and task efficiency, this collection advances our understanding of how emerging technologies can be tailored to human needs. We extend our gratitude to the authors, reviewers, and editorial team for their contributions, which pave the way for smarter, more intuitive human–machine systems.

Kaiway Li and Lu Peng
Guest Editors

applied sciences

MDPI

Editorial

Advances in Human–Machine Systems, Human–Machine Interfaces and Human Wearable Device Performance

Kai Way Li [1] and Lu Peng [2,*]

[1] Department of Industrial Management, Chung Hua University, Hsinchu 300110, Taiwan; kaiway.li@gmail.com
[2] College of Information Management, Nanjing Agricultural University, Nanjing 211800, China
* Correspondence: lupeng942@njau.edu.cn

1. Introduction

The human–machine system (HMS) and human–machine interface (HMI) are among the top factors that affect the development of advanced systems, equipment, and products. To achieve the optimal performance of systems and products and minimize system failure and disasters, human factors and ergonomics, including human-centered design [1,2], design for human use [3], user-friendly design [4], and fitting tasks to humans [5–7], need to be widely adopted. These concepts require designing products based on the capabilities and characteristics of users [8–10]. The HMS treats both the system and the humans involved as a unit. The HMI, on the other hand, focuses on the interactions between humans and the machines they operate. How users communicate with the machines they are using and whether such communications are efficient and successful affect the performance of the HMS involved. To enhance our understanding of the HMS and HMI associated with advanced systems and products, we collected submissions in these areas with particular focus given to wearable devices such as augmented reality (AR) and virtual reality (VR) devices. Our aim is to promote research and fill information gaps in these areas. This Special Issue was launched on 7 November 2022 and lasted for almost three and a quarter years. During this period, we received 21 submissions. After rigorous review processes, we finally accepted and published 10 of these works.

2. Contributions of Published Articles

Three of our collections are directly related to the issues of AR or VR. Both papers 1 ("Movement time for pointing tasks in real and augmented reality environments") and 2 ("Presenting job instructions using an augmented reality device, a printed manual, and a video display for assembly and disassembly tasks: what are the differences?") draw comparisons on the performances between human–virtual vs. human–real target interactions. The former compared the movement time of pointing tasks between pointing the real and virtual keypads of a calculator. By contrast, the latter study compared the performance of computer assembly and disassembly tasks via reading the job instructions on paper, using video images on a laptop computer, and virtual instructions generated in an AR device. The paper "Motion sickness in mixed-reality situational awareness system" (paper 4), on the other hand, discusses the motion sickness issue related to the usage of VR displays. The issues concerning visual display and the illumination of VR, including adaptation, field of view, blur, vignette, and distortion, were discussed. The results of the three above-mentioned papers are helpful for improving the design of VR/AR systems and the use of VR/AR in practice.

The third paper published is the "Impact of navigation aid and spatial ability skills on wayfinding performance and workload in indoor–outdoor campus navigation: challenges and design". This paper highlights the challenges of wayfinding on a university campus using mobile navigation apps. A smartphone was used, and the performances of the participants, together with their spatial capability using the two apps, were compared. The map app and smartphone combinations used in this study comprised a system, and the performances of the HMS were presented.

The fifth and sixth papers in this Special Issue focus on mental workload monitoring and mental fatigue assessment, respectively. Both studies in these papers focused on the development of monitoring and assessment methodologies. They tested their approaches or models using the dataset available online. In paper 5 ("Brainwaves in the cloud: cognitive workload monitoring using deep gated neural network and industrial internet of things"), a cognitive workload monitoring system was formulated by leveraging the deep gated neural network. A hybrid model integrating bidirectional long short-term memory and gated recurrent unit networks was included. Paper 6 ("Human operator mental fatigue assessment based on video: ML-driven approach and its application to HFAVD dataset") introduces an approach to assess human mental fatigue based on the postures recorded on videos. The aim of this study was to incorporate machine learning techniques in the mental fatigue assessment of human operators.

Paper 8 ("Emotional dynamics in human–machine interactive systems: effectively measuring Kuhn Poker approach with experimental validation") explores how emotional states influence decision-making behaviors in HMS. The authors used Kuhn poker as an experimental context and developed an intelligent decision support system to provide real-time strategic advice to the participants. They found that positive emotional states lead individuals to be more risk-seeking and less inclined to collaborate with the AI-driven system. Negative emotional states, on the other hand, resulted in risk-averse tendencies and higher compliance with system recommendations. Expanding on this, the paper "A wearable microphone array helmet for automotive applications" (paper 9) is a unique article concerning a wearable device. In this paper, a wearable microphone array was installed over the surface of a helmet. Instead of investigating the HMI issues of such a design, this article explores the instrumentation and applications that this specific wearable device involves. A patent request for this design has been submitted to the patent administration in the USA. The findings of papers 8 and 9 contribute to the cabin designs and improvement of the HMS, particularly for the automobile industry.

Paper 7 ("Enhancing word embeddings for improved semantic alignment") introduces a method for the improvement of word vectors. The approach proposed in this study leverages supervised learning methods and thus enhances the quality of word embeddings. Paper 10 ("SE-WiGR: A WiFi gesture recognition approach incorporating the Squeeze–Excitation mechanism and VGG16") proposes a gesture recognition method that may be used in the Internet of Things (IoT) and smart home techniques. The authors claimed that their method achieves a recognition accuracy of 94% across multiple scenarios. Such an improvement confirms that the proposed method allows hand gesture recognition even when background noise is present.

3. Focus on Future Research

As mentioned earlier, our aim is to promote publications on the HMS and HMI associated with advanced systems and products to enhance our understanding of these issues. We tried to publish more papers related to the technical and ergonomic issues of wearable devices. However, only four of the published papers (papers 1, 2, 4, and 9) met our expectations directly. The other papers discuss either technical or ergonomic issues in a broad HMS domain. We encourage scientists and scholars to increase their efforts in the study of HMS and HMI associated with advanced systems and products that incorporate new techniques.

The recent emergence of artificial intelligence (AI) tools is undoubtedly significant for both the scientific community and our society. The power of AI tools, especially generative AI, to complete human jobs involving information searching, filtering, and organizing presents an efficient alternative. Products with embedded AI functions are becoming one of the major designs used for personal products. Before these products become prevalent, many human–AI interaction issues need to be resolved [11–13]. Future research should focus on the interaction between human-wearable devices and the integration of AI functions. The performance and efficiency of these devices under various scenarios and levels of AI functioning should be explored. In addition, issues concerning long-term wearability, biocompatibility, and cybersecurity to ensure user safety and trust should also be assessed.

Conflicts of Interest: The authors declare no conflict of interest.

References

1. Ferris, T.L.J. User-Centered Design: An Integrated Approach. *IEEE Trans. Prof. Commun.* **2004**, *47*, 75–77. [CrossRef]
2. Baber, C. Intelligent Adaptive Systems: An interaction-centred design perspective. *Ergonomics* **2017**, *60*, 1458–1459. [CrossRef] [PubMed]
3. Chapanis, A.; Garner, W.R.; Morgan, C.T. *Applied Experimental Psychology: Human Factors in Engineering Design*; John Wiley & Sons Inc.: Hoboken, NJ, USA, 1949. [CrossRef]
4. Kay, A. User interface: A personal view. In *The Art of Human-Computer Interface Design*; Addison-Wesley Publishing: Boston, MA, USA, 1990; pp. 191–207.
5. Grandjean, E.; Kroemer, K.H.E. *Fitting the Task to the Human: A Textbook of Occupational Ergonomics*; CRC Press: Boca Raton, FL, USA, 1997.
6. Bentley, T.; Green, N.; Tappin, D.; Haslam, R. State of science: The future of work—Ergonomics and human factors contributions to the field. *Ergonomics* **2020**, *64*, 427–439. [CrossRef] [PubMed]
7. Richardson, M.; Thatcher, A. State of science: Refitting the human to nature. *Ergonomics* **2023**, *67*, 582–596. [CrossRef] [PubMed]
8. Fitts, P.M. *Human Engineering for an Effective Air-Navigation and Traffic-Control System*; National Research Council, Division of Anthropology and Psychology Committee on Aviation Psychology: Washington, DC, USA, 1951.
9. Greig, M.A.; Village, J.; Dixon, S.M.; Salustri, F.A.; Neumann, W.P. Assessing human factors and ergonomics capability in organisations—The Human Factors Integration Toolset. *Ergonomics* **2019**, *62*, 1254–1272. [CrossRef] [PubMed]
10. Niu, H.; Van Leeuwen, C.; Hao, J.; Wang, G.; Lachmann, T. Multimodal natural human–computer interfaces for computer-aided design: A review paper. *Appl. Sci.* **2022**, *12*, 6510. [CrossRef]
11. Peng, L.; Li, D.; Zhang, Z.; Zhang, T.; Huang, A.; Yang, S.; Hu, Y. Human-AI collaboration: Unraveling the effects of user proficiency and AI agent capability in intelligent decision support systems. *Int. J. Ind. Ergon.* **2024**, *103*, 103629. [CrossRef]
12. Ruifeng, Y.; Xu, X.; Peng, S. The impact of individual AI proficiency on human–agent collaboration: Higher sensitivity to discern the comprehension ability of intelligent agents for users with higher AI proficiency levels. *Int. J. Ind. Ergon.* **2025**, *107*, 103745. [CrossRef]
13. Markus, A.; Pfister, J.; Carolus, A.; Hotho, A.; Wienrich, C. Effects of AI understanding-training on AI literacy, usage, self-determined interactions, and anthropomorphization with voice assistants. *Comput. Educ. Open* **2024**, *6*, 100176. [CrossRef]

applied sciences

MDPI

Article

SE-WiGR: A WiFi Gesture Recognition Approach Incorporating the Squeeze–Excitation Mechanism and VGG16

Fenfang Li [1,2,*], Chujie Weng [2] and Yongguang Liang [2]

1 School of Information Engineering, Gansu Minzu Normal University, Hezuo 747000, China
2 School of Computer Science and Engineering, Northwest Normal University, Lanzhou 730070, China; wengchujie@m.scnu.edu.cn (C.W.); liang172117@163.com (Y.L.)
* Correspondence: lifenfang@nwnu.edu.cn; Tel.: +86-18189556489

Abstract: With advancements in IoT and smart home tech, WiFi-driven gesture recognition is attracting more focus due to its non-contact nature and user-friendly design. However, WiFi signals are affected by multipath effects, attenuation, and interference, resulting in complex and variable signal patterns that pose challenges for accurately modeling gesture characteristics. This study proposes SE-WiGR, an innovative WiFi gesture recognition method to address these challenges. First, channel state information (CSI) related to gesture actions is collected using commercial WiFi devices. Next, the data is preprocessed, and Doppler-shift image data is extracted as input for the network model. Finally, the method integrates the squeeze-and-excitation (SE) mechanism with the VGG16 network to classify gestures. The method achieves a recognition accuracy of 94.12% across multiple scenarios, outperforming the standalone VGG16 network by 4.13%. This improvement confirms that the SE module effectively enhances gesture feature extraction while suppressing background noise.

Keywords: WiFi; gesture recognition; channel state information; squeeze–excitement mechanism; VGG16

1. Introduction

With the advancement of Internet of Things (IoT) technology, smart homes, virtual reality, innovative healthcare, intelligent transportation, and other cutting-edge technologies have become indispensable. Researchers continue exploring more intuitive interaction methods to improve user experience and reduce the learning curve. Among these, gesture recognition—the most natural form of interaction—has emerged as a key focus in Human–Computer Interaction (HCI) research [1]. By recognizing different gestures, devices can be operated remotely or interactively. Additionally, personalized recommendations can be generated by analyzing users' movement characteristics through gesture recognition. However, low recognition accuracy may lead to operational errors, while poor real-time performance can negatively impact user experience.

In recent years, gesture recognition systems have primarily utilized computer vision technology [2], infrared sensors [3], wearable devices [4], and millimeter-wave radar [5] to detect human movements. Computer vision-based approaches employ standard cameras to capture gesture videos, then extract gesture images through frame-splitting techniques for high-precision recognition. However, their effectiveness is limited by lighting conditions. In contrast, infrared technology works by sensing infrared light reflected off hand movements to create images. While it remains effective under varying lighting conditions, the potentially recorded private details in captured images raise data protection issues.

This significantly restricts its deployment in scenarios requiring stringent confidentiality. Wearable-device solutions use sensors to collect pressure and acceleration data for action recognition. Although these systems perform well, they face practical challenges such as user discomfort from continuous wear and limited battery life. Millimeter-wave radar, grounded in the principles of Frequency Modulated Continuous Wave (FMCW) technology, achieves high recognition accuracy through detailed analysis of echo signal characteristics. Nevertheless, its widespread adoption is hindered by high hardware costs.

The emergence of WiFi technology has offered a promising solution to overcome the limitations of the aforementioned approaches. The operation of WiFi-based detection depends mainly on two indicators: Received Signal Strength Indicator (RSSI) and Channel State Information (CSI). Historically, the RSSI has been utilized to assess the strength and quality of wireless connections by quantifying the power level of incoming signals. In WiFi sensing applications, RSSI enables device presence detection and localization. By comparing RSSI values from multiple devices, the system can estimate their relative distances from the access point (AP).

In compliance with the IEEE 802.11n standard [6], Orthogonal Frequency Division Multiplexing (OFDM) is employed to modulate WiFi signals into multiple subcarriers, thereby enabling the extraction of CSI [7]. Unlike the RSSI, CSI provides a more fine-grained characterization of wireless channels, encapsulating detailed propagation characteristics such as amplitude variations, phase shifts, and frequency response characteristics. This enables precise monitoring of signal variations during transmission, yielding more accurate wireless environment sensing and analysis. CSI demonstrates exceptional sensitivity to environmental changes. Beyond detecting major movements like walking or running, it can capture subtle motions, including breathing and chewing, in static environments. Several studies have leveraged these properties for advanced gesture recognition: WiSee [8] utilizes two living room access points to extract comprehensive gesture information from wireless signals, enabling recognition of nine home-based gestures. WiMorse [9] utilizes CSI to detect finger-generated Morse code by presenting a training-free signal conversion framework that automatically maps positional changes to standardized gesture patterns, thereby enabling accurate character recognition. In Wi-SL [10], features of CSI are combined with a support vector machine enhanced by K-means clustering and the Bagging algorithm to classify twelve distinct sign language gestures. The WiReader system [11] proposes a WiFi-based adaptive framework for in-air handwriting recognition. It employs activity factors for data segmentation, applies CSI ratio-based signal transformation, generates energy feature matrices, and incorporates them into an LSTM architecture to accurately classify handwriting gestures.

This study proposes an innovative recognition method to address the limitations of traditional WiFi gesture recognition techniques in complex environments. The approach incorporates a squeeze–excitation mechanism and the VGG16 network. First, gesture action CSI data are collected using commercial WiFi devices; second, the data are preprocessed, and the short-time Fourier transform is applied to extract Doppler frequency shift images as features; then, the CSI signals are enhanced using a squeeze–excitation mechanism to reinforce gesture-related features while effectively suppressing background noise; and finally, the robust feature extraction and classification capabilities of the VGG16 network facilitate gesture recognition.

This research develops a novel WiFi-enabled gesture recognition framework that addresses the critical shortcomings of existing vision and sensor-based approaches in cluttered environments. By utilizing channel state information (CSI) and deep neural networks, robust gesture recognition is achieved. The proposed solution integrates a squeeze–excitation mechanism with the VGG16 network architecture, implementing the following processing pipeline:

(1) Data Acquisition: Commercial WiFi devices capture CSI data corresponding to gesture actions.
(2) Preprocessing: The acquired CSI measurements are processed using STFT-based spectral analysis to derive Doppler shift signatures in the time-frequency domain.
(3) Feature Enhancement: The squeeze–excitation mechanism amplifies relevant gesture features while effectively suppressing background noise in the CSI signals.
(4) Classification: The enhanced features are processed by the VGG16 network, leveraging its robust feature extraction and classification capabilities for final gesture recognition.

The rest of this paper is organized as follows. Section 2 reviews related work on WiFi-based gesture recognition, analyzing previous methods and their limitations. Section 3 details the proposed SE-WiGR method, covering data acquisition, preprocessing, and the integration of squeeze and excitation networks with VGG16. Section 4 presents the experimental design and results, including deployment, data collection, performance evaluations of different algorithms, robustness tests, and ablation studies. Section 5 concludes the study by summarizing contributions, discussing limitations, and suggesting future directions. To enhance the readability of technical terms, the following Table 1 lists the main acronyms used in this paper with their full names and explanations:

Table 1. Comparison of explanations of terms for the core abbreviations in this study.

Abbreviation	Full Title
HCI	Human–Computer Interaction
RSSI	Received Signal Strength Indicator
CSI	Channel State Information
AP	Access Point
SVMs	Support Vector Machines
CNNs	Convolutional Neural Networks
ABLSTM	Attention-based Bidirectional LSTM
GRU	Gated Recurrent Unit
SE	Squeeze-and-Excitation
NICs	Network Interface Cards
SENet	Squeeze-and-Excitation Network

2. Related Work

The ubiquitous deployment of WiFi infrastructure in indoor environments has accelerated advancements in WiFi-based gesture recognition technology. Among available signal metrics, RSS measures the power level of received signals. Several studies have demonstrated RSS-based gesture recognition systems: Abdelnasser et al. [12] developed WiGest, an RSS—based system that classifies movements by establishing separate gesture categories. Gu et al. [13] designed an activity recognition framework leveraging WiFi RSS measurements. This framework can identify directional gestures (Left/Right, Up/Down) and symbolic motions (infinity signs, open/close actions). The latter framework combines classification trees with K-nearest neighbor algorithms, achieving 96% recognition accuracy even in non-line-of-sight conditions. However, despite its ease of collection and implementation, RSS exhibits inherent limitations for gesture recognition. The metric fails to capture fundamental signal variations caused by human movements and demonstrates instability even in static environments [14], significantly constraining its practical application.

CSI serves as a critical physical-layer parameter in WiFi systems, quantitatively describing the multipath propagation properties of wireless channels. It empowers transmitters and receivers to execute intelligent adaptive strategies during wireless data transmission. Recent research has demonstrated various CSI-based gesture recognition approaches: He et al. [15] developed WiG, which identifies distinctive patterns in anomalous CSI sequences to represent different gestures, employing Support Vector Machines (SVMs) for classification. WiGeR [16] employs CSI diffusion patterns to achieve accurate gesture recognition, utilizing a customized segmentation and windowing algorithm grounded in wavelet analysis and short-time energy computations. WiAG [17] implements a DWT-based feature extraction pipeline coupled with KNN classification. AirFi [18] proposes a domain-generalized learning framework that distills environment-invariant CSI features through multi-domain training, eliminating the need for scenario-specific calibration. Isack Bulugu [19] proposed a cross-domain CSI method that extracts domain-invariant features for reliable recognition in diverse environments. The system recognizes six gesture types (push/pull, sweep, hand clap, slide, circling, and sawtooth) in known and unknown domains without additional training data from new environments.

Feature extraction represents a crucial stage in gesture recognition systems. While traditional manual feature extraction methods are both time-consuming and prone to omitting important implicit patterns, recent advances in deep learning have enabled automated extraction of key Channel State Information (CSI) features. Several notable approaches demonstrate this capability: Zou et al. [20] developed FreeGesture, a WiFi CSI-based system that employs Convolutional Neural Networks (CNNs) to learn and extract the features from CSI frames. Kong et al. [21] introduced FingerPass, a system utilizing LSTM to process temporal gesture patterns, with the derived recognition results further employed for identity verification applications. Research findings [22] demonstrate that LSTM-based approaches outperform conventional machine learning methods in CSI-based activity recognition tasks. Chen et al. [23] implemented an Attention-based Bidirectional LSTM (ABLSTM) architecture, utilizing both forward and backward layers to enhance human activity recognition. Zhou et al. [24] achieved gait identification through a Gated Recurrent Unit (GRU) network incorporating mean pooling layers.

3. SE-WiGR Gesture Recognition Method

Smart home WiFi-based gesture recognition technology offers an intuitive and convenient control approach, significantly enhancing user interaction with the home environment. In this study, we define six common gesture actions: Enlarge, Narrow, Up, Down, Left, and Right, as depicted in Figure 1. These gestures were carefully selected to represent fundamental motion primitives, balancing practical usability and computational complexity. The choice of six gestures simplifies the initial model training and validation process while still capturing a diverse range of motion characteristics. The "Enlarge" and "Narrow" gestures focus on fine finger adjustments, while the "Up", "Down", "Left", and "Right" gestures involve larger arm and wrist movements, covering both vertical and horizontal spatial dimensions.

The proposed SE-WiGR framework is designed with scalability as a key consideration. The modular architecture, especially the integration of the squeeze-and-excitation (SE) module with the VGG16 network, endows the system with the capability to effectively extract features and classify a broader range of gesture categories. The SE module's unique functionality lies in its ability to adaptively recalibrate channel-wise feature responses. By doing so, it enables the model to prioritize discriminative features, thereby mitigating potential performance degradation as the gesture set expands.

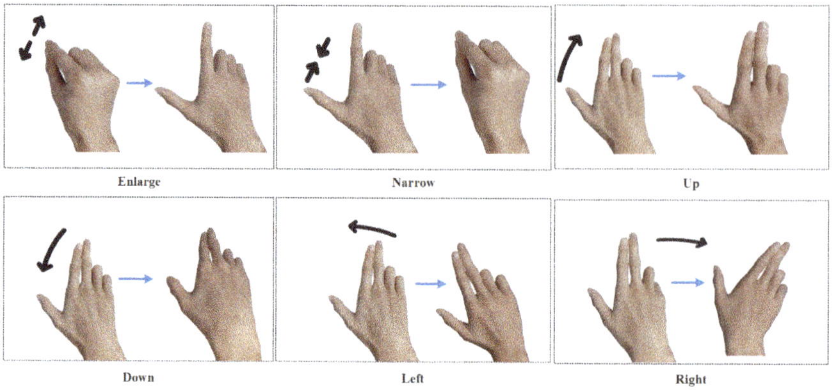

Figure 1. Gesture action design.

The structure of SE-WiGR, the gesture recognition method proposed in this study, is illustrated in Figure 2. The framework consists of three phases: data acquisition, data preprocessing, and classification and recognition. The upper half of Figure 2 covers the data acquisition and preprocessing phases, where the collected CSI data are transformed into micro-Doppler maps. The lower half depicts the gesture classification module, which is primarily based on a VGG16 network enhanced with an SE (squeeze-and-excitation) module. The training of the SE-VGG16 network involves several key configurations to ensure optimal performance. The model leverages the standard VGG16 architecture with 13 convolutional layers and 3 fully connected layers, augmented with SE modules inserted after the final layer of each convolutional block (totaling 4 SE modules). Each SE module employs global average pooling for the squeeze operation to generate channel descriptors, followed by two fully connected layers (with a reduction ratio of 16) and a Sigmoid activation for excitation, enabling adaptive channel-wise feature recalibration. The network is trained with a Softmax output layer suitable for the 6-class gesture classification task. These configurations enable the model to effectively learn discriminative gesture patterns from CSI-derived micro-Doppler maps.

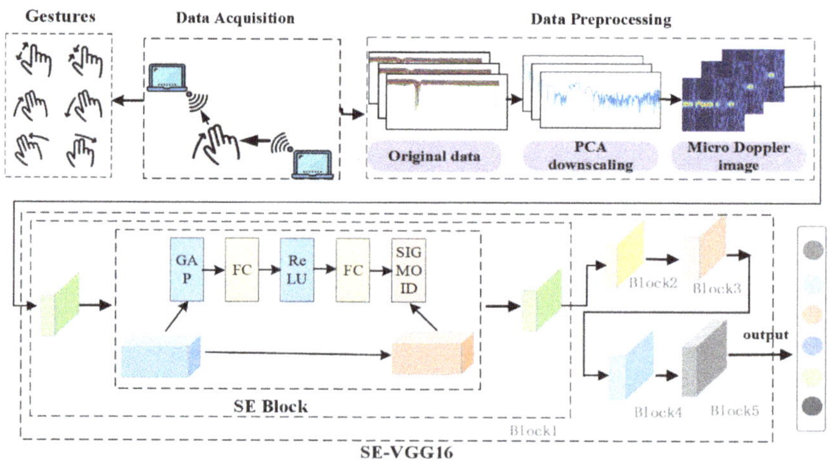

Figure 2. Structure of SE-WiGR gesture recognition method.

3.1. Data Acquisition

CSI quantifies the propagation characteristics of wireless signals through individual transmission paths, characterizing key channel phenomena including multipath scattering, environmental attenuation effects, and path loss variations. It characterizes how the wireless channel varies over time, carrier frequency, and spatial distribution [25]. In this study, we employ Intel 5300 network interface cards (NICs) (Intel, Santa Clara, CA, USA) to capture CSI measurements corresponding to various gesture states. In this study, the WiFi signals are modulated using OFDM, where the channel-received signal can be represented as

$$Y = HX + \text{Noise} \tag{1}$$

where X and Y represent the complex-valued transmitted and received signal vectors, respectively; H denotes the channel state matrix describing the wireless propagation effects; and Noise corresponds to additive white Gaussian noise. Thus, CSI is defined as the matrix H, encapsulating the channel characteristics across all subcarriers.

$$H = [H_1, H_2, \ldots, H_i, \ldots, H_N]^T \tag{2}$$

where $i \in [1, N]$ and N represent the number of subcarriers. A subcarrier H_i is represented in complex form, so the channel state information can be expressed as

$$H_i = |H_i| e^{-j \times \angle H_i} \tag{3}$$

where $|H_i|$ and $\angle H_i$ denote the amplitude and phase of the ith carrier, respectively.

3.2. Data Preprocessing

The collected CSI signals are susceptible to noise interference due to environmental factors and equipment operation. To enhance the extraction of gesture-induced CSI variations, this study employs a three-step denoising approach. First, a Hampel filter [26] removes outliers that deviate significantly from the mean and standard deviation range. Next, a Butterworth low-pass filter [27] with cutoff frequencies of 1.5 Hz (lower) and 60 Hz (upper) eliminates noise outside the gesture signal's frequency range. Finally, sym8 wavelets are applied for data smoothing. To optimize gesture recognition, PCA is used to reduce data dimensionality while retaining key information. The number of PCA components is crucial: too few may discard essential gesture features, reducing accuracy, while too many retain redundant data, increasing complexity and risking overfitting. Through experiments and cross-validation, an optimal number of components was chosen to balance information retention and dimensionality reduction, enhancing the efficiency and generalization of the recognition model. Figure 3 demonstrates the processing results: (a) and (b) show the raw data from 30 subcarriers corresponding to Narrow and Right gesture signals, while (c) and (d) display the filtered data with principal component features extracted using PCA [28]. The comparison clearly illustrates the significant improvement in signal quality after processing.

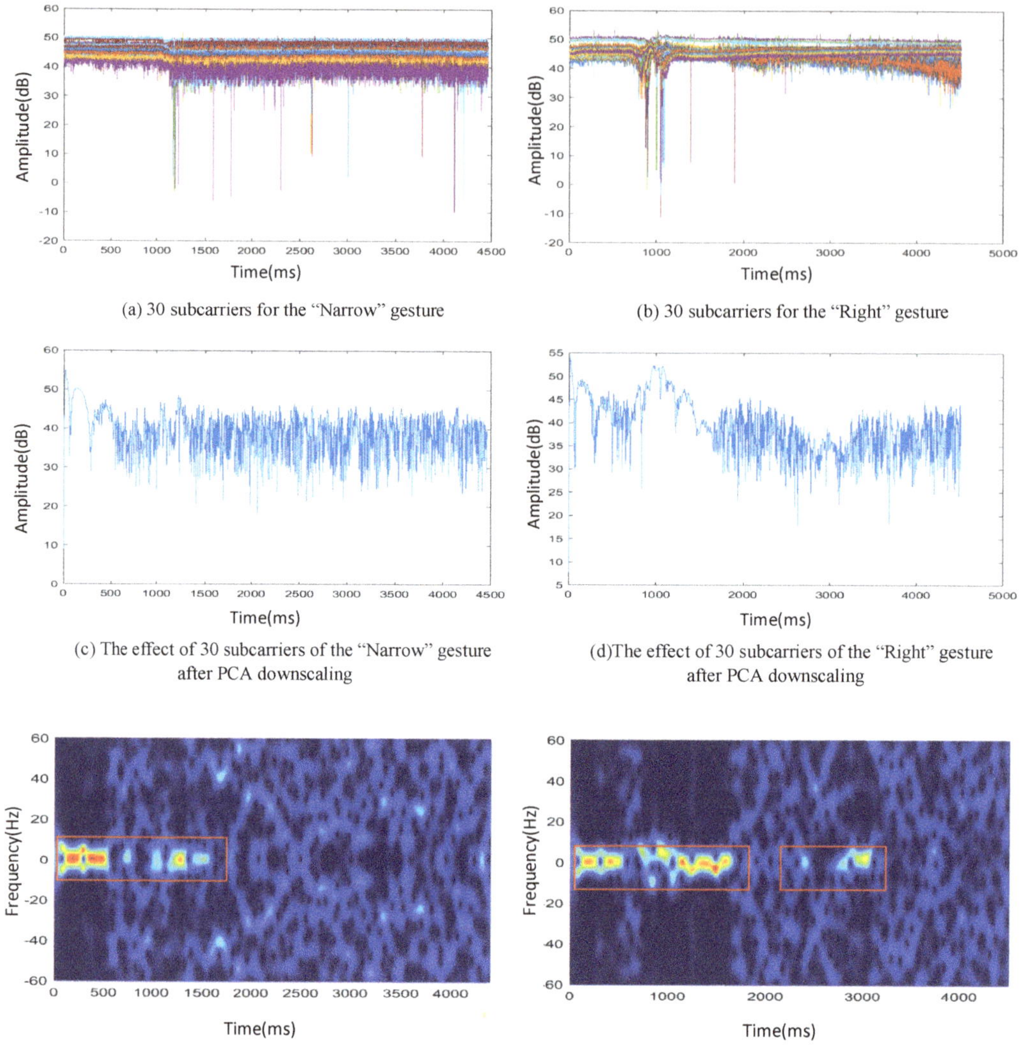

(a) 30 subcarriers for the "Narrow" gesture

(b) 30 subcarriers for the "Right" gesture

(c) The effect of 30 subcarriers of the "Narrow" gesture after PCA downscaling

(d)The effect of 30 subcarriers of the "Right" gesture after PCA downscaling

(e) Doppler spectrogram of the "Narrow" gesture

(f) Doppler spectrogram of the "Right" gesture

Figure 3. Example of the data processing process.

This study employs a sliding window approach to calculate local variance, enabling effective differentiation between inactive and motion-activated periods for precise extraction of gesture-induced signal data. The extracted signals are then processed using short-time Fourier transform (STFT) to generate Doppler shift spectrograms, as illustrated in Figure 3e,f, the red rectangular box represents the features displayed after processing the current gesture into a Doppler image. The WiFi signal's Doppler shift resulting from gesture movements can be expressed as

$$f_D = -\frac{1}{\lambda}\frac{d}{dt}d(t) \tag{4}$$

where λ is the signal wavelength, and $d(t)$ is the transmit path length. The Doppler spectrogram [29] primarily displays the signal characteristics in the frequency domain, and the Doppler spectrum of CSI is obtained by video analysis:

$$H(f,t) = h_s(f) + \sum_{k \in D} a_k(t) B(f_{D_k}(t)) \quad (5)$$

3.3. Gesture Recognition Based on SE and VGG-16 Model Fusion

The SE-VGG16 method proposed in this study focuses on data acquisition and preprocessing. The generated Doppler spectrogram data are input into the SE-VGG16 for gesture recognition. The architecture of SE-VGG16 improves upon VGG16 by incorporating the SENet module. The overall structure of SE-VGG16 is illustrated in the lower half of Figure 2. The network structure is detailed in Table 2, where SE modules are strategically embedded after specific convolutional layers to adaptively recalibrate channel-wise feature responses.

Table 2. SE-VGG16 network architecture.

Layer Type	Configuration Parameters	Output Size (H × W × C)	SE Module	Remarks
Input Layer	-	224 × 224 × 3	-	RGB image input
Convolution + ReLU	64 kernels, 3 × 3, stride = 1, pad = 1	224 × 224 × 64	-	-
Convolution + ReLU	64 kernels, 3 × 3, stride = 1, pad = 1	224 × 224 × 64	SE-64	First SE module embedding
Max Pooling	2 × 2, stride = 2	112 × 112 × 64	-	-
Convolution + ReLU	128 kernels, 3 × 3, stride = 1, pad = 1	112 × 112 × 128	-	-
Convolution + ReLU	128 kernels, 3 × 3, stride = 1, pad = 1	112 × 112 × 128	SE-128	-
Max Pooling	2 × 2, stride = 2	56 × 56 × 128	-	-
Convolution + ReLU (3 layers)	256 kernels, 3 × 3, stride = 1, pad = 1	56 × 56 × 256	-	First two layers without SE
Convolution + ReLU	256 kernels, 3 × 3, stride = 1, pad = 1	56 × 56 × 256	SE-256	Third layer with SE module
Max Pooling	2 × 2, stride = 2	28 × 28 × 256	-	-
Convolution + ReLU (3 layers)	512 kernels, 3 × 3, stride = 1, pad = 1	28 × 28 × 512	-	First two layers without SE
Convolution + ReLU	512 kernels, 3 × 3, stride = 1, pad = 1	28 × 28 × 512	SE-512	Third layer with SE module
Max Pooling	2 × 2, stride = 2	14 × 14 × 512	-	-
Convolution + ReLU (3 layers)	512 kernels, 3 × 3, stride = 1, pad = 1	14 × 14 × 512	-	First two layers without SE
Convolution + ReLU	512 kernels, 3 × 3, stride = 1, pad = 1	14 × 14 × 512	SE-512	Third layer with SE module
Max Pooling	2 × 2, stride = 2	7 × 7 × 512	-	-
Fully Connected + ReLU	4096 units	1 × 1 × 4096	-	Input after flattening
Fully Connected + ReLU	4096 units	1 × 1 × 4096	-	-
Fully Connected + Softmax	6 units (gesture categories)	1 × 1 × 6	-	Output probability for 6 gesture classes

3.3.1. VGG16 Network Architecture

VGGNet is a CNN structure network, which includes an input layer, multiple convolutional and pooling layers, fully connected layers, and an output layer. Specifically, the network architecture comprises 13 convolutional layers utilizing 3 × 3 filters for spatial feature learning, each coupled with ReLU activations to introduce nonlinearity, alongside three fully connected layers. Max pooling operations utilize a 2 × 2 kernel with a stride of 2, effectively performing down-sampling. The network architecture proceeds as follows: First, Block1 performs two 64-kernel convolutions followed by pooling. Block2 then applies two 128-kernel convolutions with pooling. Subsequently, Block3 executes three 256-kernel convolutions before pooling. This pattern continues in Block4 and Block5, each performing two 512-kernel convolutions followed by pooling. The structure concludes with three fully connected layers—the first two containing 4096 neurons each, and the final layer comprising 1000 neurons. These FC layers transform the convolutional feature maps into class probabilities. The Softmax classifier then processes the final layer's output to generate a probability distribution across categories, enabling classification. The complete VGG16 architecture is illustrated in Figure 4.

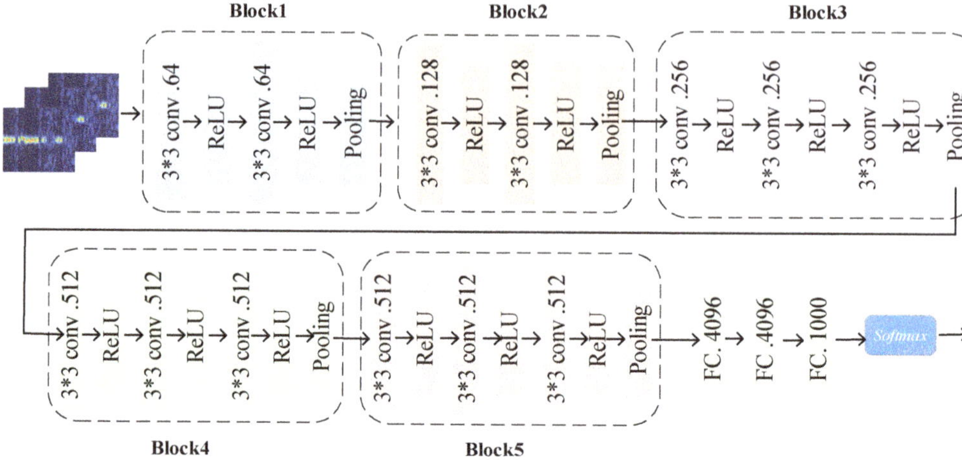

Figure 4. The structure of the VGG16 network.

3.3.2. Squeeze and Excitation Networks

The squeeze-and-excitation networks (SENets) are an innovative neural network architecture that introduces the squeeze-and-excitation (SE) block as its core component. The SE block automatically learns and determines the importance of individual feature channels, explicitly modeling inter-channel dependencies. Figure 5 shows the schematic structure of the SE block. In contrast, the excitation operation uses fully connected layers to learn nonlinear channel-wise relationships and performs feature recalibration—selectively enhancing informative features while suppressing less useful ones. This recalibration strategy allows SENet to leverage global context more effectively, optimizing the network's attention mechanism for improved performance.

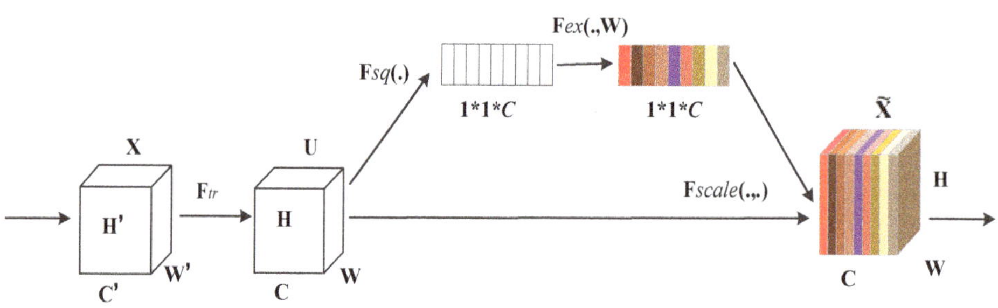

Figure 5. The structure of the squeeze–excitement (SE) module.

A squeeze and excitation block is a computational unit whose core function is to map the input data x, $x \in R^{H' \times W' \times C'}$ to the features U,$U \in R^{H \times W \times C}$ by the transformation F_{tr}. In this computation, we employ F_{tr} as the operator of the convolution operation and define a collection of filter kernels V, $V = [v_1, v_2, ..., v_C]$, and set a parameter for each filter, defined as v_c, which are learned and optimized during training. Thus, the output can be written as $U = [u_1, u_2, \ldots, u_C]$, where

$$u_c = v_c \times X = \sum_{s=1}^{C'} v_c^s * x^s \tag{6}$$

where $*$ denotes the convolution, $v_c = \left[v_C^1, v_C^2, \ldots v_C^{C'}\right]$, $X = \left[x^1, x^2, \ldots, x^{C'}\right]$ and $u_C \in R^{H \times W}$. v_C^S is a two-dimensional spatial kernel denoting the individual channels v_c acting on the corresponding channels of X. The representation is based on a two-dimensional space.

(1) Squeezing mechanism

To address the challenge of leveraging inter-channel dependencies, we begin by examining the feature distribution across individual channels in the output. Due to the locality of each filter's receptive field, the resulting activation units lack awareness of broader spatial context beyond their own regions.

To overcome this limitation, we introduce a strategy that condenses global spatial cues into a compact channel-wise representation. This is accomplished through the application of global average pooling, which computes statistical features for each channel. Formally, the statistic $Z \in R^C$ is generated by reducing U to its spatial dimension $H \times W$, so that the cth element of z is computed using Equation (7):

$$z_x = F_{sq}(U_c) = \frac{1}{H \times W} \sum_{i=1}^{H} \sum_{j=1}^{W} u_c(i, j) \tag{7}$$

(2) Excitation Mechanism

An excitation mechanism is introduced to fully utilize the information aggregated during the squeeze operation, enabling effective modeling of channel-wise dependencies. This mechanism employs a simple gating function with sigmoid activation.

$$s = F_{ex}(z, W) = \sigma(g(z, W)) = \sigma(W_2 \delta(W_1 z)) \tag{8}$$

where δ denotes the ReLU function, $W_1 \in R^{\frac{C}{r} \times C}$ and $W_2 \in R^{C \times \frac{C}{r}}$.

Fully connected layers are placed on both sides of the nonlinear region, aiming to tune the gating mechanism, constrain the model complexity, and at the same time enhance the generalization performance. Specifically, this bottleneck structure comprises: (1) a dimensionality reduction layer with a scaling ratio r to decrease parameter count and computational load, (2) a ReLU activation function introducing nonlinearity, and (3) a dimensionality restoration layer that matches the channel dimension of the transformed output U:

$$\widetilde{X}_c = F_{scale}(u_c, s_c) = s_C u_C \tag{9}$$

where $\widetilde{X} = \left[\widetilde{x}_1, \widetilde{x}_2, \ldots, \widetilde{x}_C\right]$ and $F_{scale}(u_c, s_c)$ is the channel multiplication between the metrics s_c and the feature map $U_c \in R^{H \times W}$.

4. Experimentation and Analysis of Results

4.1. Experimental Deployment

During the experimental stage, this study utilized an Intel 5300 wireless network card (Intel, Santa Clara, CA, USA) and employed the Linux 802.11n CSI-Tool to serve as the receiver (Rx), responsible for capturing signals and initiating communication with the transmitter (Tx) at a predetermined rate. The configuration includes a single antenna on the transmission side and three antennas on the reception side, with 30 subcarriers per channel and a central frequency set to 2.4 GHz. In MATLAB 2020, the collected ".dat" format data were processed to extract and visualize Doppler frequency shift information. The actual system setup is shown in Figure 6, which demonstrates the processing of raw gesture data and the recognition of different gestures, the red rectangular box represents the features displayed after processing the current gesture into a Doppler image. Model training was conducted in PyCharm 2023 using PyTorch 2.6.0, with the SE-VGG16 optimized via configurations detailed in Table 3. An Adam optimizer (learning rate = 0.001, betas = (0.9, 0.999)) ensures stable gradient descent by adapting learning rates for each parameter, combined with a stepwise learning rate decay strategy (initial LR = 0.001, 10× decay every 30 epochs) to avoid local minima. A batch size of 32 balances GPU memory efficiency, and the model is trained for 100 epochs with early stopping—terminating training if validation loss stagnates for 5 consecutive epochs to prevent overfitting.

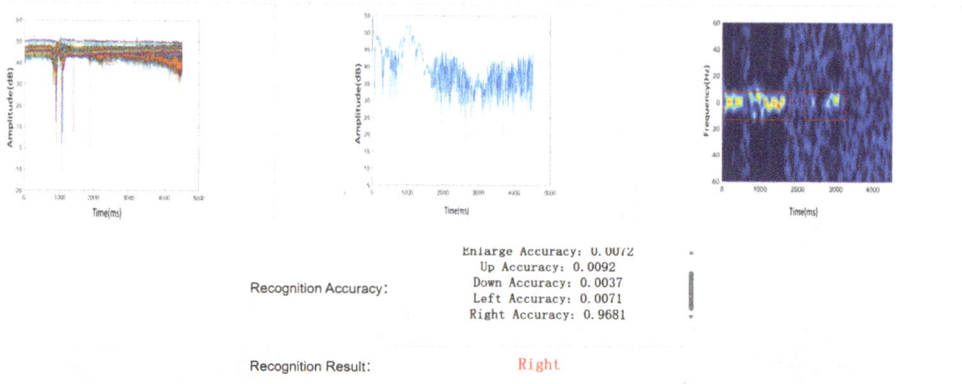

Figure 6. WiFi gesture recognition system demonstration.

Table 3. Training parameter configuration.

Parameter Name	Value/Setting
Optimizer	Adam (learning rate = 0.001, betas = (0.9, 0.999))
Initial Learning Rate	0.01, 10× decay every 30 epochs
Loss Function	CrossEntropy Loss
Batch Size	32
Number of Epochs	100
Early Stopping Condition	Terminate if validation loss does not improve for 5 consecutive epochs

To further mitigate overfitting, the model integrates ReLU activation functions, a Dropout rate of 0.5 (applied to fully connected layers to reduce neuron co-adaptation), and the aforementioned learning rate decay and early stopping strategies. This combined regularization approach significantly reduces the train–test accuracy gap, as quantitatively validated in Table 4, where the inclusion of early stopping decreases the accuracy discrepancy by 4.5% on average across environments.

Table 4. Inhibition effect of regularization strategies on overfitting.

Environment	Strategy	Optimal Validation Accuracy	Convergence Epoch	Train/Test Gap
Conference Room	Without Early Stop	91.60%	95	+6.4%
Conference Room	**With Early Stop**	**94.26%**	**88**	**+4.2%**
Office	Without Early Stop	89.73%	97	+8.7%
Office	**With Early Stop**	**93.95%**	**93**	**+3.2%**

4.2. Experimental Data

This study conducted gesture recognition experiments in two distinct environments: an office and a conference room. Figure 7 illustrates the experimental setup and equipment configuration. We recruited ten participants (five male and five female) aged 23–28 years for the experiments. The experimental protocol was designed as follows: (1) Each gesture was performed for a duration of 5 s; (2) Rest periods were implemented before and after each gesture trial; (3) Each participant executed six different gestures; and (4) Every gesture was repeated 20 times. This protocol yielded a total of 2400 samples for analysis. The dataset was divided using a stratified random split: 80% (1920 samples) for training, 10% (240 samples) for validation, and 10% (240 samples) for testing. This partition ensures that the distribution of gesture categories and environmental conditions is preserved across subsets, enabling robust model evaluation.

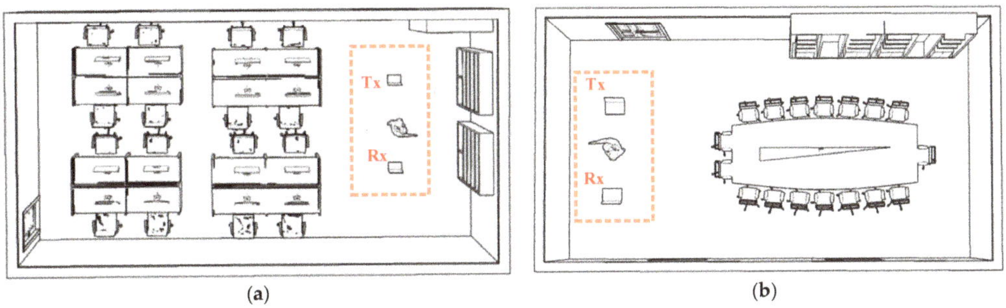

Figure 7. Experimental scenario and equipment deployment diagram: (**a**) office and (**b**) conference room.

4.3. Experimental Results and Analysis

4.3.1. Performance of Different Classification Algorithms

(1) Comparison of the accuracy of VGG16, SE-VGG16, ResNet18, ResNet34, ResNet50, MobileNetV3, and EfficientNetB0 for different gestures.

In order to evaluate the performance of the SE-VGG16 WiFi gesture recognition method, the recognition accuracy of seven models, including VGG16, SE-VGG16, ResNet18, ResNet34, ResNet50, MobileNetV3, and EfficientNetB0, is compared and analyzed in this study. The results are shown in Figure 8.

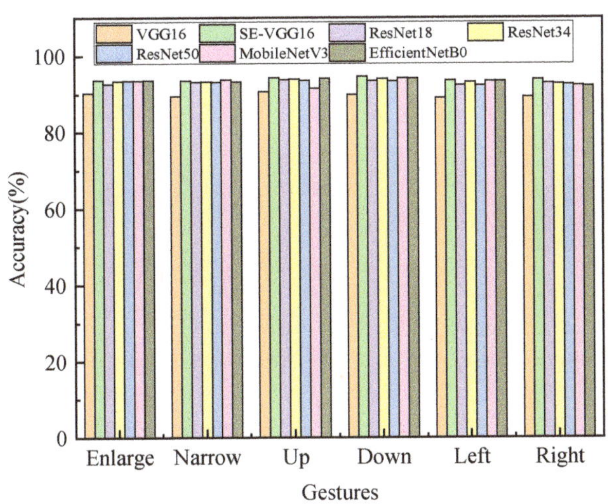

Figure 8. Gesture recognition results with different classification models.

Figure 8 compares the gesture recognition accuracy of seven deep-learning models. Among these models, SE-VGG16 has the highest overall accuracy, with ResNet variants ranking closely behind. SE-VGG16's superiority stems from its squeeze–excitation (SE) mechanism, which recalibrates channel-wise features, improving accuracy by 4.13% over VGG16. ResNet models use residual connections for better performance, but trail SE-VGG16 in some challenging scenarios. Lightweight models MobileNetV3 and EfficientNetB0 trade accuracy for computational efficiency; their depthwise separable convolutions and reduced capacity limit performance on intricate gestures. Further analysis of per-gesture accuracy reveals that SE-VGG16's advantage is most pronounced for Enlarge/Narrow gestures, where its attention mechanism effectively isolates subtle finger movements from noise. The results collectively affirm SE-VGG16 as a state-of-the-art solution, particularly for environments requiring high precision across diverse gesture types.

(2) Confusion Matrix of Different Gestures in SE-VGG16 Model

To further assess the SE-VGG16 model's gesture-recognition performance, confusion matrices for office and conference room environments are analyzed. These matrices detail misclassifications among Enlarge, Narrow, Up, Down, Left, and Right gestures.

In the office environment (Figure 9a), the SE-VGG16 model shows excellent recognition performance. The confusion matrix shows that the recognition rate of the "up" gesture is as high as 94.31%, while the "zoom in" and "zoom out" gestures with complex motion patterns also maintain a high accuracy of 93.69% and 93.56%, respectively. Despite a small number of misclassifications, the overall misclassification rate is very low, which indicates that the model can effectively distinguish the unique signal characteristics of each type of gesture and maintains stable performance even in the presence of environmental interference. In addition, in the validation of the conference room scenario (Figure 9b), the model also performs well, with a recognition rate of 94.29% for the "down" gesture, and other gesture types also maintain a high recognition accuracy. The overall low misclassification rate fully demonstrates the robustness of the model.

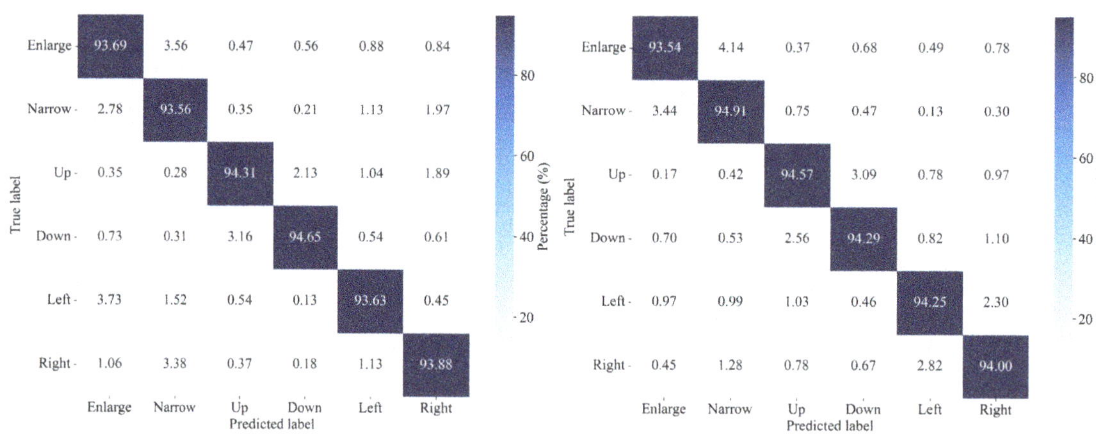

Figure 9. Confusion matrix of different gestures in SE-VGG16 mode. (**a**) Confusion matrix for gesture recognition in an office environment. (**b**) Confusion matrix for gesture recognition in a conference room.

The results of the confusion matrix clearly demonstrate the effectiveness of the SE-VGG16 model. It is worth noting that the signal uniformity due to the open space of the conference room and the signal variability due to the complex layout of the office have some comparative value, while the model maintains an overall high accuracy rate in different environments. This performance fully demonstrates the model's excellent environmental adaptability and robustness characteristics.

(3) Comparison of VGG16 and SE-VGG16 performance

To evaluate the performance enhancement achieved by incorporating the squeeze–excitation mechanism into VGG16, we compared the training and validation accuracy trajectories of both VGG16 and SE-VGG16 across successive epochs. The results are presented in Figure 10. Figure 10 demonstrates that both models exhibit progressively improving accuracy with increasing epoch count until reaching stabilization. Notably, the validation accuracy marginally exceeds the training accuracy in both architectures.

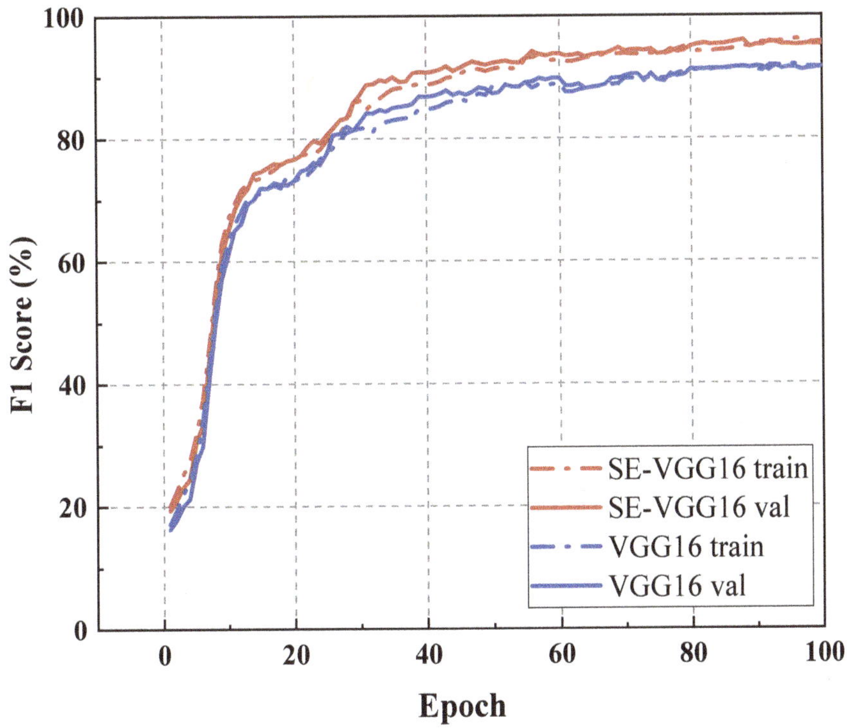

Figure 10. Comparison of F1 score between VGG16 and SE-VGG16 in training and validation sets under different Epochs.

(4) Comparison of SE-WiGR with other gesture recognition systems

To comprehensively verify the actual performance of the proposed method, we selected two typical indoor scenarios (conference room and office environment) and conducted systematic comparisons with mainstream gesture recognition algorithms under identical experimental conditions, the experimental results are shown in Table 5. In gesture recognition tasks, class imbalance (e.g., limited samples for fine-grained gestures like "Enlarge/Narrow") and inter-class feature similarity (e.g., directional gestures "Left/Right" and "Up/Down") pose challenges for traditional metrics like Accuracy. As a harmonic mean of Precision (TP/(TP + FP)) and Recall (TP/(TP + FN)), the F1 score mitigates bias from dominant gesture categories and provides a balanced assessment of minority classes, while also sensitively capturing misclassifications between similar gestures with subtle feature differences—avoiding the oversimplification of single Accuracy metrics. Additionally, in practical applications like smart homes, the F1 score's penalty of both false positives and negatives aligns with real-world reliability requirements.

Table 5. Results of SE-WiGR with other gesture recognition systems.

Method	Office			Conference Room		
	Precision	Recall	F1-Score	Precision	Recall	F1-Score
WiGest	68.21%	67.34%	67.54%	67.56%	65.44%	65.78%
HMM	73.62%	73.24%	73.31%	70.34%	69.41%	70.02%
LSTM	82.17%	81.04%	81.38%	83.75%	82.04%	82.34%
GRU	81.09%	80.25%	80.45%	81.58%	80.2%	80.27%
ABLSTM	92.13%	91.56%	91.82%	92.64%	91.88%	92.31%
Our (SE-WiGR)	**94.24%**	**93.89%**	**93.95%**	**94.78%**	**93.96%**	**94.26%**

Our experimental results demonstrate significant performance differences among the evaluated approaches. Shallow learning models relying on hand-crafted features (WiGest and HMM) achieve relatively low accuracy (approximately 70%), as manual feature extraction frequently fails to capture critical discriminative characteristics. In comparison, deep learning-based approaches (LSTM and GRU) show superior performance by automatically learning optimal feature representations. Notably, the ABLSTM method outperforms standard LSTM through its bidirectional architecture and attention mechanism. Among all benchmarked methods, our proposed SE-WiGR technique achieves the highest recognition accuracy across all gesture categories. This performance advantage stems from the effective integration of the SE module with the VGG16 network, which synergistically combines their complementary strengths for robust feature extraction and classification.

4.3.2. Robustness Experiments

(1) Evaluation of User Diversity and Cross-User Generalization

This method evaluates its robustness through a multi-faceted validation framework. First, a dual mechanism assesses:

Inter-User Variability: The recognition effect of the same gesture executed by different users (Figure 11a), where users 1–6 show minimal accuracy differences despite variations in finger width and execution postures. This indicates the method is insensitive to physiological traits like gender, weight, and height.

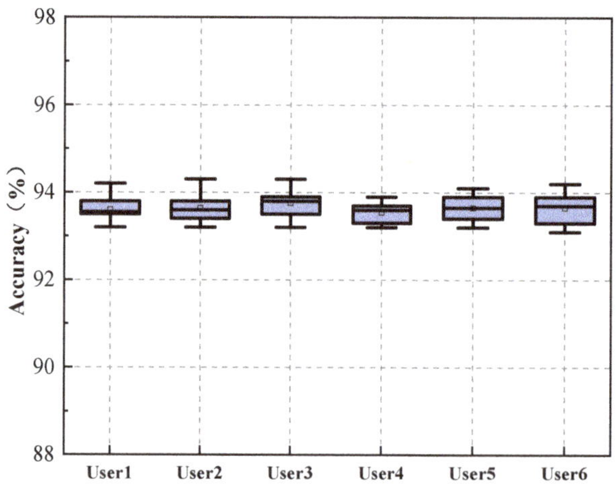

(a)

Figure 11. *Cont.*

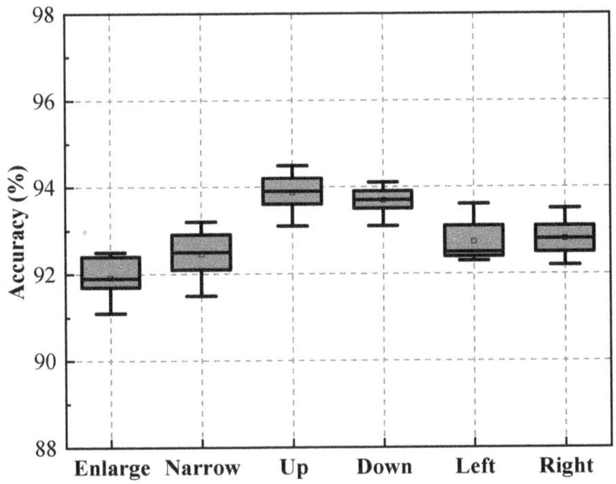

(b)

Figure 11. Recognition results of different users in two environments. (**a**) Recognition results of the same action by different users. (**b**) Recognition results of different actions of the same user.

Intra-User Consistency: The recognition stability of different gestures from the same user (Figure 11b), where accuracy fluctuations (within 3.2%) due to gesture amplitude variations do not compromise overall performance.

To quantitatively validate generalization to unseen users, a leave-one-user-out cross-validation strategy was adopted. Data from 9 users were used for training, with the remaining 1 user serving as the test set (repeated 10 times). The results (Table 6) show average accuracies of 93.95% (office) and 94.26% (conference room), with most users deviating <±1.0% from the mean. Notable exceptions (e.g., U7's 91.57% in the conference room) likely reflect environmental interference rather than user diversity, demonstrating the model's resilience to outliers.

Table 6. Cross-user generalization performance.

Test User	Office Accuracy (%)	Deviation from Mean	Conference Room Accuracy (%)	Deviation from Mean
U1	93.43%	−0.52%	94.13%	−0.13%
U2	92.76%	−1.19%	94.67%	+0.41%
U3	93.85%	−0.10%	94.32%	+0.06%
U4	93.79%	−0.16%	93.89%	−0.37%
U5	94.57%	+0.62%	94.98%	+0.72%
U6	94.28%	+0.33%	93.64%	−0.62%
U7	93.56%	−0.39%	91.57%	−2.69%
U8	93.44%	−0.51%	95.80%	+1.54%
U9	94.88%	+0.93%	94.27%	+0.01%
U10	94.98%	+1.03%	95.32%	+1.06%
Average	**93.95%**	—	**94.26%**	—

The above experiments confirm the robustness of SE-WiGR between inter-user physiological differences and intra-user gesture variations, making it suitable for real-world deployments with different user groups.

(2) Impact of speed diversity

Speed is a crucial perceptual dimension when evaluating the impact of WiFi-based devices on gesture recognition, as the same gesture displays unique speed characteristics at various execution speeds. To comprehensively verify the stability and reliability of the gesture recognition system under different speed conditions, in this study, the gesture execution duration of 4 s, as defined in the training dataset, is adopted as the baseline. To assess the system's robustness across temporal variations, a set of controlled experiments is conducted at three distinct gesture speeds: rapid (2 s), standard (4 s), and slow (6 s). The results of these experiments are shown in Table 7.

Table 7. Accuracy of speed on gesture recognition (%).

Gestures	Standard (4 s)	Fast (6 s)	Slow (2 s)
Enlarge	93.69%	88.57%	91.36%
Narrow	93.56%	87.72%	92.54%
Up	94.31%	89.78%	92.98%
Down	94.65%	91.75%	93.13%
Left	93.63%	89.63%	93.37%
Right	93.88%	90.53%	93.46%
Average	**94.12%**	**89.66%**	**92.80%**

As shown in Table 7, gesture speed significantly affects recognition accuracy. The average accuracy is 89.66% for fast gestures, 93.95% for normal-speed gestures, and 92.80% for slow gestures. Normal-speed gestures yield the highest accuracy, with Up and Down gestures reaching 94.31% and 94.65%, respectively. In contrast, fast-speed Enlarge and Narrow gestures have the lowest accuracy at 88.57% and 87.72%, while the Down gesture performs best at this speed (91.75%). For slow gestures, Left and Right gestures achieve the highest accuracy (93.37% and 93.46%, respectively), whereas Enlarge accuracy drops to 91.36%. At normal speeds, gestures exhibit stable and easily detectable signal variations, resulting in higher accuracy. Faster gestures, however, introduce pronounced signal changes with additional noise and interference, complicating recognition. Slow gestures produce smoother signal transitions, but their subtlety may also reduce recognizability. The Enlarge and Narrow gestures involve more intricate multi-finger coordination compared to simple linear motions. This complexity leads to lower accuracy, especially at extreme speeds where signal variability and uncertainty further increase recognition difficulty.

(3) Effect of positional diversity

The experiments investigated the attenuation and multipath reflection phenomena of electromagnetic signals as the gesture's position relative to the device varied. These variations cause differences in the received signals' amplitude and phase, leading to discrepancies in gesture features related to distance and angle. The impact of these feature variations on gesture recognition accuracy requires further validation. We tested 28 different positions, covering distances (D1 + D2) ranging from 1 m to 4 m and angles (θ) of 0°, 30°, 60°, and 90°. At each position, 10 datasets were collected for each gesture in both office and conference room environments, yielding a total of 560 experimental samples. Using this data, we evaluate the effectiveness of gesture recognition at varying distances and angles to analyze how feature differences affect recognition accuracy. The experimental setup is illustrated in Figure 12, and the results are shown in Table 8.

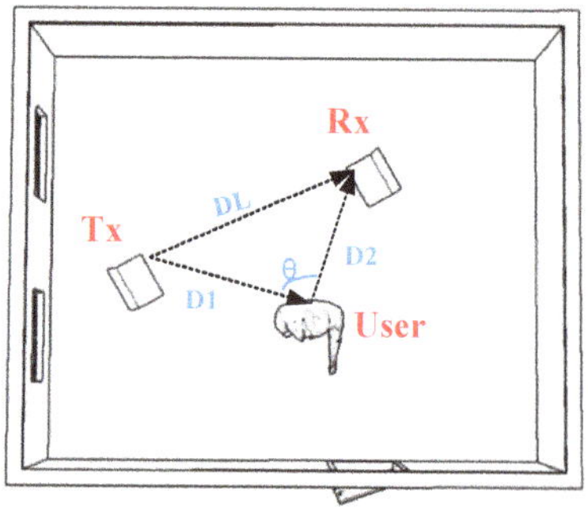

Figure 12. Graphical representation of device deployment.

Table 8. F1 score of location diversity (%).

D1+D2 D (Rx, Tx)	1.0 m 0.7 m	1.5 m 1 m	2.0 m 1.41 m	2.5 m 1.77 m	3.0 m 2.12 m	3.5 m 2.47 m	4.0 m 2.83 m
90°	91.38%	89.78%	88.53%	86.23%	85.24%	81.23%	78.75%
60°	92.49%	91.54%	89.42%	87.58%	84.45%	81.36%	77.52%
30°	92.17%	91.73%	91.77%	87.69%	84.05%	82.11%	74.36%
0°	94.25%	92.07%	93.03%	89.42%	88.39%	84.28%	80.54%

Based on the data in Table 8, it is evident that recognition accuracy varies significantly with changes in object position and angle, with overall recognition efficiency ranging from 74.36% to 94.25%. Specifically, recognition accuracy peaks at 92.17% when the object is positioned at a 0° angle and a distance of 1 m. In contrast, when the object is placed at a 90° angle and the distance increases to 4 m, recognition accuracy drops to a minimum of 78.75%. Further examination indicates that, under a fixed angular condition, recognition accuracy tends to decline progressively as the distance increases. Likewise, when distance remains constant, variations in the angle also lead to a slight decrease in accuracy, though the extent of this reduction is relatively limited. It is worth emphasizing that although recognition accuracy tends to decline as the distance increases, in practical applications such as smart home interaction or virtual reality systems, the user typically operates within a moderate range from the device. Therefore, the observed reduction in accuracy has a limited impact on the overall user experience.

In the experimental design, when the angle θ between the person and the transceiver device remains constant, the geometry of the signal path stays stable, with the only change being an increase in the total distance (D1 + D2). This alteration results in an elongated signal propagation path, increasing the probability of signal attenuation and interference.

On the contrary, assuming that D1 + D2 remains constant, the accuracy decreases slightly as the θ angle gradually increases from 0° to 90°. This occurs because, at $\theta = 0°$, the signal travels directly along a straight line, making the path the simplest, and the multipath effect has less influence, resulting in the highest recognition accuracy. However, as the θ angle increases, the signal must take a detour, leading to greater diversity and complexity in the propagation path, significantly enhancing the multipath effect. Although D1 + D2 is constant, path variation due to angle changes may cause minor signal strength fluctuations, but these are less significant than those from distance changes.

For scenarios where a secondary person performs random movements behind the primary user during gesture execution, the proposed SE-WiGR framework is well-equipped to maintain accurate recognition. The squeeze–excitation (SE) module embedded within the network adaptively recalibrates channel-wise feature responses in the CSI data, enabling the model to prioritize the signal patterns associated with the primary user's intended gestures. Meanwhile, the hierarchical feature extraction capabilities of the VGG16 network effectively filter out interference from background activities. The robustness demonstrated in previous experiments against environmental noise and occlusions further supports the framework's ability to handle such complex, real-world interactions.

In addition, combining the data in Table 8, we can further analyze the mechanism of distance and multipath effect on the recognition accuracy: at closer ranges (e.g., D1 + D2 = 1 m, $\theta = 0°$), the direct signal path dominates with weak multipath interference, enabling CSI to clearly reflect gesture dynamics—"up/down" recognition accuracy reaches 94.25%. At 4 m, signal reflection via walls/furniture causes ~20 dB CSI amplitude attenuation and fuzzy features due to multipath phase superposition, reducing "enlarge/narrow" accuracy to 78.75%, consistent with wireless channel theory.

4.3.3. Model Ablation Experiments

(1) Squeezing operator

In the SE module section, this study compares the effects of two pooling strategies, max-pooling and average-pooling, used in the squeezing operation, with the specific results presented in Table 9.

Table 9. F1 score of using different squeezing operators in SE-VGG16 (%).

Squeeze	F1
Max pooling	93.18%
Avg pooling	**94.12%**

From Table 9, we can see that both pooling methods demonstrate exemplary performance. However, average pooling performs slightly better, which may be due to its greater robustness in handling anomalous or outlier data compared to maximum pooling. It can also better capture the statistical characteristics and patterns of the data. Additionally, this experiment verifies the high robustness of the SE block, as the impact of incorporating the SE block on the performance of the final model remains relatively stable and high, with no significant difference, regardless of the pooling method used.

(2) Different activation functions

The activation functions play a crucial role in neural networks. Their primary function is to introduce nonlinear factors, allowing neural networks to learn and model complex data patterns. Regardless of its depth, such a network would be severely limited in its expressive power, capable only of approximating linear functions or linear combinations, and unable to capture the intricate nonlinear relationships within the data. Therefore, this study investigates the effect of the activation function used for the excitation operator in SE-VGG16 on the results presented in Table 10. The experimental outcomes in Table 10 indicate that employing the ReLU activation function yields greater accuracy, making SE-VGG16 more effective.

Table 10. Accuracy of using different nonlinear activation functions for the excitation operator in SE-VGG16 (%).

Activation Functions	VGG16	SE-VGG16
ReLU	**89.82%**	**94.12%**
tanh	88.64%	92.25%
sigmoid	88.98%	92.17%

(3) VGG16 integrates SE blocks at different stages

We integrated SE blocks into each stage of VGG16 sequentially to investigate their impact at different levels of the network. As illustrated in Table 11, integrating SE blocks at different stages within the VGG16 architecture leads to varying degrees of performance enhancement across the model. This indicates that the benefits produced by SE blocks at various stages are complementary and can collectively enhance the network's overall performance. Additionally, it is essential to note that while the number of model parameters increases with the addition of SE blocks, this growth remains relatively limited and does not significantly increase model training costs.

Table 11. F1 score (%) and number of parameters of integrating SE with VGG16 at different stages on gesture recognition.

Stage	F1	Parameters
VGG-16	89.82%	141.3 M
SE_Stage_2	93.21%	141.3 M
SE_Stage_3	93.33%	141.5 M
SE_Stage_4	93.35%	143.1 M
SE_All	**94.12%**	**145.3 M**

(4) Impact of SE: VGG16-SE, NoSqueese, VGG16

To verify the performance of the SE module, we first remove the pooling layer. The original fully connected (FC) layer is replaced with a 1 × 1 convolutional layer. We refer to the structure after the replacement as NoSqueeze, and the experimental results are shown in Table 12.

Table 12. F1 score (%) and parameters of squeeze operation.

	F1	Parameters
VGG-16	89.82%	141.3 M
NoSqueese	89.82%	145.3 M
SE-VGG-16	**94.12%**	**145.3 M**

From the detailed data in Table 12, we can see that the F1 score of the model improves by 4.13% when the SE module is incorporated into the VGG16 network architecture. This improvement is reflected in the increased recognition accuracy and the enhanced learning ability of the model for complex features. Notably, although we introduced the SE module to boost the network's representation capability, the model's number of parameters did not substantially increase. This indicates that we successfully optimized the performance of VGG16 while maintaining computational efficiency. This finding is significant for real-world applications, as it allows us to enhance the model's performance by introducing advanced network modules without sacrificing computational efficiency.

In summary, the data in Table 12 confirms the significant effect of SE modules on the performance enhancement of the VGG16 network. It also shows that this enhancement is achieved while keeping the model parameters streamlined and ensuring efficient utilization of computational resources. This provides new ideas and directions for us to continue exploring and optimizing deep learning models in the future.

5. Conclusions

This study proposes SE-WiGR, a WiFi gesture recognition method that fuses the squeeze-and-excitation (SE) module with VGG16. The process relies on acquired CSI signals, which are filtered and analyzed using principal component analysis and other preprocessing operations to generate Doppler spectrograms. These spectrograms are subsequently fed into the SE-VGG16 network for gesture recognition. The SE module effectively enhances gesture features by combining external excitation signals with squeezing operations while suppressing ambient noise, thereby improving recognition accuracy. Simultaneously, the powerful feature extraction capability of the VGG network further enhances recognition performance. Quantitative results demonstrate that across office and conference room scenarios, SE-WiGR achieves an average accuracy of 94.12% for six gestures (Enlarge, Narrow, Up, Down, Left, and Right), outperforming the standalone VGG16 network by 4.3%, surpassing traditional methods like WiGest (67.5%) and HMM (73.3%) by 26.62% and 20.82%, respectively, and outperforming deep learning baselines such as LSTM (81.3%) and GRU (80.5%). However, despite the SE-WiGR method's successes in gesture recognition, it still faces challenges such as multipath effects, device differences, and gesture diversity in complex environments—issues that will be the focus of future research.

Author Contributions: Conceptualization, F.L. and C.W.; methodology, F.L.; software, Y.L.; validation, F.L.; formal analysis, F.L.; investigation, C.W.; resources, F.L.; data curation, C.W.; writing—original draft preparation, F.L.; writing—review and editing, C.W.; visualization, Y.L.; supervision, Y.L. All authors have read and agreed to the published version of the manuscript.

Funding: This research was funded by the National Natural Science Foundation of China (grant nos. 62162056 and 62262061) and the Natural Science Foundation of Gansu Province (24JRRA127).

Institutional Review Board Statement: The Ethics Review Committee of Gansu Minzu Normal University confirms that the project involving the use of WiFi sensing technology to recognize human gestures does not require further ethical review. This project does not involve invasive procedures on participants, nor does it involve the collection or processing of sensitive data. Therefore, in accordance with the Declaration of Helsinki and other applicable ethical guidelines, the design of this research project meets ethical standards.

Informed Consent Statement: Informed consent was obtained from all subjects involved in the study.

Data Availability Statement: The data presented in this study are available on request from the corresponding author. The data are not publicly available due to privacy reasons.

Conflicts of Interest: The authors declare no conflicts of interest.

References

1. Li, F.; Liang, Y.; Liu, G.; Hao, Z.; Weng, C.; Luo, P. Wi-TCG: A WiFi gesture recognition method based on transfer learning and conditional generative adversarial networks. *Eng. Res. Express* **2024**, *6*, 045253. [CrossRef]
2. Chiu, P.; Kim, C.; Oda, H. Recognizing gestures on projected button widgets with an RGB-D camera using a CNN. In Proceedings of the 2018 ACM International Conference on Interactive Surfaces and Spaces, Tokyo, Japan, 25–28 November 2018; pp. 369–374.
3. Hao, S.; Wu, Y.; Ma, X.; He, T.; Wang, F. Matching of visible and infrared images based on Cy-cleGAN-SIFT. *Opt. Precis. Eng.* **2022**, *30*, 602–614. [CrossRef]
4. Dang, X.; Ke, W.; Hao, Z.; Jin, P.; Deng, H.; Sheng, Y. mm-TPG: Traffic policemen gesture recognition based on millimeter wave radar point cloud. *Sensors* **2023**, *23*, 6816. [CrossRef] [PubMed]
5. Ramanujam, E.; Perumal, T.; Padmavathi, S. Human activity recognition with smartphone and wearable sensors using deep learning techniques: A review. *IEEE Sens. J.* **2021**, *21*, 13029–13040. [CrossRef]
6. *IEEE 802.11n-2009*; IEEE Standard for Information Technology—Local and Metropolitan Area Networks—Specific Requirements—Part 11: Wireless LAN Medium Access Control (MAC) and Physical Layer (PHY) Specifications Amendment 5: Enhancements for Higher Throughput. IEEE: New York, NY, USA, 2009.
7. Tong, L.; Ma, H.; Lin, Q.; He, J.; Peng, L. A novel deep learning Bi-GRU-I model for real-time human activity recognition using inertial sensors. *IEEE Sens. J.* **2022**, *22*, 6164–6174. [CrossRef]
8. Pu, Q.; Gupta, S.; Gollakota, S.; Patel, S. Whole-home gesture recognition using wireless signals. In Proceedings of the 19th Annual International Conference on Mobile Computing & Networking—MobiCom'13, Miami, FL, USA, 30 September–4 October 2013; p. 27.
9. Niu, K.; Zhang, F.; Jiang, Y.; Xiong, J.; Lv, Q.; Zeng, Y.; Zhang, D. WiMorse: A contactless morse code text input system using ambient WiFi signals. *IEEE Internet Things J.* **2019**, *6*, 9993–10008. [CrossRef]
10. Hao, Z.; Duan, Y.; Dang, X.; Liu, Y.; Zhang, D. Wi-SL: Contactless fine-grained gesture recognition uses channel state information. *Sensors* **2020**, *20*, 4025. [CrossRef] [PubMed]
11. Guo, Z.; Xiao, F.; Sheng, B.; Fei, H.; Yu, S. WiReader: Adaptive air handwriting recognition based on commercial WiFi signal. *IEEE Internet Things J.* **2020**, *7*, 10483–10494. [CrossRef]
12. Abdelnasser, H.; Youssef, M.; Harras, K.A. Wigest: A ubiquitous WiFi-based gesture recognition system. In Proceedings of the 2015 IEEE Conference on Computer Communications (INFOCOM), Hong Kong, 26 April–1 May 2015; IEEE: Piscataway, NJ, USA, 2015; pp. 1472–1480.
13. Gu, Y.; Quan, L.; Ren, F. Wifi-assisted human activity recognition. In Proceedings of the 2014 IEEE Asia Pacific Conference on Wireless and Mobile, Bali, Indonesia, 28–30 August 2014; IEEE: Piscataway, NJ, USA, 2014; pp. 60–65.
14. Cui, W.; Li, B.; Zhang, L.; Chen, Z. Device-free single-user activity recognition using diversified deep ensemble learning. *Appl. Soft Comput.* **2021**, *102*, 107066. [CrossRef]
15. He, W.; Wu, K.; Zou, Y.; Ming, Z. WiG: WiFi-based gesture recognition system. In Proceedings of the 2015 24th International Conference on Computer Communication and Networks (ICCCN), Las Vegas, NV, USA, 3–6 August 2015; IEEE: Piscataway, NJ, USA, 2015; pp. 1–7.
16. Mohammed, A.Q.; Li, F. WiGeR: WiFi-based gesture recognition system. *ISPRS Int. J. Geo Inf.* **2016**, *5*, 92.
17. Virmani, A.; Shahzad, M. Position and orientation agnostic gesture recognition using WiFi. In Proceedings of the 15th Annual International Conference on Mobile Systems, Applications, and Services, Niagara Falls, NY, USA, 19–23 June 2017; pp. 252–264.
18. Wang, D.; Yang, J.; Cui, W.; Xie, L.; Sun, S. AirFi: Empowering WiFi-based passive human gesture recognition to unseen environment via domain generalization. *IEEE Trans. Mob. Comput.* **2022**, *23*, 1156–1168. [CrossRef]
19. Bulugu, I. Gesture recognition system based on cross-domain CSI extracted from Wi-Fi devices combined with the 3D CNN. *Signal Image Video Process.* **2023**, *17*, 3201–3209. [CrossRef]

20. Zou, H.; Zhou, Y.; Yang, J.; Jiang, H.; Spanos, C.J. WiFi-enabled device-free gesture recognition for smart home automation. In Proceedings of the 2018 IEEE 14th International Conference on Control and Automation (ICCA), Anchorage, AK, USA, 12–15 June 2018; IEEE: Piscataway, NJ, USA, 2018; pp. 476–481.
21. Kong, H.; Lu, L.; Yu, J.; Chen, Y.; Tang, F. Continuous authentication through finger gesture interaction for smart homes using WiFi. *IEEE Trans. Mob. Comput.* **2020**, *20*, 3148–3162. [CrossRef]
22. Yousefi, S.; Narui, H.; Dayal, S.; Ermon, S.; Valaee, S. A survey on behavior recognition using WiFi channel state information. *IEEE Commun. Mag.* **2017**, *55*, 98–104. [CrossRef]
23. Chen, Z.; Zhang, L.; Jiang, C.; Cao, Z.; Cui, W. WiFi CSI based passive human activity recognition using attention based BLSTM. *IEEE Trans. Mob. Comput.* **2018**, *18*, 2714–2724. [CrossRef]
24. Zhou, Z.; Liu, C.; Yu, X.; Yang, C.; Duan, P.; Cao, Y. Deep-WiID: WiFi-based contactless human identification via deep learning. In Proceedings of the 2019 IEEE SmartWorld, Ubiquitous Intelligence & Computing, Advanced & Trusted Computing, Scalable Computing & Communications, Cloud & Big Data Computing, Internet of People and Smart City Innovation (SmartWorld/SCALCOM/UIC/ATC/CBDCom/IOP/SCI), Leicester, UK, 19–23 August 2019; IEEE: Piscataway, NJ, USA, 2019; pp. 877–884.
25. Alhamdi, M.J.; Lopez-Guede, J.M.; AlQaryouti, J.; Rahebi, J.; Zulueta, E.; Fernandez-Gamiz, U. AI-based malware detection in IoT networks within smart cities: A survey. *Comput. Commun.* **2025**, *233*, 108055. [CrossRef]
26. Nozari, S.; Haleem, H.; Garibotto, C.; Sciarrone, A.; Bisio, I.; Lavagetto, F. CrowdWatch: Privacy-Preserving Monitoring Leveraging Wi-Fi Multiple Access Information. *IEEE Internet Things J.* **2025**, *12*, 15238–15248. [CrossRef]
27. Daniş, F.S. Live rssi filtering for indoor positioning with bluetooth low-energy. In Proceedings of the 2022 IEEE 12th International Conference on Indoor Positioning and Indoor Navigation (IPIN), Beijing, China, 5–7 September 2022; IEEE: Piscataway, NJ, USA, 2022; pp. 1–8.
28. Yang, J.; Chen, X.; Zou, H.; Lu, C.X.; Wang, D.; Sun, S.; Xie, L. SenseFi: A library and benchmark on deep-learning-empowered WiFi human sensing. *Patterns* **2023**, *4*, 100703. [CrossRef] [PubMed]
29. Wang, H.; Wang, Q.; Huang, L.; Shi, C.; Zhang, R. A PCA acceleration algorithm for WiFi sensing and its hardware implementation. In Proceedings of the 2024 IEEE International Symposium on Circuits and Systems (ISCAS), Singapore, 19–22 May 2024; IEEE: Piscataway, NJ, USA, 2024; pp. 1–5.

applied sciences

MDPI

Article

A Wearable Microphone Array Helmet for Automotive Applications

Daniel Pinardi [1], Andrea Toscani [2,*], Marco Binelli [1], Angelo Farina [1] and Jong-Suh Park [3]

[1] Department of Engineering for Industrial Systems and Technologies, University of Parma, 43124 Parma, Italy; daniel.pinardi@unipr.it (D.P.); marco.binelli@unipr.it (M.B.); angelo.farina@unipr.it (A.F.)

[2] Department of Engineering and Architecture, University of Parma, 43124 Parma, Italy

[3] Automotive R&D Division, Hyundai Motor Company, Namyang-eup, Hwaseong-si 18280, Gyeongi-do, Republic of Korea; jspark@hyundai.com

* Correspondence: andrea.toscani@unipr.it

Featured Application: Development of immersive infotainment and ANC systems in the automotive industry with Ambisonics audio.

Abstract: Growing interest in microphone array technology has been observed in the automotive industry and in this work, specifically, for Active Noise Control (ANC) systems. However, the human presence always limits the usage of microphone arrays in driving conditions at the driver's seat. This is often the most important position of the car cabin; a wearable microphone array is particularly interesting. In this paper, a wearable helmet microphone array is presented featuring 32 microphones arranged over the surface of a helmet, which also integrates a specially designed Analog-to-Digital (A/D) converter, delivering digital signals over the Automotive Audio Bus (A^2B). Digital signals are collected using a control unit located in the passenger compartment. The control unit can either deliver digital signals to a personal computer or analog signals to an external acquisition system, by means of Digital-to-Analog (D/A) converters. A prototype was built and acoustically characterized to calculate the beamforming filter matrix required to convert the recordings (pressure signals) into Ambisonics signals (a spatial audio format). The proposed solution was compared to the reference spherical microphone array of the last decade, demonstrating better performance in sound source localization at low frequencies, where ANC systems are mostly effective.

Keywords: active noise control (ANC); Ambisonics; Automotive Audio Bus (A^2B); beamforming; immersive audio; microphone array; spatial audio; wearable helmet

1. Introduction

Microphone arrays are more and more employed for several purposes, such as electronic devices (e.g., dereverberation, speech enhancement and recognition), industries (e.g., machine condition monitoring), and even civil applications (e.g., structural health monitoring). Among these applications, the automotive industry particularly benefited from this technology, which is now used in a variety of problems. In [1], a uniform circular microphone array featuring seven capsules (of which one in the center) was employed to localize the speaker in autonomous shared vehicles. In [2,3], a spherical array with 32 capsules was used to auralize the car sound system of a large sedan. The same array was also employed in [4] to spatially evaluate the sound quality inside a car cabin, while in [5] the authors did a similar work with the Bruel and Kjaer WA-1565, a rigid spherical microphone array of 195 mm diameter, equipped with 36 capsules and 12 cameras. In [6], a uniform planar Micro electro-mechanical system (MEMS) array featuring 150 capsules was used to detect pedestrians in an autonomous emergency braking system. Finally, K. Vasudev et al. attempted to enhance hands-free communication by integrating a microphone array into the seat belt [7]. The usage of microphone arrays for hands-free communication inside cars is widely studied and offers a large volume of the literature, among which three relevant references worth to be cited are [8–10].

Active Noise Control (ANC) is another field where microphone arrays find relevant practical applications in the automotive industry. In [11], linear and circular arrays with 4, 8, 16, or 32 capsules were employed to enhance noise cancelling inside a car cabin by means of the virtual sensing technique. A comprehensive review of virtual sensing algorithms for ANC can be found in [12]. In [13], a head-shaped array with 32 capsules and 8 cameras was used to spatially assess the effectiveness of an ANC system installed in a segment-D car. In [14], two linear arrays of 6 microphones were each used to enhance the performance of an ANC system by means of the noise source separation technique. From the previous examples, it is evident that geometry plays a crucial role when designing a microphone array. In fact, it determines the portion of space where the system exhibits optimal accuracy, such as a line, a plane, a half-space, or the full sphere. As a result, microphone arrays with a variety of shapes were observed. In [15], S. J. Patel et al. discussed the optimal design for linear arrays, while in [16] X. Wang et al. did the same for planar arrays and in [17] J. Trevino et al. for the cylindrical ones. Nevertheless, most of the microphone arrays are spherical, namely the spherical microphone array (SMA), constituted by a rigid shell; however, in some applications, open frame SMA are also employed. An in-depth analysis of the design of open frame SMA is discussed in [18,19], while four of the most relevant papers discussing design, analysis, and processing of rigid SMA are [20–23].

The dimension of the array is another key factor for the design, since the larger the array, the lower the minimum frequency where beamforming is effective [24]. Therefore, large arrays, either real or virtual [25], are required to localize noise sources at low frequency. This is particularly important when dealing with ANC systems for cars, which are mostly effective at very low frequencies, e.g., 50 Hz–400 Hz for road noise as referred in [26] and up to 1 kHz for tonal applications such as engine noise order cancelling, as referred in [27].

Another fundamental feature for microphone array design is the number of capsules, as the closer the microphones, the higher the maximum beamforming frequency, as demonstrated in [28] by B. Rafaely. Consequently, a trend toward the continuous appearance on the market of systems with an ever-increasing number of capsules was observed. The first microphone array was launched in 1978 by K. Farrar with the name "Soundfield" [29,30]. It featured four capsules positioned in the vertices of a tetrahedron; several other microphone arrays featuring four capsules have come to the market in the following decades, the last of them being the Saramonic Sr-VR in 2024. In 2009, Eingemike32 (also employed in this work) was launched by G. Elko [31], i.e., a SMA featuring 32 capsules; in 2017, the Octomic by Core Sound, featuring 8 capsules and the Zylia, another SMA featuring 19 MEMS capsules, were launched; and finally in 2023, the new Eingemike64, featuring 64 capsules was launched. In 2015, a prototype featuring 252 capsules was presented by Sakamoto et al. [32], while in 2021, Carsten et al. showed a portable array with 512 MEMS microphones [33].

The layout of the capsules over the surface of the array is at least as important. A heuristic approach was used for planar arrays in [34], while several mathematical approaches were developed for SMA, with the aim of optimizing the spatial sampling over the sphere, as well treated in [35]. In the case of equiangular sampling, the number N of required microphones is given by (1), while in the case of nearly uniform sampling, it is given by (2)

$$N = 4(O+1)^2 \tag{1}$$

$$N = 1/2(O+1)^2 \tag{2}$$

where O is the Ambisonics order. The Ambisonics theory, consisting of a Spherical Harmonics (SHs) expansion of the sound field at the recording point, was first presented by M. Gerzon in 1975 [36,37] and nowadays is widely employed for delivering spatial audio. More on the SH can be found in [38], while an exhaustive discussion of SH decomposition with SMA can be found in [39]. In the case of nearly uniform sampling, a mathematical approach called Spherical Design is commonly employed to optimize the distribution of the capsules over the surface of a rigid SMA. A spherical design is a finite set of N points on the d-dimensional unit sphere S_d such that the average value of any polynomial f of degree t or less equals the average value of f on the whole sphere [40–42]. Such a set is often also called the Spherical T-design. In [43,44], many spherical designs and their properties are presented. Other interesting distributions of points are based on regular polyhedrons, such as tetrahedron, dodecahedron, and icosahedron, with N = 4, 12, and 20, respectively. Eigenmike32, widely recognized as the reference system of the last decade and employed as a comparative target also in the presented work, features 32 capsules arranged in a truncated icosahedron over a rigid sphere of 84 mm diameter.

When selecting the type of capsule, the choice is between analog and digital transducers. Analog transducers can provide high-quality audio signals, but they come with some drawbacks, such as cumbersome wiring and susceptibility to noise interference, especially when long cables are used to connect the capsules to the Analog-to-Digital (A/D) converters. Affordable digital MEMS microphones are less affected by electrical interference, though this often comes at the expense of lower acoustic performance, such as reduced dynamic range and signal-to-noise ratio. A review of MEMS microphones can be found in [45].

In this work, a wearable microphone array is presented for the development and performance assessment of ANC systems installed on vehicles. The array solves a common limitation encountered in measuring automotive ANC systems with microphone arrays: the presence of the driver in the front left seat during road tests. The array was built by using a normal rigid helmet as a frame. The dimension of the helmet, which is larger than the microphone arrays currently available on the market, allows for shifting the frequency range of beamforming toward low frequencies, where ANC systems are mostly effective. With the shape of the helmet being almost spherical, the nearly uniform sampling theory was applied to optimize the distribution of the capsules over the surface of the array. A Spherical Design of degree t = 7 with N = 32 points was employed. The proposed solution features electret capsules connected to a miniaturized A/D converter incorporated in the helmet. The A/D converter also integrates an A^2B transceiver (see Section 2.3), allowing it to deliver digital signals over an Ethernet cable, thus ensuring immunity to electromagnetic disturbances, which are always present in a car cabin.

The proposed solution is flexible and can be used in many scenarios, such as internal combustion engines or electric vehicles, different road conditions (smooth, rough, very rough asphalt, with/without bumps are common tests performed by car makers), and speed. The only limitation is encountered with cabriolet and spider cars due to the aerodynamic noise and the unavailability of a windshield for the array helmet. If equipped with internal loudspeakers, the array helmet would find another interesting application in motor sport for the communication between the pilot and the co-pilot in rally competitions. Thanks to the beamforming capability, the array helmet can focus on the speaker voice, rejecting the noise and enhancing the intelligibility of the instructions provided by the co-pilot. This application is known as Audio Augmented Reality (AAR) or augmented listening. Despite no previous works being found in motor sports applications, in [46], several wearable microphone arrays, both rigid and flexible, were built and tested for AAR.

The paper is arranged as follows: Section 2 describes the development of the helmet (Section 2.1), the acoustic characterization (Section 2.2), and the architecture of the electronics (Section 2.3); Section 3 illustrates the main findings obtained within this work; in Section 4 the results are discussed; and finally, Section 5 summarizes the conclusions, the limitations, and possible future developments.

2. Materials and Methods

2.1. Array Helmet Design

The first step of the development consisted of the choice of the helmet. It must be as spherical as possible, with a smooth surface without holes, grids, vents, knobs, and protruding parts to avoid any practical limitation for capsule mounting. This led to the choice of the model X3000 by the Italian manufacturer AGV (Vicenza, Italy).

The microphonic capsule chosen to build the array is AOM-5024L-HD-R from PUI Audio (Fairborn, OH, USA), whose technical specifications are reported in Table 1. It has a very low self-noise of just 14 dB(A) and more than 90 dB(A) of dynamic range. In Figure 1 (provided by the manufacturer), one can note that the frequency response is almost flat in the interested frequency range, which is 20 Hz–1 kHz.

Table 1. Microphonic capsule specification (AOM-5024L-HD-R).

Characteristic	Value
Diameter	9.7 mm
Height	5 mm
Type	Analogue, electret
Directivity	Omnidirectional
Freq. range	20 Hz–10 kHz
Self-noise	14 dB (A)
Max SPL	110 dB

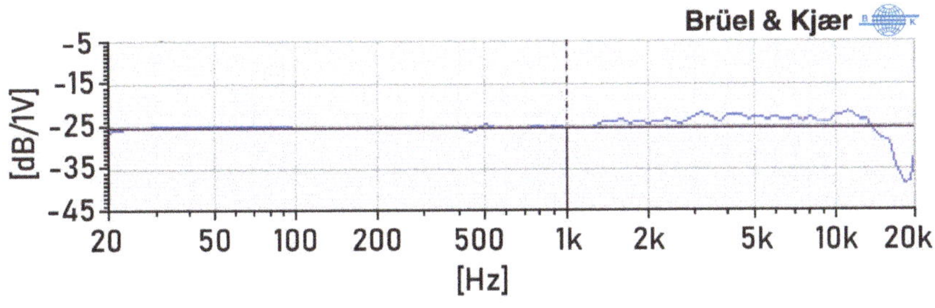

Figure 1. Frequency response of the PUI Audio AOM-5024L-HD-R capsule.

A 3D model of the helmet was acquired with the laser scanning technique by employing a Shining 3D EinScan H2. The result was stored as a stereolithography file, namely stl (also referred to as standard triangle language or standard tessellation language). In this way, it was possible to easily import and manipulate the geometry in MATLAB (R2024b). To define the optimal distribution of the capsules, the nearly uniform sampling was adopted. First, the geometrical center of the model was calculated, then 32 straight lines were propagated from the center in the same directions of a spherical design of degree $t = 7$, with N = 32 points. This configuration, as well as many others up to $t = 21$ and $N = 240$, can be found in [47]. The bundle of lines was rotated to minimize the intersections with the areas where it was not possible to mount any capsules, namely the visor and the hole for the neck. The minimum number of unallowed points was six: two on the visor and four on the neck. These points were manually repositioned. Finally, 10 mm diameters holes were drilled by using a computer numerical control (CNC) machine and the capsules were fixed in positions by using a special adhesive modeling paste.

2.2. Acoustics Characterization of the Array Helmet

Once assembled, the helmet was characterized in the acoustics laboratory at the University of Parma, Parma (Italy), with the dimensions 5 × 3 × 8 m (W × H × L). The measurement procedure made use of a two-axis turntable and a loudspeaker (studio monitor type 8351a, by Genelec, Iisalmi, Finland) in a fixed position (Figure 2). Since the helmet must be used in a small environment, the measurement was performed at a very short distance, d = 0.5 m, to consider the near-field effect provided by the curvature of the sound waves. In fact, the commonly employed analytical beamformers rely on the plane wave hypothesis, which is only true in a far field condition, as demonstrated in [48]. This assumption is not true inside a car cockpit, due to its limited size compared to the wavelengths interested in an ANC system. The near-field beamforming [49,50] allows us to overcome the far field limitation of analytical beamformers. The test signal employed for the measurement was an Exponential Sine Sweep (ESS) [51], pre-equalized for flatting the spectrum and to reduce the time domain impulse response (IR).

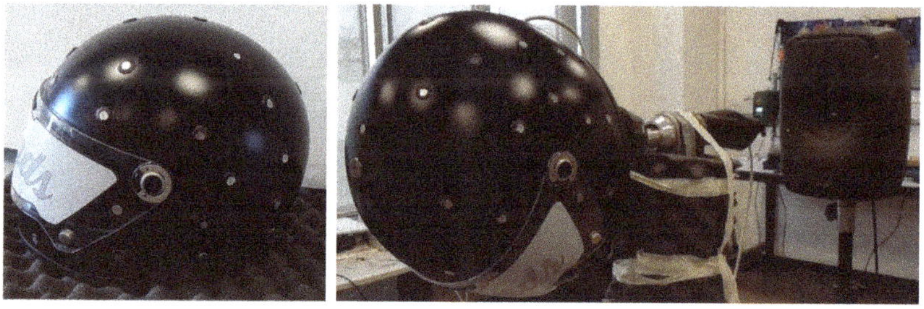

Figure 2. Side view of the array helmet after assembling the capsules (**left**). A view of the array helmet mounted on the two-axis turntable while performing the acoustic characterization (**right**).

Since the loudspeaker was positioned in the center of the laboratory, the time of arrival t_1 of the first reflections (floor and ceilings) can be easily calculated as

$$t_1 = \frac{\sqrt{H^2 + d^2}}{c} \tag{3}$$

where H = 3 m is the height of the room, d = 0.5 m is the distance between the microphone array and the loudspeaker, and c = 343 m/s is the sound celerity in air at 20 °C. Note that the numerator of (3) is the total travel path of the first reflections, which is about 3.04 m, and hence $t_1 \cong 8.9$ ms or 427 samples at the standard audio sampling frequency f_s = 48 kHz. As can be seen in Figure 3, the effective number of samples between the direct sound (first peak of the IR, occurring after 70 samples) and the first reflections (second peak of the IR, occurring after 491 samples) corresponds almost exactly to the analytically calculated time of flight. Therefore, the entire direct sound can be isolated before the first reflections occur, and this technique is referred as a "pseudo-anechoic chamber" [52]. It can also be seen that the delay of the first peak in samples, s_d, perfectly matches the source–microphone distance, as follows:

$$s_d = \frac{d}{f_s} \cdot c \quad (4)$$

with $d = 0.5$ m, $f_s = 48$ kHz, and $c = 343$ m/s, it results in $s_d = 70$ samples, as expected.

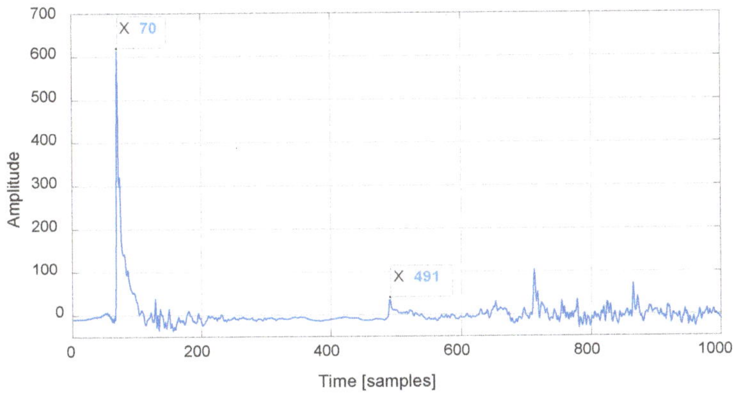

Figure 3. Pre-equalized loudspeaker IR.

In Figure 4, the spectrum of the IR direct sound is shown in the frequency range of interest, that is 20 Hz–1 kHz. Taking the lowest IR level in this range as an acceptance threshold, i.e., 85.6 dB, we can assess that the measurement can be used above 40 Hz (black dot in Figure 4). However, this can be further improved by measuring the array in a true anechoic chamber, where the IR can be kept longer thanks to the absence of reflections. In that case, without the pre-equalization filter, the lower limit would be given by the frequency response of the loudspeaker (32 Hz according to manufacturer's datasheet).

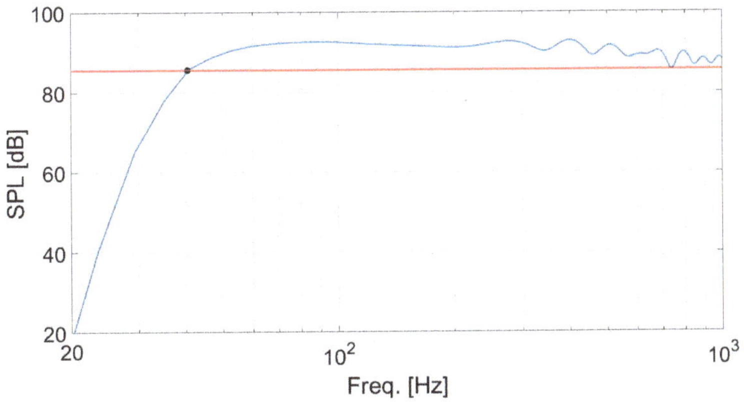

Figure 4. Spectrum of the direct sound of the pre-equalized IR.

At each repetition of the measurement, the helmet is automatically rotated by the turntable in a specific direction, and the test signal is recorded by the N capsules of the array. This procedure also addresses the problem of spatial sampling of a sphere, since the choice of the measurement directions has a strong impact on the quality of the beamforming filters. This problem was already treated by one of the same authors in [53], where it is shown that a nearly uniform sampling is again a wise choice. A set of $D = 240$ directions was used to characterize the acoustic response of the array helmet, corresponding to a spherical design of degree $t = 21$.

The measurement procedure provides the matrix C of Equation (5), which consists of the regularized Kirkeby inversion [54], a linear processing employed to compute a beamforming matrix of Finite Impulse Response (FIR) filters. In the frequency domain, it is defined as

$$H = [C^{*} \cdot C + \beta \cdot I]^{-1} \cdot \left[C^{*} \cdot A \cdot e^{-j\pi k}\right] \tag{5}$$

where C is the frequency response of the array for the N capsules, D directions, and K frequencies, * is the Hermitian transpose, $\cdot$ is the scalar product, β is a parameter that varies with frequency for regularized inversion [55], I is the identity matrix, $^{-1}$ is the pseudo-inverse, $e^{-j\pi k}$ introduces a delay to guarantee the causality of the filters, and k is the frequency. Matrix A represents the desired directivity patterns of the virtual microphones encoded by the beamformer. It does not depend on the frequency and must be defined by the user for each measured direction of the matrix C and for all the N capsules. In our case, A must define the first 16 Spherical Harmonics (SHs) employed by the Ambisonics format at order 3. The 3D polar patterns of these SH are shown in Figure 5, while (6) provides the relation between Ambisonics order O and the number V of SH, as follows:

$$V = (O + 1)^2 \tag{6}$$

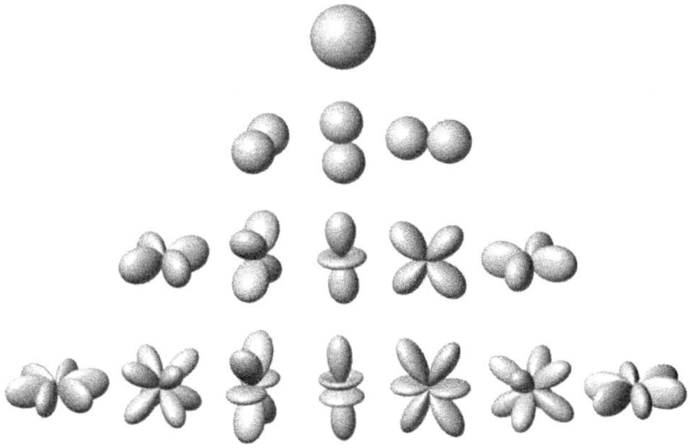

Figure 5. Ideal 3D directivity of the first $v = 16$ Spherical Harmonics (Ambisonics 3rd order).

The virtual microphones, i.e., SH, are then obtained by means of convolution, which is the multiplication in frequency domain, as follows:

$$A' = C \cdot H \tag{7}$$

Note that only in the ideal case of perfect reconstruction of the SH will $A' = A$ result at all frequencies, while in the real case, A' will always approximate A.

2.3. Electronics Design

The system, whose scheme is depicted in Figure 6, is composed of a helmet and a digital-to-analog (D/A) box. The helmet houses microphones and the electronic board to collect the signals, while the D/A box is connected either to a personal computer (PC) or an external acquisition system. When a PC is employed, the entire system (after A/D conversion) is full-digital, while a double A/D and D/A conversion is introduced when an external recording system is employed. However, this solution guarantees a robust signal-to-noise (S/N) ratio even inside a car cockpit, which is always electrically noisy.

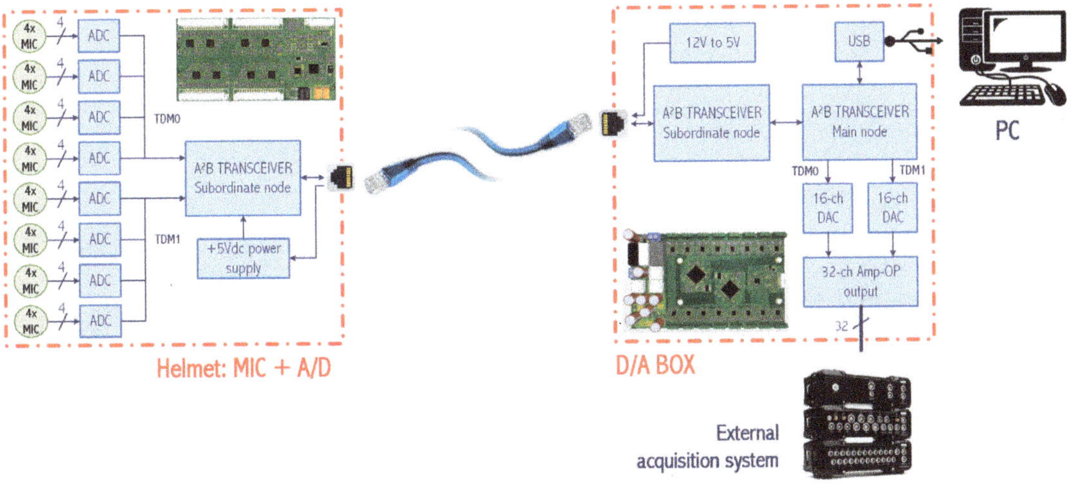

Figure 6. Block scheme of entire system: A/D board installed inside the helmet (**left**), D/A box (**right**).

The system was designed to use the A^{2}B bus by Analog Devices [56], which can handle up to 32 channels on a single Unshielded Twisted Pair (UTP), up to 15 m distance. An A^{2}B network consists of a main node and up to 10 subordinate nodes arranged in daisy-chain. Dedicated transceivers handle bus access and data flow, eliminating the need for extra devices that would increase the cost and complexity of the system. An A^{2}B network provides a bandwidth of 50 Mbit/s, enabling the transmission of up to 32 channels at the standard sample rate of 48 kHz with 16-, 24-, or 32-bits resolutions. This digital bus is particularly suitable for transmitting audio signals in both microphone arrays [57] and loudspeaker arrays [58], since it guarantees a deterministic latency of just two samples at 48 kHz and the synchronization between all the channels [59].

The electronic board installed in the helmet is provided with eight 4-channels, 16 bits, audio A/D converters, and type TLV320ADC6140 by Burr-Brown, and it also integrates the A^{2}B transceiver to acquire and deliver the data via A^{2}B bus. The A/D converters (specification are reported in Table 2) are connected to the transceiver via two Time-Division Multiplexing (TDM) 4 digital lines (TDM0 and TDM1). The signals are transferred from the helmet to the D/A box unit over an Ethernet cable that also carries the 5 Vdc power supply for the board installed inside the helmet. The ethernet cable was chosen as the physical layer of the A^{2}B bus, being cheap and common, with robust connectors that are easy to insert and remove. Note that a standard Ethernet cable includes four UTP cables, of which one is used for the A^{2}B bus and one for the power supply.

The D/A box receives the digital signals from the A^{2}B bus and transmits them to a PC via Universal Serial Bus (USB) or to an external acquisition system, after the D/A conversion. This feature is of particular interest for Noise, Vibration, and Harshness (NVH) applications since standard automotive equipment, like the SCADA by Siemens, are usually required by car makers. In addition, when such systems are employed, additional sensors like accelerometers or microphones can be recorded synchronously with the helmet array, e.g., opening to the implementation of more advanced ANC algorithms. The D/A box is made of two boards: an A^{2}B main board and an A^{2}B subordinate board. The A^{2}B main board generates the A^{2}B network and provides the clock to the A^{2}B subordinate board and the USB connection to deliver digital data to a PC. The A^{2}B subordinate board features two 16-channels, 16 bits, D/A converters, type ADAU1966 by Analog Devices (specifications are resumed in Table 2), and it is provided with 32 BNC connectors to connect to external acquisition systems. In addition, the D/A box produces all the supply voltages for the system, starting from the 12 V on-board battery of the vehicle. To isolate the system from disturbances of the vehicle's electrical system, usually noise, all the regulators are designed to ensure full galvanic insulation.

Table 2. A/D (TLV320ADC6140) and D/A (ADAU1966) converter specification.

A/D Characteristics	Value
Number and type of input channels	4-channels, analog
Dynamic range	123 dB
Total harmonic distortion + noise (THD + N)	−98 dB
Input voltage range	2 Vrms
Sample rate	8 to 768 kHz
Power consumption	9.2 mW/ch @ 48 kHz
Audio serial data interface	TDM or I^2S
D/A Characteristics	**Value**
Number of output channels	16-channels, single ended
Resolution	24 bits
Dynamic range	110 dB (A)
Total harmonic distortion + noise (THD + N)	−97 dB
Sample rate	32 to 192 kHz

3. Results

At first, the beamforming capability of the proposed array helmet was compared to the Eigenmike32. The comparison was made by using a well-established metric for microphone array assessment [60], which consists of evaluating the deviation of the directivity A' with respect to the ideal directivity defined by A. This can be performed by relying on two parameters, the Spatial Correlation (SC), which sets the limit of the beamforming at high frequency and is the defined in (8), and the Level Difference (LD), which sets the limit of the beamforming at low frequency and is defined in (9) as follows:

$$\mathrm{SC} = \frac{\sum_d (A')^T \cdot A}{\sum_d \left(\sqrt{(A')^T \cdot A'} \cdot \sqrt{A^T \cdot A} \right)} \tag{8}$$

$$\mathrm{LD} = \frac{1}{D} \sum_d \frac{A^2}{A' \cdot (A')^*} \cdot [\mathrm{dB}] \tag{9}$$

The parameters are calculated for each frequency, each direction, and each virtual microphone; then, the D directions are summed, while the virtual microphones are averaged among those belonging to the same Ambisonics order. Hence, from 1 to 4 in the first order, from 5 to 9 in the second order, and from 10 to 16 in the third order. In the ideal case of perfect reconstruction of the SH, it will be SC = 1 and LD = 0 dB. However, a certain amount of deviation is always present in the real case; hence, the upper and lower frequency limits of the beamforming are usually defined by considering two thresholds: SC > 0.9 and LD > −1 dB. The results are shown in Figure 7 for Eigenmike32 and in Figure 8 for the array helmet.

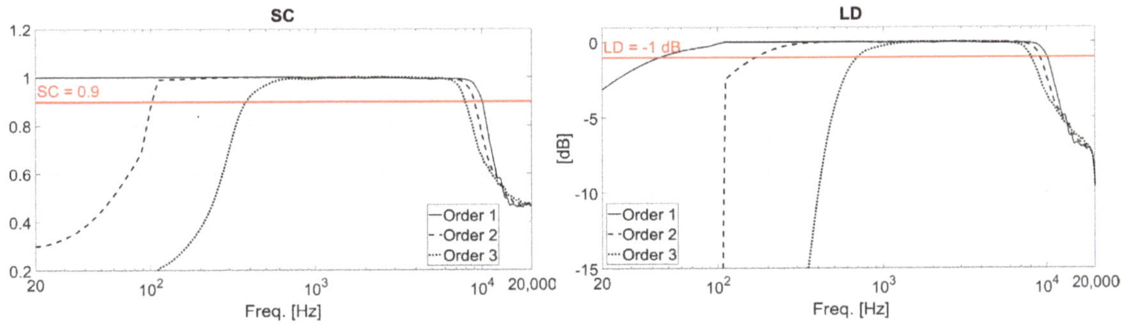

Figure 7. SC (**left**) and LD (**right**) calculated for the Eigenmike32.

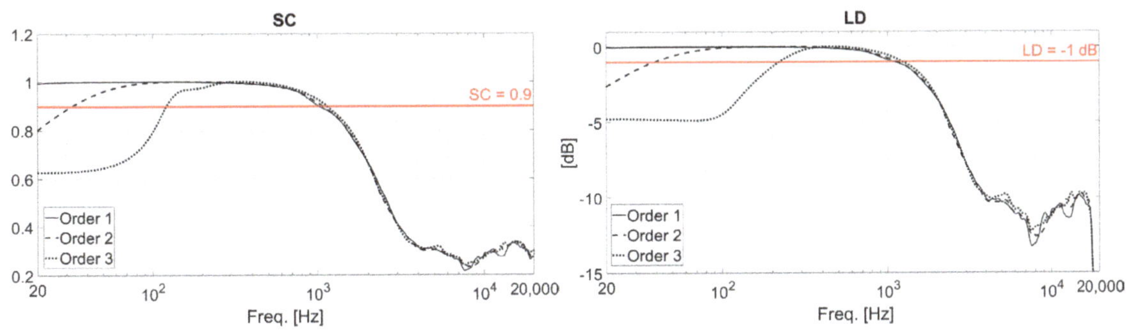

Figure 8. SC (**left**) and LD (**right**) calculated for the array helmet.

Then, a localization test was performed in the acoustic laboratory at the University of Parma, Parma (Italy), using the sound color mapping technique [61,62]. The array helmet was mounted on a microphone stand, in front of the Genelec studio monitor 8351a playing pink noise. A 30 s signal was recorded and converted into Ambisonics third order. Then, the array helmet was replaced with a dual lenses camera to take a 360° picture of the environment, which is used as the background of the sound color maps. The analysis was performed in the octave bands centered at 31.5 Hz, 63 Hz, 125 Hz, 250 Hz, 500 Hz, and 1 kHz. The quantity being mapped is the Sound Pressure Level (SPL), calculated with a resolution of 1 degree. The pseudo-color maps are generated through graphical interpolation, with a color scale that goes from blue (lowest SPL value) to red (highest SPL value). The results are shown in Figure 9.

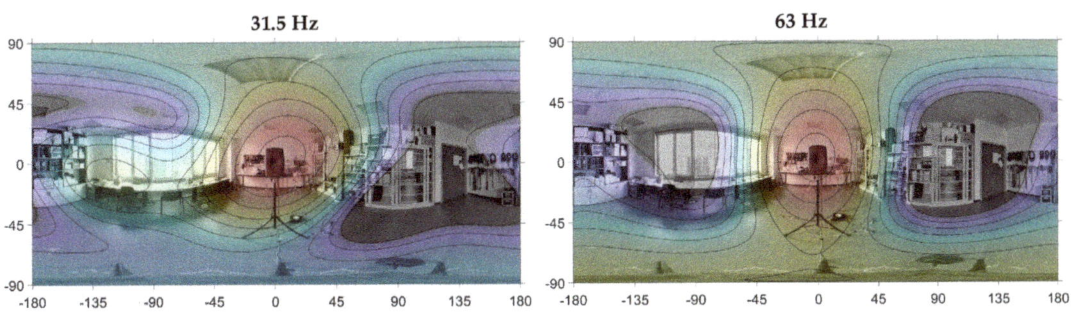

Figure 9. *Cont.*

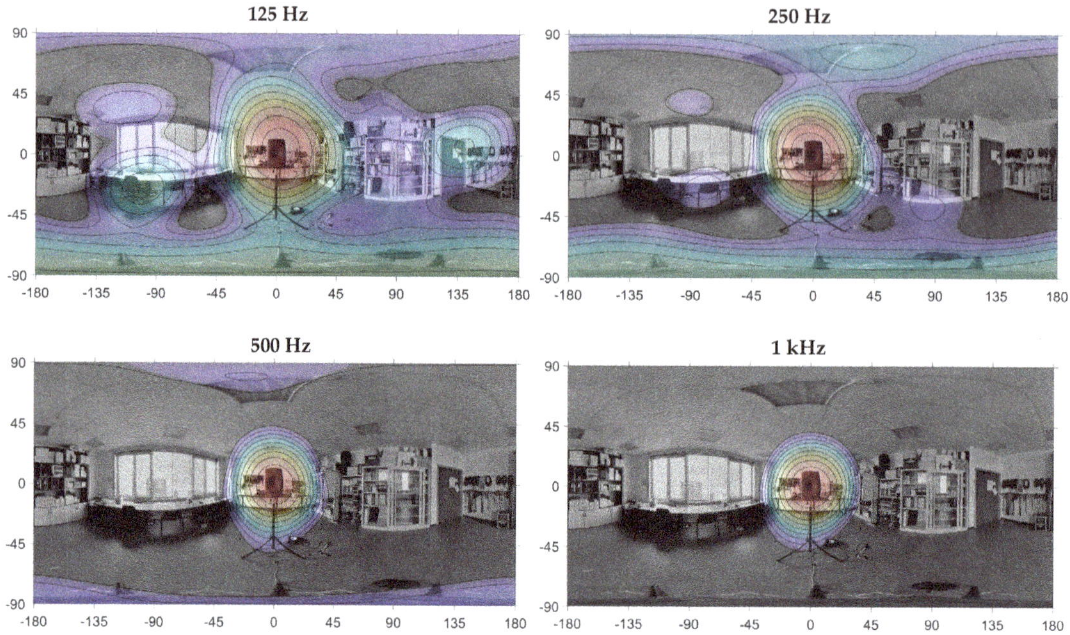

Figure 9. Sound color map analysis of a pink noise source located in (0°; 0°) at various octave bands.

4. Discussion

The SC and LD charts of Figures 7 and 8 can be used to precisely define the frequency limits of beamforming at each Ambisonics order. This is obtained by applying the previously stated thresholds, SC > 0.9 and LD > −1 dB. Each of them intersects the chart for each Ambisonics order in two points, thus identifying a low and a high acceptable frequency. These results are shown in Table 3, where it can be seen that the LD is always more restrictive at low frequency, while the SC is at high frequency. Therefore, the frequency limits provided by the two metrics must be combined, taking the most restrictive condition at low and at high frequency for each Ambisonics order (Table 4).

Table 3. Acceptable low and high frequency limits at various Ambisonics orders for SC and LD metrics.

	Eigenmike32				Array Helmet			
	SC		LD		SC		LD	
Ambisonics Order	Low Freq. [Hz]	High Freq. [Hz]	Low Freq. [Hz]	High Freq. [Hz]	Low Freq. [Hz]	High Freq. [Hz]	Low Freq. [Hz]	High Freq. [Hz]
1	-	10,000	45	10,100	-	1050	-	1100
2	100	9000	170	9100	35	1050	40	1100
3	380	7950	700	8050	120	1050	220	1100

The maximum beamforming frequency for the array helmet is 1 kHz at all orders. Despite being lower than the Eigenmike32 (10 kHz at 1st order, 9 kHz at 2nd and 8 kHz at 3rd), it is still acceptable for ANC applications. As referred to in [28] by R. Boaz, the maximum theoretical beamforming frequency, f_{max}, can be calculated as

$$f_{max} = \frac{1}{2} \cdot \frac{c}{d_{min}} \tag{10}$$

where $c = 343$ m/s is the celerity of the sound wave and d_{min} is the minimum distance between the capsules of the array. The results shown in Table 5 are obtained by replacing the value of d_{min} in (10) for the Eigenmike32 and for the array helmet.

Table 4. Combined frequency limits at various Ambisonics orders.

	Eigenmike32		Array Helmet	
Ambisonics Order	**Min. Freq. [Hz]**	**Max. Freq. [Hz]**	**Min. Freq. [Hz]**	**Max. Freq. [Hz]**
1	45	10,000	20 *	1050
2	170	9000	40	1050
3	700	7950	220	1050

* Theoretical value. The current practical limit is 40 Hz due to the pseudo-anechoic measurement; further improvement can be made with a true anechoic measurement.

Table 5. Theoretical maximum beamforming frequencies for Eigenmkie32 and the array helmet.

	d_{min} [mm]	Theoretical f_{max} [kHz]
Eigenmike32	17	10
Array helmet	170	1

The main advantage offered by the proposed solution consists of a significant shift in the beamforming capability toward low frequencies, as follows:

- First order Ambisonics at 20 Hz (array helmet) instead of 45 Hz (Eingemike32);
- Second order Ambisonics at 40 Hz (array helmet) instead of 170 Hz (Eingemike32);
- Third order Ambisonics at 220 Hz (array helmet) instead of 700 Hz (Eingemike32).

Usually, the comparison between microphone arrays is also performed in terms of octave bands. As can be seen in Table 6, the array helmet allowed gaining one Ambisonics order at the octave bands from one to five. The two solutions have the same performance at the sixth octave band (centered at 1 kHz), while only the Eigenmike32 operates at the octave bands centered at 2, 4, and 8 kHz. A particularly useful improvement for ANC applications is obtained at the first octave band (centered at 31.5 Hz), at which the Eigenmike32 does not provide any beamforming, while the array helmet offers a first order localization.

Table 6. Ambisonics orders at different octave bands.

Octave Bands Center Frequency [Hz]	Highest Available Ambisonics Order	
	Eigenmike32	**Array Helmet**
31.5	-	1
63	1	2
125	1	2
250	2	3
500	2	3
1000	3	3
2000	3	-
4000	3	-
8000	3	-
16,000	-	-

The second laboratory experiment, related to the localization of a noise source through sound color mapping, is now analyzed. The acoustic center of noise source was positioned coincident with the optical center of the panoramic camera. Therefore, in polar coordinates (azimuth, a, and elevation, e) the acoustic center is in $a = 0°\ ; e = 0°$. Table 7 shows the estimated position of the direction of arrival (DoA) of the direct sound, in polar coordinates. The DoA of the direct sound was estimated in correspondence of the maximum SPL, which is the center of the red spots in the color maps of Figure 9.

Table 7. Polar coordinates of the DoA of the direct sound at different octave bands for the array helmet.

Octave Bands Center Frequency [Hz]	Estimated DoA of Direct Sound Located at (0°; 0°)	
	Azimuth [°]	Elevation [°]
31.5	0	1
63	−2	3
125	3	4
250	−1	1
500	1	−1
1000	0	−1

As can be seen, an average absolute error of 1.2° is made along the azimuth and 1.8° along the elevation. The array helmet correctly localized the sound source (Genelec studio monitor) even at the first octave band, where only the first order Ambisonics is available. At 125 Hz, when the second order Ambisonics becomes available, some reflections can also be seen, e.g., the door (135°; 0°) and the desk (−100°; −30°). Above 250 Hz, the third order Ambisonics provides its own contribution too, and the spot on the noise source becomes narrower as the frequency increases.

5. Conclusions

A wearable helmet microphone array featuring 32 electret capsules has been developed, built, and characterized. It integrates a miniaturized 32-channel A/D converter, which avoids the bulky wiring that would occur with an analog solution and ensures a high S/N ratio thanks to the usage of the A^2B digital bus. In addition, this solution allows only one Ethernet cable to come out of the helmet, maintaining good comfort and ease of use. The A^2B signal is received by an external D/A box, which can deliver digital data to the PC via USB or convert the data back to the analog domain by means of two D/A converters, allowing us to record them with an external acquisition unit.

The array helmet was compared with the Eigenmike32, a spherical microphone array that widely considered the reference equipment for spatial audio recording during the last decade. By analyzing two metrics for Ambisonics performance evaluation, namely Spatial Correlation and Level Difference, it was possible to assess that the proposed system shifted toward low frequencies, with the Ambisonics orders by one or even two octaves, making it accurate even at the lowest octave band, centered at 31.5 Hz, where the Eigenmike32 is not effective at all. These results have been proved with a laboratory test consisting of a noise source localization problem, making use of the sound color mapping technique. The array helmet was demonstrated to correctly localize the noise source at all frequencies, with a trend of increasing accuracy with frequency, as expected. The highest valid frequency for beamforming has been reduced to 1 kHz, due to the large size of the helmet compared to the number of capsules. However, such a frequency is still above the maximum frequency at which Active Noise Control systems for cars are effective. In conclusion, the proposed solution can be employed for the assessment and the development of road noise cancelling or engine order cancelling systems at the driver seat in driving condition.

The presented array has two limitations. Most of the rolling noise from the wheels comes from below; therefore, the highest microphone density would be desired in the lower part of the helmet to increase the spatial resolution in the interested portion of space. However, this is not possible due to the presence of the neck. The second limitation consists in the movement of the driver's head, which is naturally unstable when driving a car, in particular while cornering. Therefore, when the recording is played back, a mismatch between the orientation of the listener's head and the orientation of the sound field may occur. However, this problem will be overcome in a future update of the helmet by installing a head-tracking system to record the quaternion of spatial rotation synchronously with the pressure signals. Such a solution will allow compensating the head's movements in post-processing, by counter-rotating the sound field after Ambisonics conversion.

6. Patents

A patent request was submitted for the presented work, application number US 18/519,990.

Author Contributions: Conceptualization, A.F. and J.-S.P.; methodology, A.F.; software, D.P.; validation, D.P. and A.T.; formal analysis, A.F.; investigation, A.T. and D.P.; resources, J.-S.P.; data curation, D.P.; writing—original draft preparation, D.P. and A.T.; writing—review and editing, A.F., A.T., M.B. and J.-S.P.; visualization, D.P.; supervision, A.F.; project administration, A.F. and J.-S.P.; funding acquisition, A.F. All authors have read and agreed to the published version of the manuscript.

Funding: This research received no external funding.

Institutional Review Board Statement: Not applicable.

Informed Consent Statement: Not applicable.

Data Availability Statement: The datasets presented in this article are not readily available because they are subjected to industrial secrecy. Requests to access the datasets should be directed to jspark@hyundai.com.

Acknowledgments: The authors are grateful to Hyundai Motor Company for the economic support that made this research possible.

Conflicts of Interest: Author Jong-Suh Park was employed by the company Hyundai Motor Company. The remaining authors declare that the research was conducted in the absence of any commercial or financial relationships that could be construed as a potential conflict of interest.

Abbreviations

The following abbreviations are used in this manuscript:

A^2B	Automotive Audio Bus
A/D	Analog-to-Digital
AAR	Audio Augmented Reality
ANC	Active Noise Control
CNC	Computer Numerical Control
D/A	Digital-to-Analog
DC	Direct Current
DoA	Direction of Arrival
DSP	Digital Signal Processing
ESS	Exponential Sine Sweep
IR	Impulse Response
FIR	Finite Impulse Response
LD	Level Difference
MEMS	Micro Electro-mechanical Systems
NVH	Noise, Vibration and Harshness
PC	Personal Computer

PWD	Plane Wave Decomposition
S/N	Signal-to-Noise
SC	Spatial Correlation
SH	Spherical Harmonics
SMA	Spherical Microphone Array
SPL	Sound Pressure Level
TDM	Time-division multiplexing
USB	Universal Serial Bus
UTP	Unshielded Twisted Pair

References

1. Marques, I.; Sousa, J.; Sá, B.; Costa, D.; Sousa, P.; Pereira, S.; Santos, A.; Lima, C.; Hammerschmidt, N.; Pinto, S.; et al. Microphone Array for Speaker Localization and Identification in Shared Autonomous Vehicles. *Electronics* **2022**, *11*, 766. [CrossRef]
2. Pinardi, D.; Farina, A.; Park, J.-S. Low Frequency Simulations for Ambisonics Auralization of a Car Sound System. In Proceedings of the 2021 Immersive and 3D Audio: From Architecture to Automotive, I3DA 2021, Bologna, Italy, , 8–10 September 2021. [CrossRef]
3. Pinardi, D.; Riabova, K.; Binelli, M.; Farina, A.; Park, J.-S. Geometrical Acoustics Simulations for Ambisonics Auralization of a Car Sound System at High Frequency. In Proceedings of the 2021 Immersive and 3D Audio: From Architecture to Automotive, I3DA 2021, Bologna, Italy, 8–10 September 2021. [CrossRef]
4. Pinardi, D.; Ebri, L.; Belicchi, C.; Farina, A.; Binelli, M. *Direction Specific Analysis of Psychoacoustics Parameters Inside Car Cockpit: A Novel Tool for NVH and Sound Quality*; SAE Technical Paper 2020-01-1547; SAE International: Warrendale, PA, USA, 2020; Available online: https://saemobilus.sae.org/papers/direction-specific-analysis-psychoacoustics-parameters-inside-car-cockpit-a-novel-tool-nvh-sound-quality-2020-01-1547#citation (accessed on 12 March 2025). [CrossRef]
5. Jin, J.; Cheng, H.; Xie, T.; Lu, H. Interior Sound Field Subjective Evaluation Based on the 3D Distribution of Sound Quality Objective Parameters and Sound Source Localization. *Materials* **2021**, *14*, 429. [CrossRef] [PubMed]
6. Izquierdo, A.; del Val, L.; Villacorta, J.J. Feasibility of Using a MEMS Microphone Array for Pedestrian Detection in an Autonomous Emergency Braking System. *Sensors* **2021**, *21*, 4162. [CrossRef]
7. Rajan, V.K.; Krini, M.; Rodemer, K.; Schmidt, G. Signal processing techniques for seat belt microphone arrays. *EURASIP J. Adv. Signal Process.* **2016**, *2016*, 35. [CrossRef]
8. Nordholm, S.E.; Low, S.Y.; Dam, H.Q. A Microphone Array Solution for Duplex Hands-Free Communication Systems. In Proceedings of the 2007 Digest of Technical Papers International Conference on Consumer Electronics, Las Vegas, NV, USA, 10–14 January 2007; IEEE: New York, NY, USA, 2007; pp. 1–2. [CrossRef]
9. Oh, S.; Viswanathan, V. Microphone Array for Hands-Free Voice Communication in a Car. In *Modern Methods of Speech Processing*; Springer: Boston, MA, USA, 1995; pp. 351–375. [CrossRef]
10. Grenier, Y. A microphone array for car environments. In Proceedings of the ICASSP-92: 1992 IEEE International Conference on Acoustics, Speech, and Signal Processing, San Francisco, CA, USA, 23–26 March 1992; IEEE: New York, NY, USA, 1992; Volume 1, pp. 305–308. [CrossRef]
11. Ahamed, P.S.S.; Duraiswamy, P. Virtual Sensing Active Noise Control System with 2D Microphone Array for Automotive Applications. In Proceedings of the 2019 6th International Conference on Signal Processing and Integrated Networks (SPIN), Noida, India, 7–8 March 2019; IEEE: New York, NY, USA, 2019; pp. 151–155. [CrossRef]
12. Moreau, D.; Cazzolato, B.; Zander, A.; Petersen, C. A Review of Virtual Sensing Algorithms for Active Noise Control. *Algorithms* **2008**, *1*, 69–99. [CrossRef]
13. Pinardi, D. A Human Head Shaped Array of Microphones and Cameras for Automotive Applications. In Proceedings of the 2021 Immersive and 3D Audio: From Architecture to Automotive, I3DA 2021, Bologna, Italy, 8–10 September 2021. [CrossRef]
14. Iwai, K.; Kinoshita, S.; Kajikawa, Y. Multichannel feedforward active noise control system combined with noise source separation by microphone arrays. *J. Sound Vib.* **2019**, *453*, 151–173. [CrossRef]
15. Patel, S.J.; Grant, S.L.; Zawodniok, M.; Benesty, J. On the Design of Optimal Linear Microphone Array Geometries. In Proceedings of the 2018 16th International Workshop on Acoustic Signal Enhancement (IWAENC), Tokyo, Japan, 17–20 September 2018; IEEE: New York, NY, USA, 2018; pp. 501–505. [CrossRef]
16. Wang, X.; Li, M.; Zhao, Y.; Wang, J.; Tan, X. Design of Planar Differential Microphone Array Beampatterns with Controllable Mainlobe Beamwidth and Sidelobe Level. *Sensors* **2023**, *23*, 3733. [CrossRef]
17. Trevino, J.; Koyama, S.; Sakamoto, S.; Suzuki, Y. Mixed-order Ambisonics encoding of cylindrical microphone array signals. *Acoust. Sci. Technol.* **2014**, *35*, 174–177. [CrossRef]

18. Du, C.; Leclere, Q.; Li, B. Design And Evaluation of Open Spherical Microphone Arrays. In Proceedings of the 24th International Congress on Sound And Vibration (ICSV24), London, UK, 23–27 July 2017; HAL Open Science: London, UK, 2017.
19. Yang, B.; Shi, S.; Yang, D. Acoustic source localization using the open spherical microphone array in the low-frequency range. *MATEC Web Conf.* **2019**, *283*, 04001. [CrossRef]
20. González, R.; Pearce, J.; Lokki, T. Modular design for spherical microphone arrays. In Proceedings of the AES International Conference, Tokyo, Japan, 7–9 August 2018; p. 1. Available online: http://www.aes.org/e-lib/browse.cfm?elib=19701 (accessed on 12 March 2025).
21. Fernandez-Grande, E. Sound field reconstruction using a spherical microphone array. *J. Acoust. Soc. Am.* **2016**, *139*, 1168–1178. [CrossRef]
22. Rafaely, B. Analysis and design of spherical microphone arrays. *IEEE Trans. Speech Audio Process.* **2005**, *13*, 135–143. [CrossRef]
23. Gover, B.N.; Ryan, J.G.; Stinson, M.R. Designing a Spherical Microphone Array for the Directional Analysis of Reflections and Reverberation. In Proceedings of the Audio Engineering Society Convention 115, New York, NY, USA, 10–13 October 2003; Available online: http://www.aes.org/e-lib/browse.cfm?elib=12447 (accessed on 12 March 2025).
24. Sachar, J.M.; Silverman, H.F.; Patterson, W.R. Large vs small aperture microphone arrays: Performance over a large focal area. In Proceedings of the 2001 IEEE International Conference on Acoustics, Speech, and Signal Processing. Proceedings (Cat. No.01CH37221), Salt Lake City, UT, USA, 7–11 May 2001; IEEE: New York, NY, USA, 2001; pp. 3049–3052. [CrossRef]
25. Yang, B.; Gao, Y.; Guo, Q.; Shi, S. A Low Frequency Noise Source Localization and Identification Method Based on a Virtual Open Spherical Vector Microphone Array. *Appl. Sci.* **2023**, *13*, 4368. [CrossRef]
26. Tripodi, C.; Costalunga, A.; Ebri, L.; Vizzaccaro, M.; Cattani, L.; Ugolotti, E.; Nili, T. Experimental Results on Active Road Noise Cancellation in Car Interior. In Proceedings of the Audio Engineering Society Convention 144, Milan, Italy, 23–26 May 2018; Available online: http://www.aes.org/e-lib/browse.cfm?elib=19493 (accessed on 12 March 2025).
27. Ferrari, C.L.; Cheer, J.; Mautone, M. Investigation of an engine order noise cancellation system in a super sports car. *Acta Acust.* **2023**, *7*, 1. [CrossRef]
28. Rafaely, B. Fundamentals of Spherical Array Processing. In *Springer Topics in Signal Processing*; Springer International Publishing: Cham, Switzerland, 2019. [CrossRef]
29. Farrar, K. Soundfield microphone 2—Detailed functioning of control unit. *Wirel. World* **1979**, *85*, 99–103.
30. Farrar, K. Soundfield microphone—Design and development of microphone and control unit. *Wirel. World* **1979**, *85*, 48–50.
31. Meyer, J.; Elko, G. A highly scalable spherical microphone array based on an orthonormal decomposition of the soundfield. In Proceedings of the IEEE International Conference on Acoustics Speech and Signal Processing, Orlando, FL, USA, 13–17 May 2002; IEEE: New York, NY, USA, 2002; pp. II-1781–II-1784. [CrossRef]
32. Sakamoto, S.; Hongo, S.; Okamoto, T.; Iwaya, Y.; Suzuki, Y. Sound-space recording and binaural presentation system based on a 252-channel microphone array. *Acoust. Sci. Technol.* **2015**, *36*, 516–526. [CrossRef]
33. Spehr, C.; Ernst, D.; Raumer, H.-G. MEMS microphone intensity array for cabin noise measurements. In Proceedings of the INTER-NOISE and NOISE-CON Congress and Conference Proceedings, Washington, DC, USA, 1–5 August 2021; Volume 263, pp. 3023–3034. [CrossRef]
34. Kodrasi, I.; Rohdenburg, T.; Doclo, S. Microphone position optimization for planar superdirective beamforming. In Proceedings of the 2011 IEEE International Conference on Acoustics, Speech and Signal Processing (ICASSP), Prague, Czech Republic, 22–27 May 2011; IEEE: New York, NY, USA, 2011; pp. 109–112. [CrossRef]
35. Ciric, D.; Djordjevic, A.; Licanin, M. Analysis of effects of spherical microphone array physical parameters using simulations. *Facta Univ.-Ser. Electron. Energet.* **2013**, *26*, 107–119. [CrossRef]
36. Gerzon, M. Ambisonics. Part two: Studio techniques. *Studio Sound* **1975**, *17*, 24–30. Available online: https://www.michaelgerzonphotos.org.uk/articles/Ambisonics%202.pdf (accessed on 12 March 2025).
37. Gerzon, M.A. The Design of Precisely Coincident Microphone Arrays for Stereo and Surround Sound. In Proceedings of the Audio Engineering Society Convention 50, London, UK, 4–7 March 1975; Available online: http://www.aes.org/e-lib/browse.cfm?elib=2466 (accessed on 12 March 2025).
38. Ferres, N.M. *An Elementary Treatise on Spherical Harmonics and Subjects Connected with Them*; Cornell University Library: Ithaca, NY, USA, 1877.
39. Battista, G.; Chiariotti, P.; Castellini, P. Spherical Harmonics Decomposition in inverse acoustic methods involving spherical arrays. *J. Sound Vib.* **2018**, *433*, 425–460. [CrossRef]
40. Colbourn, C.J.; Dinitz, J.H. (Eds.) *Handbook of Combinatorial Designs*; Chapman and Hall/CRC: Boca Raton, FL, USA, 2006. [CrossRef]
41. Bannai, E. Rigid spherical t-designs and a theorem of Y. Hong. *J. Fac. Sci. Univ. Tokyo* **1987**, *34*, 485–489.
42. Bannai, E.; Bannai, E. A survey on spherical designs and algebraic combinatorics on spheres. *Eur. J. Comb.* **2009**, *30*, 1392–1425. [CrossRef]
43. Bannai, E.; Damerell, R.M. Tight spherical designs, I.J. *Math. Soc. Jpn.* **1979**, *31*, 199–207. [CrossRef]

44. Bannai, E.; Okuda, T.; Tagami, M. Spherical designs of harmonic index *t*. *J. Approx. Theory* **2015**, *195*, 1–18. [CrossRef]
45. Zawawi, S.A.; Hamzah, A.A.; Majlis, B.Y.; Mohd-Yasin, F. A Review of MEMS Capacitive Microphones. *Micromachines* **2020**, *11*, 484. [CrossRef]
46. Corey, R.M. Microphone Array Processing for Augmented Listening. Ph.D. Thesis, University of Illinois, Urbana, IL, USA, 2019.
47. Hardin, R.H.; Sloane, N.J.A. McLaren's improved snub cube and other new spherical designs in three dimensions. *Discret. Comput. Geom.* **1996**, *15*, 429–441. [CrossRef]
48. Shen, J.; Jia, X. Diffraction of a plane wave by an infinitely long circular cylinder or a sphere: Solution from Mie theory. *Appl. Opt.* **2013**, *52*, 5707–5712. [CrossRef]
49. Ryan, J.G.; Goubran, R.A. Near-field beamforming for microphone arrays. In Proceedings of the 1997 IEEE International Conference on Acoustics, Speech, and Signal Processing, Munich, Germany, 21–24 April 1997; IEEE Computer Society Press: Los Alamitos, CA, USA, 1997; pp. 363–366. [CrossRef]
50. Pinardi, D. Spherical Wave Diffraction for Microphone Arrays Operating in Near Field. In Proceedings of the 2023 Immersive and 3D Audio: From Architecture to Automotive, I3DA 2023, Bologna, Italy, 5–7 September 2023.
51. Otero, J.; Felis, I. Measurement Transducer Impulse Response Using an Exponential Sine Sweep Method. *Proceedings* **2019**, *4*, 53. [CrossRef]
52. Denk, F.; Kollmeier, B.; Ewert, S.D. Removing Reflections in Semianechoic Impulse Responses by Frequency-Dependent Truncation. *J. Audio Eng. Soc.* **2018**, *66*, 146–153. [CrossRef]
53. Pinardi, D.; Farina, A.; Binelli, M. Transducer Distribution on Spherical Arrays for Ambisonics Recording and Playback. In Proceedings of the 2023 Immersive and 3D Audio: From Architecture to Automotive (I3DA), Bologna, Italy, 5–7 September 2023; IEEE: New York, NY, USA, 2023; pp. 1–9. [CrossRef]
54. Kirkeby, O.; Orduna, F.; Nelson, P.A.; Hameda, H. Inverse Filtering in Sound Reproduction. *Meas. Control* **1993**, *26*, 261–266. [CrossRef]
55. Tokuno, H.; Kirkeby, O.; Nelson, P.A.; Hamada, H. Inverse Filter of Sound Reproduction Systems Using Regularization. *IEICE Trans Fundam. Electron. Commun. Comput. Sci.* **1997**, *E80-A*, 809–820. Available online: https://www.melaudia.net/zdoc/kirkebyInverseFilter.PDF (accessed on 12 March 2025).
56. Devices, A. A2B Audio Bus: An Easier, Simpler Solution for Audio Design. Available online: https://www.analog.com/en/applications/technology/a2b-audio-bus.html (accessed on 12 March 2025).
57. Pinardi, D.; Toscani, A.; Binelli, M.; Saccenti, L.; Farina, A.; Cattani, L. Full-Digital Microphone Meta-Arrays for Consumer Electronics. *IEEE Trans. Consum. Electron.* **2023**, *69*, 640–648. [CrossRef]
58. Rocchi, N.; Toscani, A.; Pinardi, D.; Binelli, M.; Chiesi, L.; Farina, A.; Bonomi, E.; Tronchin, L. A Modular, Low Latency, A^2B-based Architecture for Distributed Multichannel Full-Digital Audio Systems. In Proceedings of the 2021 Immersive and 3D Audio: From Architecture to Automot (I3DA), Bologna, Italy, 8–10 September 2021; IEEE: New York, NY, USA, 2021; pp. 1–8. [CrossRef]
59. Rocchi, N.; Toscani, A.; Chiorboli, G.; Pinardi, D.; Binelli, M.; Farina, A. Transducer Arrays over A^2B Networks in Industrial and Automotive Applications: Clock Propagation Measurements. *IEEE Access* **2021**, *9*, 118232–118241. [CrossRef]
60. Bertet, S.; Daniel, J.; Moreau, S. 3D Sound Field Recording with Higher Order Ambisonics—Objective Measurements and Validation of Spherical Microphone. In Proceedings of the Audio Engineering Society Convention 120, Paris, France, 20–23 May 2006. Available online: http://www.aes.org/e-lib/browse.cfm?elib=13661 (accessed on 12 March 2025).
61. Fredianelli, L.; Pedrini, G.; Bolognese, M.; Bernardini, M.; Fidecaro, F.; Licitra, G. Features for Evaluating Source Localization Effectiveness in Sound Maps from Acoustic Cameras. *Sensors* **2024**, *24*, 4696. [CrossRef]
62. Csóka, B.; Fiala, P.; Rucz, P. Tracking Sound Sources with Microphone Arrays and Beamforming Algorithms. In Proceedings of the Proceedings of the 10th Convention of the European Acoustics Association Forum Acusticum 2023, Torino, Italy, 11–15 September 2023; European Acoustics Association: Turin, Italy, 2024; pp. 1115–1122. [CrossRef]

applied sciences

MDPI

Article

Emotional Dynamics in Human–Machine Interactive Systems: Effectively Measuring Kuhn Poker Approach with Experimental Validation

Haoran Yang [1] **and Shu Liang** [1,2,*]

[1] Department of Control Science and Engineering, College of Electronic and Information Engineering, Tongji University, 4800 Caoan Road, Shanghai 201804, China; 2232956@tongji.edu.cn
[2] Shanghai Research Institute for Intelligent Autonomous Systems, Tongji University, 55 Chuanhe Road, Shanghai 201210, China
* Correspondence: sliang@tongji.edu.cn

Abstract: The impact of immediate emotions on human decision-making has been extensively studied; however, their role within human–machine interactive systems remains underexplored. This study investigates how emotional states influence decision-making behaviors in such systems by introducing a novel two-tiered progressive inquiry model—environment–emotion and emotion–decision. Using Kuhn poker as an experimental context, we developed an intelligent decision support system based on Counterfactual Regret Minimization algorithms to provide real-time strategic advice to participants. Our findings reveal that positive emotional states lead individuals to be more risk-seeking and less inclined to collaborate with the AI-driven system, whereas negative emotional states result in risk-averse tendencies and higher compliance with system recommendations. By systematically uncovering the causal mechanisms through which environmental factors regulate emotions and subsequently affect decision-making, this research provides critical theoretical and empirical insights for optimizing human–machine interaction design.

Keywords: immediate emotions; two-tiered progressive model; counterfactual regret minimization; dynamic emotion regulation; risk-taking tendency; cooperation level

1. Introduction

The influence of emotions on human decision-making has emerged as a critical area of research in both decision science and artificial intelligence [1,2]. Particularly, the role of immediate emotions in dynamically regulating individual risk preferences and cooperation tendencies [3] has garnered increasing attention [4]. Immediate emotions refer to an individual's transient emotional responses in specific contexts, characterized by their short-term nature and dynamic variability. Research indicates that these emotions can alter subjective perceptions of risk and reward, influencing decision-making tendencies [5]. For example, Loewenstein et al. [6] found that positive emotion (PE) generally improves individuals' propensity to take risks, while negative emotion (NE) prompts more conservative choices. Riedl and Paul [7] further noted that emotional states significantly modulate cooperative intentions, impacting team collaboration efficiency and decision outcomes.

In human–machine collaboration, immediate emotions may determine the level of trust individuals place in machine-generated recommendations and their willingness to adopt such advice. This is crucial for designing effective and user-friendly interactive decision systems [8,9]. Therefore, understanding the dynamic mechanisms of immediate emotions and integrating emotion sensing and regulation technologies into human–machine interactive decision systems have become key challenges in this field [10,11]. However, existing research has several limitations. First, traditional decision models, such as Expected Utility Theory and Prospect Theory, primarily focus on the impact of anticipated emotions on long-term decisions, neglecting the immediate effects of immediate emotions in short-term decision contexts [12,13]. For example, in investment scenarios, investors' immediate emotions triggered by market fluctuations often lead to irrational behaviors, such as selling assets at low points or blindly increasing investments at high points, rather than making rational choices based on long-term profit maximization goals [14,15]. Similarly, in medical decision-making, patients' fears or anxieties may cause them to overly rely on doctors' advice or even reject reasonable treatment plans, highlighting the need to understand the mechanisms of immediate emotions better.

The choice of Kuhn poker as the experimental context was motivated by its ability to simulate decision-making under incomplete information, which aligns closely with the study's objective of investigating the impact of emotions on risk-taking behavior and cooperation levels. Kuhn poker is a zero-sum game typically involving two players, with a deck of only three cards. Players draw cards and make sequential decisions (e.g., betting or folding), with the final outcome determined by the relative strength of their hands. This structure, which incorporates elements of incomplete information and strategic interaction, makes Kuhn poker particularly suitable for studying how individuals adjust their decision-making behaviors based on emotional states in uncertain environments. In addition to Kuhn poker, other game-theoretic approaches, such as Bluffing Games, Rock–Paper–Scissors, and Schelling Points, were considered but ultimately deemed less suitable for this study. Compared to Bluffing Games, Kuhn poker avoids the excessive complexity and trust-related dynamics that could obscure the direct effects of emotions on decision-making. Rock–Paper–Scissors, while simple and widely understood, lacks the strategic depth and risk–reward trade-offs necessary to explore nuanced emotional influences on risk preferences and cooperation. Schelling Points, which focus on coordination and focal points in decision-making, are more relevant to cooperative scenarios rather than the competitive and risk-oriented context of this study. Therefore, Kuhn poker was selected for its balance of simplicity, strategic depth, and relevance to the study of risk preferences and cooperation under the influence of immediate emotions. Its tractable structure allows for precise control and analysis of decision-making dynamics, making it an ideal framework for examining the interplay between emotions, risk-taking, and human–machine collaboration. Nevertheless, most current experiments employ static emotion-induction methods (e.g., short videos or questionnaires), ignoring the dynamic changes in emotions during the experiment and the continuous regulatory effects of environmental factors [16]. Such experimental designs diverge from real-world dynamic emotional variations, leading to lower external validity [17,18]. Additionally, existing human–machine decision support systems are often based on the "rational agent" assumption, emphasizing algorithmically generated rational suggestions while overlooking individual behavioral deviations due to immediate emotional changes and differences in adoption willingness, thus limiting the practical applicability of these systems.

To tackle these challenges and fill the gap in the literature, in this paper, we propose an innovative approach by incorporating environmental variables into the emotion process, constructing a two-tiered progressive inquiry environment–emotion and emotion–decision model. Utilizing the classic game scenario of Kuhn poker and Counterfactual Regret Minimization (CFR) algorithms [19,20], we developed an intelligent decision support system. By dynamically regulating the experimental environment, our study systematically reveals the dynamic impact of immediate emotions on participants' risk preferences and cooperation tendencies. The core objective is to elucidate the regulatory role of immediate emotions within intelligent assistance systems through the "two-tiered progressive model", providing new insights and empirical support for the design of emotion-driven intelligent decision systems.

The main contributions of this paper are summarized in the following.

- We introduce a novel environment–emotion and emotion–decision model to systematically investigate how environmental factors affect emotion in human–computer interaction systems and hence decision-making behavior. This model addresses the limitations of traditional approaches by focusing on short-term emotional dynamics.
- We develop an intelligent decision support system based on Kuhn poker and Counterfactual Regret Minimization algorithms. This system provides real-time strategic advice and examines how immediate emotions affect participants' willingness to adopt AI recommendations, enhancing human–machine collaboration.
- Through an experiment with 80 participants, this study empirically validates the significant influence of immediate emotions on risk preferences and cooperation tendencies within human–machine interactive systems.

The remainder of this paper is organized as follows: Section 2 provides a brief review of related work and highlights the novelty of this paper. Section 3 introduces the materials and methods in Kuhn poker, incorporating the two ways of environment–emotion and emotion–decision. Simulation results are presented in Section 4, followed by discussion and conclusions in Sections 5 and 6.

2. Related Work

In recent years, the effect of emotions on decision-making behavior, especially the role of immediate emotions, has become an important area of research in cognitive psychology and behavioral economics. Isen et al. [21] showed that positive emotions increase individuals' creative problem-solving abilities and enhance the propensity for risk-taking behaviors. Raghunathan and Pham [22] found that different types of negative emotions have different effects on decision-making: Anxiety predisposes individuals to choose low-risk, low-return options, whereas sadness predisposes individuals to choose high-risk, high-return options. Lerner et al. [23] showed through a systematic review that immediate emotions affect decision-making behavior and the processing of external information by modulating the way in which an individual evaluates his or her cognition. Béroule and Gisquet-Verrier [24] proposed a computational architecture that demonstrates how emotions guide the decision-making process, emphasizing the role of emotions in fast and ambiguous decision-making channels. In addition, Kuppens et al. [25] showed that the intensity and duration of emotions have a significant effect on the decision-making process, especially in complex social decision-making situations. Fiorenzato et al. [26] investigated gender differences in the effects of emotion induction on decision-making across time and found that the induction of fearful emotions made females more inclined to choose immediately smaller rewards, whereas males' decision-making was not significantly affected by their emotional state. Although these studies revealed the important role of immediate emotions in cognitive and behavioral regulation, the mechanisms by which immediate emotions affect the level of human–computer collaboration and complex decision-making behaviors in dynamic human–computer interaction scenarios still need to be further explored.

Recent studies have underscored the growing importance of emotional dynamics in human–machine interactions, revealing that emotion-aware systems significantly enhance decision-making efficiency, improve user engagement, and foster more effective collaboration. For instance, Wang et al. [27] explored the role of emotion recognition in AI-powered musical and entertainment robotics, emphasizing how emotional awareness enables more natural and effective collaboration between humans and robots. Similarly,

Wei et al. [28] proposed CoMIF, a model that integrates complex emotional, personality, and conversational factors into human–machine interaction, showing how these elements influence decision-making and improve interaction quality. In healthcare, Jain and Jain [29] investigated the integration of emotion detection systems within AI-powered healthcare platforms, demonstrating how emotionally aware systems can improve patient interaction and overall care outcomes. Furthermore, Chung and Tan [30] focused on enhancing user experiences in smart hospitality services by integrating AI and the Internet of Things, where emotional feedback plays a critical role in adapting AI-driven systems to better serve users' emotional needs. These studies collectively emphasize the significance of emotional intelligence in human–machine collaboration, underlining its impact on decision-making, user satisfaction, and system efficiency.

Building on this understanding of emotional dynamics in HCI, the role of human–computer hybrid systems in decision-making has evolved significantly. In the research of human–computer hybrid systems and human decision-making, and Shneiderman's [31] concept of 'human-centered AI' advocates that technology should augment human capabilities rather than replace human roles. Huang et al. [32] significantly improved the efficiency and interpretability of decision-making in a safety inspection system by introducing human involvement into augmented intelligence. In addition, Sun et al. [33] investigated a personalised control system in the field of autonomous driving, optimizing the driver's decision-making behavior based on the Mixed Observable Markov Decision Process framework. In terms of applications, Xian et al. [34] proposed a hierarchical hybrid control scheme, which achieves robust performance guarantees through optimized human–machine collaboration models. In addition, H Xia et al. [35] investigated the role of human–computer collaboration in medical imaging diagnosis and proved that it can effectively improve diagnostic accuracy. Overall, the role of human–computer hybrid systems in decision-making has gradually shifted from mere assistance to deep collaboration. However, while improving the system's interpretability and transparency, how to effectively weigh the distribution of power and responsibility between humans and machines and fully consider the potential impact of emotional factors on decision-making and human–computer collaboration dynamics are still important directions for future research.

In summary, this paper differs from the existing literature by introducing a two-layer progressive exploration model of environment–emotion and emotion–decision. Specifically, we first construct an intelligent assisted decision-making system based on the CFR algorithm, and then propose a novel model for the study of emotions. The novelty of our work lies in the comprehensive integration of environmental factors, immediate emotions, and decision-making algorithms to systematically analyse the impact of emotions and environmental interactions on decision-making, which provides important theoretical support and practical guidance to improve the efficiency of human–computer collaboration and decision-making, and to enhance the adaptability of the system and the user experience, and further promotes the practical application of human–computer interaction systems in complex dynamic environments.

3. Materials and Methods

In this section, we provide a systematic overview of how to construct a human–machine interaction model under game scenarios. As shown in Figure 1, this figure illustrates the methodology presented in this chapter, as well as the experimental process and results analysis outlined in Section 4. Specifically, our discussion begins with an examination based on typical sequential game models. Subsequently, we propose the utilization of the CFR algorithm—an extensively applied iterative method in game theory and imperfect information games—to build a two-tier progressive exploration model that investigates the relationships between environment and emotion, and subsequently between emotion and decision-making.

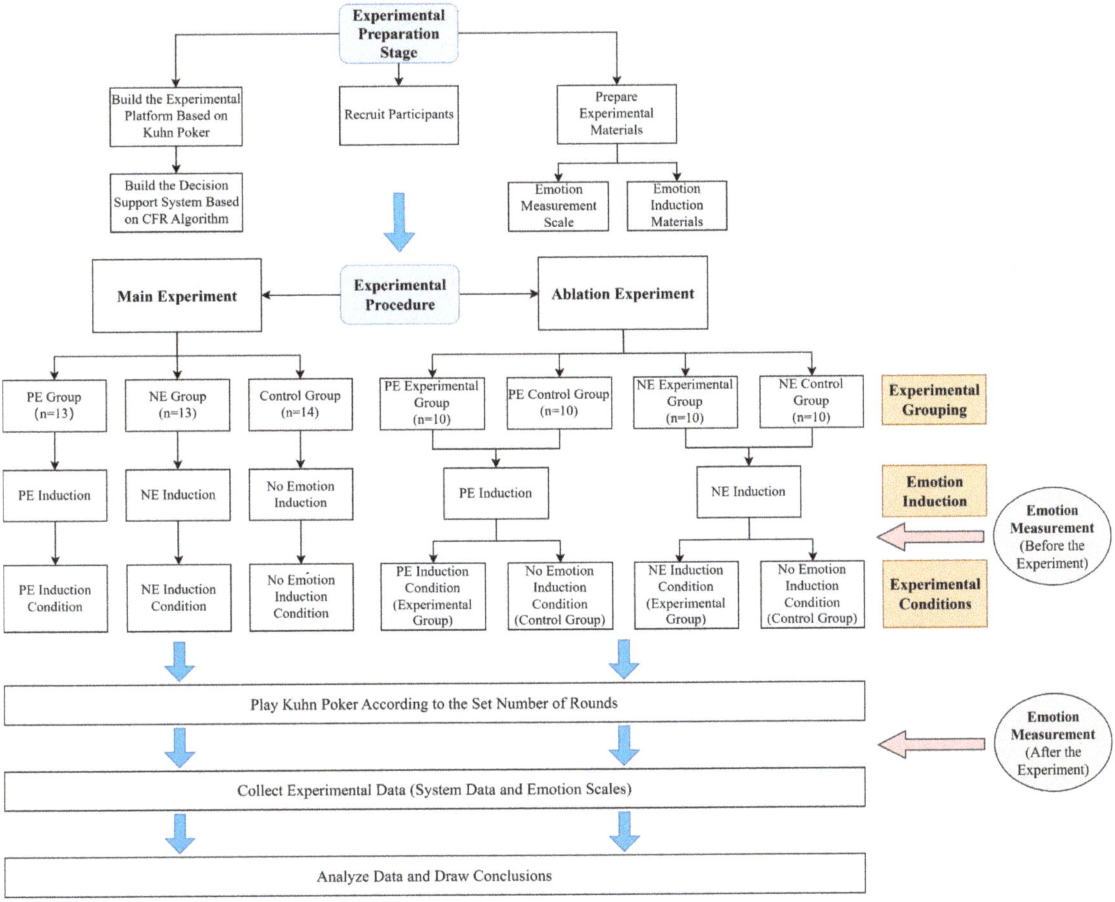

Figure 1. Flowchart of this study.

3.1. Game Scenario and System Architecture

We first constructed an intelligent system for human–computer interaction based on Kuhn poker [36,37], aiming to optimize the decision-making process by integrating environmental factors and immediate emotions. The system consists of two players (human player A and AI player B). At the beginning of the game, each player randomly draws a card from a deck of three cards (card values of 1, 2 and 3, respectively). During the game, the first player (A) chooses to 'pass' or 'bet'. The second player (B) then responds to A's choice with either 'pass' or 'bet'. If A chooses 'pass' and B chooses 'bet', then A will again have the opportunity to decide whether to 'pass' or 'bet'. Ultimately, the outcome and payoff is determined by the order of the player's actions and the value of the cards held. Figure 2 shows the complete flow of the game in the form of a decision tree, which visualizes the decision paths in each phase.

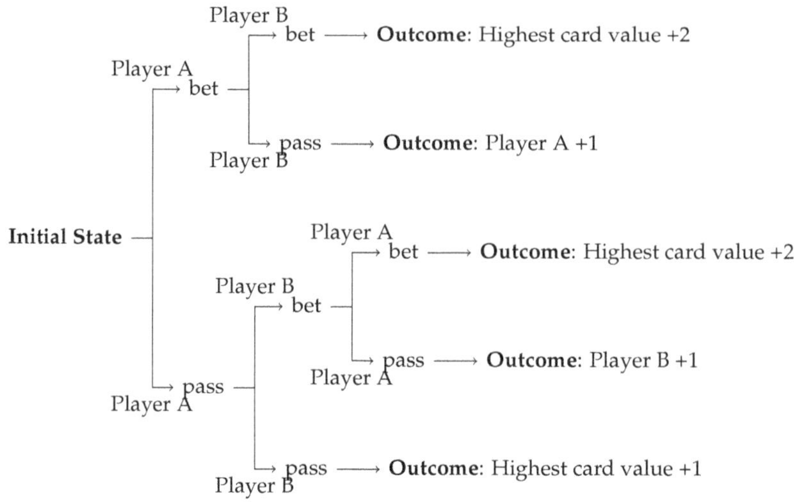

Figure 2. Decision tree for the game scenario.

The AI player employs the CFR algorithm to optimize its strategy, continuously adjusting its approach through iterative training to improve performance in future games. The overall system architecture consists of three main components: the human player module, the game engine module (responsible for managing game state, handling player actions, and determining outcomes), and the AI player module (incorporating the CFR algorithm and decision mechanisms). Through this integrated design, the system not only simulates complex game scenarios but also provides new perspectives and methods for researching human–machine interactive intelligent systems.

3.2. An Enhanced Intelligent Decision Support System That Minimizes Counterfactual Regret

The Counterfactual Regret Minimization algorithm, a widely utilized iterative technique in game theory and games of incomplete information, aims at calculating equilibrium strategies and approximations to the Nash equilibrium [38]. Unlike traditional regret minimization algorithms [39] that require exhaustive exploration of the entire strategy space, CFR operates by iteratively updating strategies based on the regrets associated with each action taken at different information sets. This research employs CFR as the cornerstone for an intelligent decision support system due to its efficiency in updating strategies. The core concept involves minimizing regret values within each information set to progressively converge towards optimal strategies.

In the context of an information set I, counterfactual regret for player i not choosing action a is defined as the difference between the utility of taking action a and the actual utility under strategy σ_t. Mathematically, it is represented as Equation (1):

$$R_t^i(a) = \sum_{h \in I} \pi^{-i}(h)[u^i(h,a) - u^i(h,\sigma_t)] \tag{1}$$

where $\pi^{-i}(h)$ denotes the probability of reaching leaf node h by players other than i, and $u^i(h,a)$ represents the utility of player i choosing action a at history state h. After T iterations, the cumulative counterfactual regret is given by Equation (2):

$$R_T^i(a) = \sum_{t=1}^{T} R_t^i(a) \tag{2}$$

Subsequently, the current strategy is updated via regret matching, as given by Equation (3):

$$\sigma_T^i(a) = \begin{cases} \frac{\max(R_T^i(a),0)}{\sum_{b \in A} \max(R_T^i(b),0)} & \text{if } \sum_{b \in A} \max(R_T^i(b),0) > 0 \\ \frac{1}{|A(I)|} & \text{otherwise} \end{cases} \quad (3)$$

where $A(I)$ denotes the set of actions that can be taken under information set I.

Through continuous iteration, the strategy converges to the Nash equilibrium. This convergence property makes CFR particularly suitable for complex games like Kuhn poker, where the number of possible strategies can be vast. Prior to experimentation, the CFR algorithm was applied to train on Kuhn poker scenarios. Specifically, 100,000 rounds of training were conducted to ensure robustness and accuracy in approximating Nash equilibrium outcomes across various information sets.

These outcomes provided foundational guidelines for the decision support system's recommendations. For example, in the Kuhn poker game, if we denote "p" as pass and "b" as bet, the detailed strategies for both the first and second players are shown in Tables 1 and 2, respectively. These tables illustrate how the probabilities of actions, i.e., passing or betting, vary depending on the specific information set encountered during gameplay. The iterative process of the strategies of the two players with different information sets is shown in Figure 3a,b. By comparison, the solution obtained by the CFR algorithm is closer to that obtained by the theory, which verifies the effectiveness of the algorithm.

Table 1. First player strategy.

Information Set	1	1pb	2	2pb	3	3pb
pass	0.73	1.00	1.00	0.39	0.18	0.00
bet	0.27	0.00	0.00	0.61	0.82	1.00

Table 2. Second player strategy.

Information Set	1b	1p	2b	2p	3b	3p
pass	1.00	0.65	0.66	1.00	0.00	0.00
bet	0.00	0.35	0.34	0.00	1.00	1.00

These strategies demonstrate the dynamic adjustment of probabilities based on the evolving game states, providing participants with informed decisions throughout the experiment.

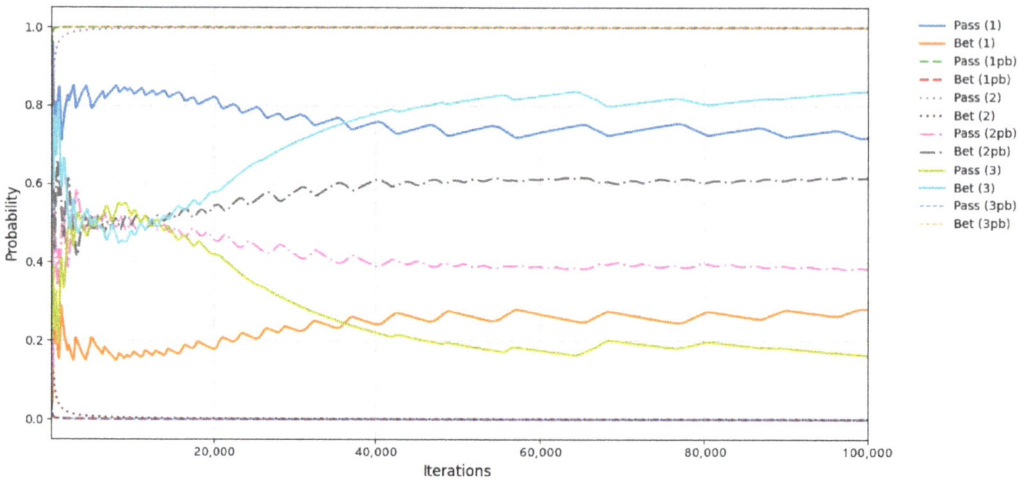

(a)

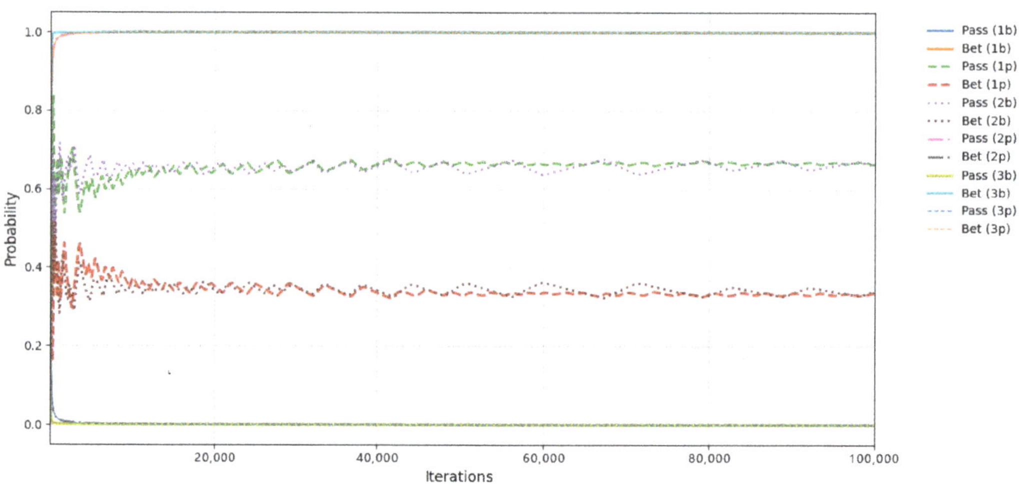

(b)

Figure 3. (**a**) shows the strategy iteration curves for different information sets of the first player, and (**b**) shows the strategy iteration curves for different information sets of the second player.

3.3. Computational Complexity

To analyze the time complexity of the CFR algorithm, we need to consider several factors, including the size of the game tree, the number of iterations, and the operations performed during each iteration. Let $|I|$ denote the number of information sets in the game tree, and let A_i represent the set of actions available at information set i. The total number of actions across all information sets can be denoted as $|A| = \sum_{i \in I} |A_i|$.

In extensive-form games, the size of the game tree is exponential in the depth of the tree. However, the CFR algorithm does not require exploring the entire game tree explicitly; instead, it traverses the game tree implicitly during each iteration.

During each iteration t, the CFR algorithm performs the following operations:

1. For each information set i:
    - Compute the counterfactual regret for each action $a \in A_i$.

- Update the strategy using regret matching.

2. Accumulate the cumulative regrets and update the average strategy.

The computation of counterfactual regret involves summing over all possible histories $h \in I$, which requires $O(|I| \cdot |A|)$ operations per iteration. Let T denote the number of iterations required for convergence. The overall time complexity of the CFR algorithm can be expressed as: $O(T \cdot |I| \cdot |A|)$. This complexity arises from performing $O(|I| \cdot |A|)$ operations during each of the T iterations.

3.4. Environment–Emotion and Emotion–Decision Model

To address the issue of emotional influence decay over time, this study introduces a two-tier progressive exploration model: environment–emotion and emotion–decision. By manipulating environmental factors such as background imagery and music during the experiment, we aimed to sustain participants' emotional states throughout the process. Specifically, three experimental conditions were established:

- **Control Condition**: No enhancements.
- **Positive Emotion Induction**: Graffiti backgrounds paired with punk rock music (e.g., Green Day's "Basket Case").
- **Negative Emotion Induction**: Minimalist blue-toned visuals accompanied by classical music pieces (e.g., Chopin's "Farewell Waltz").

Participants' emotions were measured using standardized emotional scales at the beginning, during, and after the experiment to track changes and validate the effectiveness of the environmental manipulations. An interactive human–machine platform based on Kuhn poker was developed, allowing participants to engage in gameplay while receiving real-time feedback on game dynamics, including current round details, card values, and actions taken. Participants had the option to utilize AI-generated strategic advice before making decisions by selecting "pass" or "bet". Post-game, detailed results, scores, and overall performance metrics were provided, with all data recorded for subsequent analysis. This setup ensures a controlled environment where the impact of environmental and emotional variables can be systematically studied.

4. Experimental Setup and Results

In this section, simulations are conducted to numerically evaluate the performance of emotional dynamics in human–machine interactive systems using the proposed approach. As shown in Figure 1, the overall experimental process, which includes both the methodology presented in Section 3 and the experimental process outlined in this chapter, is illustrated. This figure provides a visual summary of the key steps involved in the experiments, from the setup to the analysis phase. This section verifies and tests the effectiveness from a statistical perspective as follows.

4.1. Experimental Setup

The primary objective of this experiment is to investigate the influence of emotions on participants' willingness to adopt auxiliary decision-making support and their risk-taking behavior in a human–machine interactive intelligent system. A total of 40 participants were recruited, including 22 males and 18 females, aged between 18 and 60 years. All participants were physically and mentally healthy and had not participated in similar experiments before. Participants were randomly divided into three groups of 13, 13, and 14 individuals, ensuring an equal distribution of genders across groups. These groups were designated as the positive emotion group, negative emotion group, and control group. Research indicates that the method chosen for emotion induction significantly impacts its effectiveness. Emotion induction is a well-established technique in psychology, with visual and auditory stimuli, such as films and music, being commonly used to evoke specific emotional states. The effectiveness of these stimuli has been supported by multiple studies, including Siedlecka and Denson [40] , who reviewed various methods for inducing basic emotions. Furthermore, Ansani et al. [41] showed that combining music with film scenes induces mood-coherent emotional biases, strengthening the validity of using these methods in experimental settings. Movies are considered one of the most effective methods for inducing strong emotional experiences. Therefore, this study employed movie clips to induce positive or negative emotions. Specifically, a 5-min clip from *Mr. Bean* and a 5-min clip from *Zootopia* were used to induce positive emotions, while a 5-min clip from *Hachi: A Dog's Tale* and a 5-min clip from *Titanic* was used to induce negative emotions. Meanwhile, research has shown that the Positive and Negative Affects Scale (PANAS) [42] has high internal consistency reliability (Cronbach's alpha = 0.87) and structural validity, making it suitable for measuring emotions. Therefore, this study utilized the Chinese version of PANAS, which consists of ten positive and ten negative emotion descriptors. Participants rated these descriptors on a scale from 1 (very little or not at all) to 5 (extremely), with higher scores indicating stronger positive or negative emotions.

The specific experimental procedure was as follows:

Positive emotion group: subjects were informed that they would be paid for the experiment. Prior to the start of the experiment, a warm welcome was given to the subjects, followed by an introduction of the rules of the Khun poker game, the presentation of a small gift (snack) to the subjects, and the playing of a funny classic scene from *Mr Bean* to induce a positive emotion and the completion of an emotion measurement scale. Subjects were asked to watch a funny scene from *Zootopia* again, and then they were asked to operate the Kuhn poker platform in a positive emotion-inducing environment. Subjects were asked to complete 20 rounds of poker, and for each round they were able to see the winning and losing results at that time. After completing the 20 rounds, the subjects were asked to fill out an emotion measurement scale.

Negative emotion group: Subjects were not informed of the experimental payoff. The rules of Kuhn's poker game were introduced directly to the subjects without a welcome speech before the start of the experiment. Subjects were shown a scene from *Hachi: A Dog's Tale* in which a dog waits for its master but never gets there, and filled out an emotion measurement scale. Subjects were asked to watch the scene of the hero sinking to the bottom of the sea and saying goodbye to the heroine in *Titanic* again, and then were asked to operate the Kuhn poker game experimental platform under the negative emotion-inducing environment. Subjects were asked to complete 20 rounds of poker, and for each round they were able to see the winning or losing results at that time. After completing the 20 rounds, the subjects were asked to complete an emotion measurement scale.

Control group: We did not induce any emotion in the subjects. When the subjects arrived, we first asked them to calm down and then fill in the emotion scale. After that, we introduced the rules of Kuhn poker game to the subjects, and after the subjects understood the rules of the game, we asked them to operate the Kuhn poker game experiment platform in the control environment directly. The subjects were asked to complete 20 rounds of the game, and they could see the winning and losing results of each round at that time. After completing the 20 rounds, the subjects were asked to complete an emotion scale.

4.2. Ablation Experiment Setup

The ablation experiment aims to validate the effectiveness of the two-tier progressive model (environment–emotion and emotion–decision) in maintaining continuous emotional induction. An additional 40 participants were recruited, including 24 males and 16 females, aged between 18 and 60 years. Participants were randomly divided into two main groups of 20 each, further subdivided into two subgroups of 10. One subgroup used the experimental platform with environmental factors, while the other used the platform without environmental factors. Similar to the main experiment, movie clips were used to induce positive or negative emotions. The Chinese version of PANAS was used to measure the effectiveness and maintenance of induced emotions. Participants followed a similar procedure as in the main experiment, with the addition of environmental factors for the experimental group.

4.3. Experiment Results

In this section, we will present the results of the main experiment, aiming to explore the impact of emotions on individual decision-making behaviors in human–machine interaction systems. Through emotion induction experiments, we evaluated the specific effects of emotions on participants' decision-making processes and further analyzed how these effects manifested in different contexts. The following subsections will discuss the emotion induction effectiveness, the impact of emotions on risk-taking tendencies, and the influence of emotions on human–machine cooperation levels.

4.3.1. Emotion Induction Effectiveness

In the 40 valid datasets, the positive emotion group had 13 participants, the negative emotion group had 13 participants, and the control group had 14 participants. The PANAS scores are summarized in Tables 3–5.

Table 3. Positive emotion group scores before and after experiment.

	PE		NE	
	M	**SD**	**M**	**SD**
Before	3.99	0.33	1.28	0.17
After	4.08	0.36	1.29	0.18

Table 4. Negative Emotion group scores before and after experiment.

	PE		NE	
	M	**SD**	**M**	**SD**
Before	1.31	0.17	3.71	0.25
After	1.36	0.18	3.82	0.26

Table 5. Control group scores before and after experiment.

	PE		NE	
	M	**SD**	**M**	**SD**
Before	1.26	0.15	1.23	0.14
After	1.25	0.14	1.25	0.15

The control group showed stable emotions, while the positive and negative emotion groups exhibited significant changes, confirming the effectiveness of the emotion induction methods.

4.3.2. Risk-Taking Tendency

We used the betting ratio (number of bets/total rounds) to reflect participants' risk-taking tendency during the game. ANOVA results are shown in Tables 6 and 7.

Participants in the positive emotion group showed a higher risk-seeking tendency compared to the control group, while those in the negative emotion group exhibited risk aversion.

4.3.3. Human–Machine Cooperation Level

Cooperation ratio (number of times checking AI advice/total rounds) and cooperation but inconsistent ratio abbreviated as CNAR (number of times checking AI advice but not following/number of times checking AI advice) were used to measure human–machine cooperation. ANOVA results are shown in Tables 8–11.

Table 6. Risk-taking ratios analysis among three groups.

Group	PE (n = 13)	NE (n = 13)	Control (n = 14)	F	p
BR (Mean ± SD) (%)	60.38 ± 14.60	35.76 ± 14.79	51.43 ± 13.40	9.42	0.000 ***
95% Confidence Interval	(0.5120, 0.6957)	(0.2647, 0.4507)	(0.4371, 0.5914)		

*** $p < 0.001$.

Table 7. Effect sizes between three groups of risk-taking ratios.

	PE and NE	PE and Control	NE and Control	η^2	Cohen's f
Cohen's d	1.67	0.65	1.13	0.34	0.71

Table 8. Cooperation ratios analysis among three groups.

Group	PE (n = 13)	NE (n = 13)	Control (n = 14)	F	p
CR (Mean ± SD) (%)	35.77 ± 14.79	60.00 ± 16.98	48.93 ± 10.89	8.57	0.000 ***
95% Confidence Interval	(0.2647, 0.4507)	(0.4932, 0.7068)	(0.4240, 0.5546)		

*** $p < 0.001$.

Table 9. Effect sizes between three groups of cooperation ratios.

	PE and NE	PE and Control	NE and Control	η^2	Cohen's f
Cohen's d	1.52	1.01	0.78	0.32	0.68

Table 10. Cooperation but not acceptance ratios analysis among three groups.

Group	PE (n = 13)	NE (n = 13)	Control (n = 14)	F	p
CNAR (Mean ± SD) (%)	59.85 ± 20.81	43.23 ± 10.58	49.29 ± 8.77	4.17	0.02
95% Confidence Interval	(0.4674, 0.7295)	(0.3657, 0.4989)	(0.4403, 0.5454)		

Table 11. Effect sizes between three groups of CNAR.

	PE and NE	PE and Control	NE and Control	η^2	Cohen's f
Cohen's d	1.00	0.66	0.62	0.18	0.47

Participants in the negative emotion group showed higher levels of human–machine cooperation compared to those in the positive emotion group.

4.4. Ablation Experiment Results

In the 40 valid datasets, each group (positive emotion experimental, positive emotion control, negative emotion experimental, negative emotion control) had 10 participants. The

PANAS scores before and after the experiment for the positive emotion group and the negative emotion group are summarized in Tables 12 and 13.

Table 12. PE group scores before and after experiment.

	Experimental Group		Control Group	
	M	**SD**	**M**	**SD**
Before	3.97	0.23	3.93	0.22
After	3.90	0.31	3.37	0.26

Table 13. NE group scores before and after experiment.

	Experimental Group		Control Group	
	M	**SD**	**M**	**SD**
Before	3.87	0.24	3.82	0.20
After	4.02	0.21	3.35	0.32

To further validate the differences between the experimental and control groups, we first conducted an independent samples *t*-test on the preexperimental emotion scores of the two groups. The results of the experiment are shown in Tables 14 and 15.

Table 14. Results of *t*-test for PE group before experiment.

	Category (Mean ± SD)	**t**	***p***
Experimental Group (n = 10)	3.97 ± 0.23		
Control Group (n = 10)	3.93 ± 0.22	0.55	0.588

Table 15. Results of *t*-test for NE group before experiment.

	Category (Mean ± SD)	**t**	***p***
Experimental Group (n = 10)	3.87 ± 0.24		
Control Group (n = 10)	3.82 ± 0.20	0.52	0.609

We then conducted an independent samples *t*-test on the emotion scores of the two groups after the experiment. The results of the experiment are shown in Tables 16 and 17.

Table 16. Results of *t*-test for PE group after experiment.

	Category (Mean ± SD)	**t**	***p***
Experimental Group (n = 10)	3.90 ± 0.31		
Control Group (n = 10)	3.37 ± 0.26	4.36	0.000 ***

*** $p < 0.001$.

Table 17. Results of *t*-test for NE group after experiment.

	Category (Mean ± SD)	**t**	***p***
Experimental Group (n = 10)	4.02 ± 0.21		
Control Group (n = 10)	3.35 ± 0.32	5.83	0.000 ***

*** $p < 0.001$.

After that, we conducted independent samples *t*-tests on the emotion scores before and after the experiment for the experimental and control groups in the positive emotion group and the negative emotion group, respectively. The results of the experiment are shown in Tables 18–21.

Table 18. Results of t-test for PE experimental group pre–post experiment.

	Category (Mean ± SD)	t	p
Pre-experiment (n = 10)	3.97 ± 0.23		
Post-experiment (n = 10)	3.90 ± 0.31	0.96	0.363

Table 19. Results of t-test for PE control group pre–post experiment.

	Category (Mean ± SD)	t	p
Pre-experiment (n = 10)	3.93 ± 0.22		
Post-experiment (n = 10)	3.37 ± 0.26	7.67	0.000 ***

*** $p < 0.001$.

Table 20. Results of t-test for NE experimental pre–post experiment.

	Category (Mean ± SD)	t	p
Pre-experiment (n = 10)	3.87 ± 0.24		
Post-experiment (n = 10)	4.02 ± 0.21	−2.32	0.046

Table 21. Results of t-test for NE control group pre–post experiment.

	Category (Mean ± SD)	t	p
Pre-experiment (n = 10)	3.82 ± 0.20		
Post-experiment (n = 10)	3.35 ± 0.32	6.82	0.000 ***

*** $p < 0.001$.

The ablation experiment confirmed the effectiveness of the two-tier progressive model. Before the experiment, there were no significant differences in positive (t = 0.55, $p > 0.05$) or negative (t = 0.52, $p > 0.05$) emotion scores between the experimental and control groups, indicating consistent initial emotional states. After the experiment, significant differences were observed in both positive (t = 4.36, $p < 0.001$) and negative (t = 5.83, $p < 0.001$) emotion scores, validating the role of environmental factors in maintaining emotional stability. Before and after change analyses showed that the emotion of the positive emotion experimental group remained stable and there was no significant change in the emotion scores after the experiment (t = 0.96, $p > 0.05$), while the control group had a significant decrease in the emotion scores (t = 7.67, $p < 0.00$). Negative emotion experimental group emotion scores increased slightly after the experiment (t = −2.32, $p < 0.05$), while the control group emotion scores decreased significantly (t = 6.82, $p < 0.001$). These results also indicate that environmental regulation is an important factor influencing emotional stability, validating the effectiveness of the 'environment–emotion' pathway.

5. Discussion

This study delves into the role of immediate emotions in individual decision-making behaviors and extends this exploration into human–machine hybrid systems. Our findings reveal that immediate emotions not only significantly influence risk preferences and cooperation willingness but also play a crucial regulatory role in complex dynamic tasks. By constructing a two-tier progressive inquiry model of "environment–emotion" and "emotion–decision", and integrating classic game scenario with Counterfactual Regret Minimization algorithms, we demonstrate the regulatory impact of immediate emotions in intelligent auxiliary systems. This fills a gap in traditional decision-making theories that have overlooked the influence of immediate emotions and provides new insights for designing efficient human–machine hybrid decision systems.

However, despite these achievements, our study has several limitations. For instance, the diversity of our sample was limited, which restricts the generalizability of our results. Future research could address this limitation by incorporating a broader and more diverse sample, which would enhance the external validity of the findings. Moreover, although we discussed the technical challenges related to authority allocation and system explainability, these remain unresolved. In future studies, exploring new algorithms and models, such as ensemble learning and deep learning approaches, could enhance the transparency and usability of the system. In addition, although Kuhn poker provides us with a simplified experimental environment, its application may not fully represent the more complex real-world decision-making scenarios, such as those in medical artificial intelligence, financial risk analysis, or autonomous driving. In these practical applications, the decision-making process involves not only multivariable analysis but also a high degree of uncertainty, making the scenarios much more complex than those in the poker model. Specifically, in the field of medical AI, real-world medical decisions face additional complexities, including the emotional factors of patients, uncertainty in medical knowledge, and interactions between various stakeholders. Future research could explore how the game strategies from Kuhn poker can be applied to medical decision systems, particularly in handling incomplete information and dynamic situations, and how to effectively balance the interests of all parties. In the field of financial risk analysis, participants in the financial market often face high uncertainty and rapidly changing risk factors. Game theory models similar to Kuhn poker can be used to analyze investor behavior, thereby developing more effective risk management strategies. Future studies should focus on how these game strategies can be integrated with the ever-changing environment of financial markets to improve the accuracy of risk analysis and decision-making. For autonomous driving systems, the real-time decision-making process involves interactions between multiple traffic agents, and coordinating the risk preferences and cooperation willingness of different agents is crucial. Future research could further extend the model from this study, considering more complex traffic scenarios and dynamic decision-making processes, exploring how to effectively coordinate decisions among multiple intelligent agents. In conclusion, future research should consider how the findings from Kuhn poker can be extended to these more complex decision-making scenarios to enhance the practical significance and broader applicability of the research, thereby promoting the application of human–machine hybrid decision systems in a wider range of fields.

Furthermore, this study employs the CFR algorithm to solve the Kuhn poker approximate Nash equilibrium and uses it as the decision-making basis for the decision support system. However, as the complexity of the experimental scenarios increases, the computational complexity of the CFR algorithm will become a potential limiting factor. Specifically, the size of the tree grows exponentially as the problem scale increases, and the time complexity of the CFR algorithm is $O(T \cdot |I| \cdot |A|)$, which means that for large-scale instances, the computation time may exceed the acceptable range for practical applications. In this case, alternative algorithms may provide more effective solutions. Heuristic methods (such as genetic algorithms, simulated annealing, or local search methods) can significantly improve computational speed, although they typically sacrifice some degree of optimality. These methods are particularly useful for large-scale instances where response time is crucial. Furthermore, CFR has some scalable variants, such as CFR+, which improves the efficiency of the algorithm by optimizing the regret update process, and Deep CFR, which integrates deep learning techniques and uses neural networks to estimate the regret values, enabling it to handle more complex and large-scale state spaces. Therefore, although precise methods like CFR remain advantageous for smaller problems, particularly when accuracy is critical, heuristic methods and CFR variants (such as CFR+ and Deep CFR) provide effective alternatives for large-scale optimization tasks with more stringent time constraints.

In this study, we propose a two-tier progressive model of environment–emotion and emotion–decision, where the environment influences emotions, and emotions, in turn, affect decision-making. This model captures the interconnected nature of human decision-making in dynamic environments, where emotional states are shaped by environmental factors and subsequently influence the decision-making process. We chose not to separate the layers of environment and emotion in our analysis because emotions cannot be fully decoupled from their environmental antecedents. Emotions are inherently shaped by the environment, and attempting to isolate them for independent analysis would overlook this intrinsic connection. For example, isolating emotions as an independent variable in decision-making experiments would require disregarding their dependence on environmental factors, thus potentially distorting the understanding of how emotions truly function in decision-making. Given that emotions are an essential intermediary between the environment and decision outcomes, their role is best understood within the context of the environment that induces them. Thus, in the present work, we maintain the integrated structure of our model, which reflects the natural progression from environmental stimuli to emotional response and finally to decision-making. However, we acknowledge that further research could explore the independent influences of these factors more explicitly. Future work may investigate how the environment affects decision-making directly, with a focus on the mediating role of emotions, and whether such causal relationships can be isolated in more controlled experimental settings. This direction would provide a deeper understanding of the underlying mechanisms driving decision-making in complex, dynamic contexts.

Moreover, future research should focus on developing advanced dynamic emotion monitoring technologies to capture and analyze individuals' emotional changes in real-time during complex tasks. Additionally, cross-cultural studies are needed to explore how immediate emotions affect decision-making behaviors in different cultural contexts. Efforts should also be made to improve the explainability of human–machine hybrid systems, enabling users to better understand and trust the suggestions provided by the system.

Alongside technical advancements, another significant aspect that requires careful consideration in AI-driven decision-making is the ethical implications of emotional biases. In particular, the ethical issues surrounding emotional biases have gradually become an important topic. On one hand, AI systems can be designed to "override" the biases introduced by human emotions, especially in high-risk areas such as healthcare or financial decisions, by eliminating the influence of emotions to ensure rational and fair decisions. However, completely ignoring emotional factors may lead to AI decisions lacking empathy and failing to provide personalized support in situations that require emotional care. For example, in mental health counseling or caregiving services, AI that ignores emotional responses may not fully understand the user's emotional needs, thus affecting the quality of the service provided. On the other hand, AI can also be designed to adapt to and respond to human emotional biases, thereby optimizing decision-making personalization and emotional intelligence, improving user experience. For instance, in personalized customer service, AI adjusting based on the user's emotions may help improve service quality and enhance customer satisfaction. However, adapting to emotional biases may introduce the risk of unfairness or exacerbate emotionally driven decision-making. For example, AI might make more impulsive decisions when emotionally charged, relying excessively on emotional judgment while overlooking rational analysis. In conclusion, when designing AI, it is essential to balance the influence of emotions with ethical principles such as fairness, transparency, and accountability, ensuring that AI systems can make decisions that are both ethical and fair in different contexts. At the same time, in order to promote the widespread application and validation of the methodology proposed in this study, and to lay the foundation for our future work, we plan to develop a software package containing the experimental platform and decision support system in our subsequent research. This package will allow researchers to easily implement and extend our methodology, including the CFR-based decision support system and the game platform for Kuhn poker experiments. Through this approach, we hope to provide a reusable tool for researchers in related fields, promoting the popularization and improvement of this method and advancing further research and applications. Ultimately, these endeavors aim to further advance the field.

6. Conclusions

This study investigated the impact of immediate emotions on human–machine interactive systems using a two-stage progressive exploration model ("environment–emotion" and "emotion–decision") within the context of Kuhn poker, supported by an intelligent auxiliary decision system based on CFR algorithm. Our findings indicate that emotional states significantly influence risk-taking behaviors and cooperation levels with AI systems: participants in positive emotional states exhibited greater risk-seeking tendencies and lower cooperation with AI recommendations, while those in negative emotional states showed risk-averse behavior and higher acceptance of AI advice. The ablation experimental results validated the effectiveness of our two-stage model in maintaining emotional stability and influencing decision-making processes. These insights underscore the importance of incorporating emotion-aware design principles in human–machine interactive systems to enhance trust, acceptance, and overall decision-making performance. Future research should explore these dynamics in diverse contexts and with larger sample sizes to further validate and extend these findings, ultimately leading to more intuitive and user-friendly interactive systems.

Author Contributions: Software, formal analysis, investigation, data curation, writing—original draft preparation, H.Y.; conceptualization, supervision, resources, writing—review and editing, project administration, S.L.; methodology and validation, H.Y. and S.L. All authors have read and agreed to the published version of the manuscript.

Funding: This research received no external funding

Institutional Review Board Statement: The research was conducted in accordance with the Declaration of Helsinki and was approved by the Ethics Review Committee of Tongji University, with the approval number tjdxsr2025026.

Informed Consent Statement: Informed consent was obtained from all subjects involved in the study.

Data Availability Statement: The data presented in this study are available on request from the corresponding author.

Acknowledgments: We would like to thank all the players who voluntarily participated in this study. Their contribution was essential to the success of this research.

Conflicts of Interest: The authors declare no conflicts of interest.

References

1. Barbierato, E.; Zamponi, M.E. Shifting Perspectives on AI Evaluation: The Increasing Role of Ethics in Cooperation. *AI* **2022**, *3*, 331–352. [CrossRef]
2. Macmillan-Scott, O.; Musolesi, M. (Ir) rationality in AI: State of the Art, Research Challenges and Open Questions. *arXiv* **2023**, arXiv:2311.17165.

3. Braun, M.; Weber, F.; Alt, F. Affective Automotive User Interfaces–Reviewing the State of Driver Affect Research and Emotion Regulation in the Car. *ACM Comput. Surv.* **2021**, *54*, 137. [CrossRef]
4. Chuah, S.H.W.; Yu, J. The future of service: The power of emotion in human-robot interaction. *J. Retail. Consum. Serv.* **2021**, *61*, 102551. [CrossRef]
5. Xia, S. The role of incidental emotions in decision-making: An empirical study. *J. Behav. Econ. Policy* **2022**, *6*, 79–86.
6. Loewenstein, G.; Weber, E.U.; Hsee, C.K.; Welch, N. Risk as feelings. *Psychol. Bull.* **2001**, *127*, 267. [CrossRef]
7. Riedl, R.; Paul, P. Emotional states in decision making: An experimental investigation of the role of emotions on choice behavior. *J. Econ. Psychol.* **2010**, *31*, 651–661.
8. Brave, S.; Nass, C. Emotion in Human–Computer Interaction. In *The Human-Computer Interaction Handbook: Fundamentals, Evolving Technologies and Emerging Applications*; CRC Press: Boca Raton, FL, USA, 2002 . [CrossRef]
9. Cowie, R.; Douglas-Cowie, E.; Tsapatsoulis, N.; Votsis, G.; Kollias, S.; Fellenz, W.; Taylor, J. Emotion recognition in human-computer interaction. *IEEE Signal Process. Mag.* **2001**, *18*, 32–80. [CrossRef]
10. Prati, E.; Peruzzini, M.; Pellicciari, M.; Raffaeli, R. How to include User eXperience in the design of Human-Robot Interaction. *Robot. Comput.-Integr. Manuf.* **2021**, *68*, 102072. [CrossRef]
11. Stock-Homburg, R. Survey of Emotions in Human–Robot Interactions: Perspectives from Robotic Psychology on 20 Years of Research. *Int. J. Soc. Robot.* **2021**, *14*, 1–23. [CrossRef]
12. Loomes, G.; Sugden, R. Regret Theory: An Alternative Theory of Rational Choice Under Uncertainty. *Econ. J.* **1982**, *92*, 805–824 . [CrossRef]
13. Loomes, G.; Sugden, R. Disappointment and Dynamic Consistency in Choice under Uncertainty. *Rev. Econ. Stud.* **1986**, *53*, 271–282. [CrossRef]
14. Vieira da Costa Cunha, N. Comparison of Different Approaches to Solving Sequential Decision Problems with Applications to Poker. Ph.D. Thesis, Ludwig Maximilian University of Munich, Munich, Germany, 2022.
15. Lewis, M. Self-conscious emotions: Embarrassment, pride, shame, and guilt. *Handb. Cogn. Emot.* **2005**, *25*, 630–640.
16. Dmitrieva, E.S.; Gel'Man, V.I.; Zaĭtseva, K.; Orlov, A.M. Influence of Human Personal Features on Acoustic Correlates of Speech Emotional Intonation Characteristics. *Zhurnal Vyss. Nervn. Deiatelnosti Im. IP Pavlov.* **2009**, *59*, 538–546.
17. Frijda, N.H. *The Emotions*; Cambridge University Press: Cambridge, UK, 1986.
18. Isen, A.M. An influence of positive affect on decision making in complex situations: Theoretical issues with practical implications. *J. Consum. Psychol.* **2001**, *11*, 75–85. [CrossRef]
19. Zinkevich, M.A.; Johanson, M.; Bowling, M.H.; Piccione, C. *Regret Minimization in Games with Incomplete Information*; Curran Associates Inc.: Red Hook, NY, USA, 2007.
20. Li, H.; Wang, X.; Jia, F.; Wu, Y.; Zhang, J.; Qi, S. RLCFR: Minimize Counterfactual Regret by Deep Reinforcement Learning. *Pergamon* **2022** , *187*, 115953. [CrossRef]
21. Isen, A.M.; Daubman, K.A.; Nowicki, G.P. Positive affect facilitates creative problem solving. *J. Personal. Soc. Psychol.* **1987**, *53*, 1122. [CrossRef]
22. Pham, M. All Negative Moods Are Not Equal: Motivational Influences of Anxiety and Sadness on Decision Making. *Organ. Behav. Hum. Decis. Process.* **1999**, *79*, 56–77. [CrossRef]
23. Lerner, J.; Li, Y.; Valdesolo, P.; Kassam, K. Emotion and Decision Making. *Annu. Rev. Psychol.* **2014**, *66*, 799–823. [CrossRef]
24. Béroule, D.G.; Gisquet-Verrier, P. Decision Making guided by Emotion A computational architecture. *arXiv* **2019**, arXiv:1911.09948.
25. Peter, K.; Francis, T.; A, R.J.; Feldman, B.L. The relation between valence and arousal in subjective experience. *Psychol. Bull.* **2013**, *139*, 917–940.
26. Fiorenzato, E.; Bisiacchi, P.; Cona, G. Gender differences in the effects of emotion induction on intertemporal decision-making. *PLoS ONE* **2024**, *19*, e0299591. [CrossRef] [PubMed]
27. Wang, H.; Hughes, J.; Nonaka, T.; Abdulali, A.; Lalitharatne, T.D.; Iida, F. Editorial: AI-powered musical and entertainment robotics. *Front. Robot. AI* **2025**, *12*, 1572828. [CrossRef]
28. Chen, Y.; Wei, W.; Fan, S.; Xu, K.; Chen, D. CoMIF: Modeling of Complex Multiple Interaction Factors for Conversation Generation. In *Proceedings of the 31st International Conference on Computational Linguistics*; Rambow, O., Wanner, L., Apidianaki, M., Al-Khalifa, H., Eugenio, B.D., Schockaert, S., Eds.; Association for Computational Linguistics: Abu Dhabi, United Arab Emirates, 2025; pp. 7355–7366.
29. Jain, A.; Jain, A. AI-Based Emotion Detection System in Healthcare for Patient. In *Generative Artificial Intelligence for Biomedical and Smart Health Informatics*; John Wiley & Sons, Ltd.: Hoboken, NJ, USA, 2025; chapter 23, pp. 455–470. [CrossRef]
30. Chung, K.C.; Tan, P.J.B. Artificial intelligence and internet of things to improve smart hospitality services. *Internet Things* **2025**, *31*, 101544. [CrossRef]
31. Ben, S. Human-Centered Artificial Intelligence: Reliable, Safe and Trustworthy. *Int. J. Hum. Comput. Interact.* **2020**, *36*, 495–504.
32. Huang, Y.; Wang, X.; Zhang, Y.; Chen, L.; Zhang, H. Application of human-in-the-loop hybrid augmented intelligence approach in security inspection system. *Front. Artif. Intell.* **2025**, *8*, 1518850. [CrossRef]

33. Sun, B.; Shan, Y.; Wu, G.; Zhao, S.; Xie, F. Personalized Shared Control for Automated Vehicles Considering Driving Capability and Styles. *Sensors* **2024**, *24*, 7904. [CrossRef]
34. Xian, Y.; Huang, K.; Zhu, Z.; Zhen, S.; Chen, Y.H. Guaranteeing Performance Robust Control for Human–Machine Systems with Optimal Human Decision. *IEEE Trans. Cybern.* **2024**, *55*, 854–866. [CrossRef]
35. Xia, H.; Yu, J.; Nie, K.; Yang, J.; Zhu, L.; Zhang, S. CT Radiomics and human-machine Hybrid System for Differentiating Mediastinal Lymphomas from Thymic Epithelial Tumors. *Cancer Imaging Off. Publ. Int. Cancer Imaging Soc.* **2024**, *24*, 163. [CrossRef]
36. Kuhn, H.W. A simplified two-person poker. *Contrib. Theory Games* **1950**, *1*, 2.
37. Werf, L. Analysis of Nash Equilibria for Kuhn Poker and Its Extensions. Bachelor's Thesis, University of Twente, Enschede, The Netherlands, 2022.
38. Nash, J.F., Jr. Equilibrium points in n-person games. *Proc. Natl. Acad. Sci. USA* **1950**, *36*, 48–49. [CrossRef] [PubMed]
39. Gibson, R. Regret Minimization in Non-Zero-Sum Games with Applications to Building Champion Multiplayer Computer Poker Agents. *arXiv* **2013**, arXiv:1305.0034.
40. Siedlecka, E.; Denson, T.F. Experimental Methods for Inducing Basic Emotions: A Qualitative Review. *Emot. Rev.* **2018**, *11*, 87–97. [CrossRef]
41. Ansani, A.; Marini, M.; Poggi, I.; Mallia, L. Recognition memory in movie scenes: The soundtrack induces mood-coherent bias, but not through mood induction. *J. Cogn. Psychol.* **2023**, *35*, 59–75. [CrossRef]
42. Watson, D.; Clark, L.A.; Tellegen, A. Development and validation of brief measures of positive and negative affect: The PANAS scales. *J. Personal. Soc. Psychol.* **1988**, *54*, 1063. [CrossRef]

applied sciences

MDPI

Article

Enhancing Word Embeddings for Improved Semantic Alignment

Julian Szymański [1,*], Maksymilian Operlejn [1] and Paweł Weichbroth [2]

[1] Department of Computer Systems Architecture, Faculty of Electronics, Telecommunications and Informatics, Gdansk University of Technology, 80-233 Gdansk, Poland

[2] Department of Software Engineering, Faculty of Electronics, Telecommunications and Informatics, Gdansk University of Technology, 80-233 Gdansk, Poland; pawel.weichbroth@pg.edu.pl

* Correspondence: julian.szymanski@pg.edu.pl

Abstract: This study introduces a method for the improvement of word vectors, addressing the limitations of traditional approaches like Word2Vec or GloVe through introducing into embeddings richer semantic properties. Our approach leverages supervised learning methods, with shifts in vectors in the representation space enhancing the quality of word embeddings. This ensures better alignment with semantic reference resources, such as WordNet. The effectiveness of the method has been demonstrated through the application of modified embeddings to text classification and clustering. We also show how our method influences document class distributions, visualized through PCA projections. By comparing our results with state-of-the-art approaches and achieving better accuracy, we confirm the effectiveness of the proposed method. The results underscore the potential of adaptive embeddings to improve both the accuracy and efficiency of semantic analysis across a range of NLP.

Keywords: natural language processing; semantic ambiguity; word vector representation; Word2vec; polysemous word embedding; word sense disambiguation

1. Introduction

To effectively process text, it is necessary to represent it numerically, which allows computers to analyze a document's content. Creating an appropriate representation of a text is fundamental to achieving good results from machine learning algorithms. Traditional approaches to text representation treat documents as points in a high-dimensional space, where each axis corresponds to a word in the vocabulary. In this model, documents are represented as vectors with coefficients indicating the frequency of each word in the text (e.g., using the TF-IDF in a vector space model (VSM) [1]).

But this approach is limited by its inability to capture the semantics of words. When using word-based representations in a VSM, we cannot distinguish between synonyms (words with similar meanings but different forms) and homonyms (words with the same form but different meanings). For example, the words "car" and "automobile" may be treated as completely different entities in word space, even though they have the same meaning. Such information can be provided to the NLP system by an external database. A good example that provides such information is the WordNet dictionary [2], which is organized in the form of a semantic network that contains relationships between synsets that describe groups of words that have the same meaning [3]. Other approaches may include interactions with humans that enrich the vector representations with semantic features [4]. Nevertheless, in the context of NLP, word embeddings have a lot of potential compared to the use of semantic networks. One advantage is their ability to perform classical algebraic operations on vectors, such as addition or multiplication.

To address the problem of semantic word representation, an extension of the VSM was introduced using a method known as word embeddings. In this model, each word is represented as a vector in a multidimensional space, and the similarity between vectors reflects the semantic similarity between words. Models such as Word2Vec [5], GloVe [6], FastText [7] and others learn word representations in an unsupervised manner. The training is based on the context in which words appear in a text corpus, which allows some elementary relationships between them to be captured [8]. In the research presented in our paper, we refer to these embeddings as original embeddings (OE), which were used in the experiments as GloVe vectors.

Using vector-based representations, geometric operations become possible, enabling basic semantic inference. A classic example, proposed by Tomas Mikolov [5], is the operation king-man + woman = queen, which demonstrates how vectors can capture hierarchical relationships between words [9]. These operations enhance the modeling of language dependencies, enabling more advanced tasks like machine translation, semantic searches, and sentiment analysis.

Our study aims to explore methods for creating semantic word embeddings, namely embeddings that carry precise meanings with a strong reference to WordNet. We present methods for building improved word embeddings based on semi-supervised techniques. In our experiments, we refer to these embeddings as neural embeddings (NE). After applying the alignment we refer to them as Fine-tuned Embeddings (FE). Our approach extends the method proposed in [10], modifying word vectors so that geometric operations on them correspond to fundamental semantic relationships. We call this method geometrical embeddings (GE).

The word embeddings generated by our approach have been evaluated in terms of their usefulness for text classification, clustering, and the analysis of class distribution in the embedded representation space.

The rest of this paper is organized as follows: Section 2 describes various methods for creating word embeddings, both supervised and unsupervised. Section 3 presents our approach, including the dataset, the preprocessing methods and techniques used to create semantic word embeddings, and the evaluation methodology. Section 4 presents the experimental results. Section 5 contains a brief discussion of the overall findings, followed by Section 6 which outlines potential applications and future research directions.

2. Related Works

Word embeddings, also known as distributed word representations, are typically created using unsupervised methods. Their popularity stems from the simplicity of the training process, as only a large corpus of text is required. The embedding vectors are first constructed from sparse representations and then mapped to a lower-dimensional space, producing vectors that correspond to specific words. To provide context for our research, we briefly describe the most popular approaches to building word embeddings.

After the pioneering work of Bengio et al. on neural language models [11], research on word embeddings stalled due to limitations in computational power and algorithms being insufficient for training large vocabularies. However, in 2008, Collobert and Weston showed that word embeddings trained on sufficiently large datasets capture both syntactic and semantic properties, improving performance in subsequent tasks. In 2011, they introduced SENNA embeddings [12], based on probabilistic models that allow for the creation of word vectors. For each training iteration, an n-gram was used by combining the embeddings of all n words. The model then modified the n-gram by replacing the middle word with a random word from the vocabulary. It was trained to recognize that the intact n-gram was more meaningful than the modified (broken) one, using a hinge loss function. Shortly thereafter, Turian et al. replicated this method, but instead of replacing the middle word, they replaced the last word [13], achieving slightly better results.

These embeddings have been tested on a number of NLP tasks, including Named Entity Recognition (NER) [14], Part of Speech (POS) [15], Chunking (CHK) [16], Syntactic Parsing (PSG) [17], and Semantic Role Labeling (SRL) [18]. This approach has shown improvement in a wide number of tasks [19] and is computationally efficient, using only 200 MB of RAM.

Word2vec is a set of models introduced by Mikolov et al. [5] that focuses on word similarity. Words with similar meanings should have similar vectors, while words with different vectors do not have similar meanings. Considering the two sentences *"They like watching TV "* and *"They enjoy watching TV"*, we can conclude that the words *like* and *enjoy* are very similar, although not identical. Primitive methods like one-hot encoding would not be able to capture the similarity between them and would treat them as separate entities, so the distance between the words *{like, TV}* would be the same as the distance between *{like, enjoy}*. In the case of word2vec, the distance between *{like, TV}* would be greater than the distance between *{like, enjoy}*. Word2vec allows one to perform basic algebraic operations on vectors, such as addition and subtraction.

The GloVe algorithm [6] integrates word co-occurrence statistics to determine the semantic associations between words in the corpus, which is somewhat the opposite to Word2vec—which depends only on local features in words. GloVe implements the global matrix factorization technique, which uses a matrix to encode the presence or absence of word occurrences. Word2vec is commonly known as neural word embeddings; hence, it is a feedforward neural network model, whereas GloVe is commonly known as a count-based model, also called a log-bilinear model. By analyzing how frequently words co-occur in a corpus, GloVe captures the relationships between words. The co-occurrence likelihood can encode semantic significance, which helps improve performance on tasks like the word analogy problem.

FastText is another method developed by Joulin et al. [20] and further improved [21] by introducing an extension of the continuous skipgram model. In this approach, the concept of word embedding differs from Word2Vec or GloVe, where words are represented by vectors. Instead, word representation is based on a bag of character n-grams, with each vector corresponding to a single character n-gram. By summing the vectors of the corresponding n-grams, we obtain the full word embedding.

As the name suggests, this method is fast. In addition, since the words are represented as a sum of n-grams, it is possible to represent words that were not previously added to the dictionary, which increases the accuracy of the results [7,22]. The authors of this approach have also proposed a compact version of fastText, which, due to quantization, maintains an appropriate balance between model accuracy and memory consumption [23].

Peters et al. [24] introduced the ELMo embedding method, which uses a bidirectional language model (biLM) to derive word representations based on the full context of a sentence. Unlike traditional word embeddings, which map each token to a single dense vector, ELMo computes representations through three layers of a language model. The first layer is a Convolutional Neural Network (CNN), which produces a non-contextualized word representation based on the word's characters. This is followed by two bidirectional Long Short-Term Memory (LSTM) layers that incorporate the context of the entire sentence.

By design, ELMo is an unsupervised method, though it can be applied in supervised settings by passing pretrained vectors into models like classifiers [25]. When learning a word's meaning, ELMo uses a bidirectional RNN [26], which incorporates both preceding and subsequent context [24]. The language model learns to predict the likelihood of the next token based on the history, with the input being a sequence of n tokens.

Bidirectional Encoder Representations of Transformers, often referred to as BERT [27], is a transformer-based architecture. Transformers were first introduced relatively recently, in 2017 [28]. This architecture does not rely on recursion, as previous state-of-the-art systems (such as LSTM or ELMo) did, significantly speeding up training time. This is because sentences are processed in parallel rather than sequentially, word by word. A key unit, self-attention, was also introduced to measure the similarity between words in a sentence. Instead of recursion, positional embeddings were created to store information about the position of a word/token in a sentence.

BERT was designed to eliminate the need for decoders. It introduced the concept of masked language modeling, where 15% of the words in the input are randomly masked and predicted using positional embeddings. Since predicting a masked token only requires the surrounding tokens in the sentence, no decoder is needed. This architecture set a new standard, achieving state-of-the-art results in tasks like question answering. However, the model's complexity leads to slow convergence [27].

Beringer et al. [10] proposed an approach to represent unambiguous words based on their meaning in the context [29]. In this method, word vector representations are adjusted according to polysemy: vectors of synonyms are brought closer together in the representation space, while vectors of homonyms are separated. The supervised iterative optimization process leads to the creation of vectors that offer more accurate representations compared to traditional approaches like word2vec or GloVe.

Pretrained models can be used across a wide range of applications. For instance, the developers of Google's Word2Vec trained their model on unstructured data from Google News [30], while the team at Stanford used data from Wikipedia, Twitter, and web crawling for GloVe [31]. Competing with these pretrained solutions is challenging; instead, it is often more effective to leverage them. For example, Al-Khatib and El-Beltagy applied the fine-tuning of pretrained vectors for sentiment analysis and emotion detection tasks [32].

To further leverage the power of pretrained embeddings, Dingwall and Potts developed the Mittens tool for fine-tuning GloVe embeddings [33]. The main goal is to enrich word vectors with domain-specific knowledge, using either labeled or unlabeled specialized data. Mittens transforms the original GloVe goal into a retrofitting model by adding a penalty for the squared Euclidean distance between the newly learned embeddings and the pretrained ones. It is important to note that Mittens adapts pretrained embeddings to domain-specific contexts, rather than explicitly incorporating semantics to build semantic word embeddings.

3. Material and Methods

In this section, we describe our method for building word embeddings that incorporate elementary semantic information. We begin by detailing the dedicated dataset used in the experiments. Next, we explain our method, which modifies the hidden layer of a neural network initialized with pretrained embeddings. We then describe the fine-tuning process, which is based on vector shifts. The final subsection outlines the methodology used to evaluate the proposed approach.

3.1. Dataset

For our experiments, we constructed a custom dataset consisting of four categories: *animal*, *meal*, *vehicle*, and *technology*. Each category contains six to seven keywords, for a total of 25 words.

We also collected 250 sentences for each keyword (6250 in total) using web scraping from various online sources, including most English dictionaries. The sentences contain no more than one occurrence of the specified keyword.

The words defining each category were selected such that the boundaries between categories were fuzzy. For instance, the categories *vehicle* and *technology*, or *meal* and *animal*, overlap to some extent. Additionally, some keywords were chosen that could belong to more than one category but were deliberately grouped into one. An example is the word *fish*, which can refer to both a *food* and an *animal*. This was carried out intentionally to make it more difficult to classify the text and highlight the contribution our word embeddings aim to address.

To prepare and transform raw data into useful formats, preprocessing is essential for developing or modifying word embeddings [34,35]. The following steps were taken to prepare the data for further modeling:

1. Change all letters to lowercase;
2. Remove all non-alphabetic characters (punctuation, symbols, and numbers);
3. Remove unnecessary whitespace characters: duplicated spaces, tabs, or leading spaces at the beginning of a sentence;

4. Remove selected stop words using a function from the Gensim library: articles, pronouns, and other words that will not benefit the creation of embeddings;
5. Tokenize, i.e., divide the sentence into words (tokens);
6. Lemmatize, bringing variations of words into a common form, allowing them to be treated as the same word [36]. For example, the word *cats* would be transformed into *cat*, and *went* would be transformed into *go*.

3.2. Method of Semantic Word Embedding Creation

In this section, we describe in detail our method for improving word embeddings. This process involves three steps: training a neural network to obtain an embedding layer, tuning the embeddings, and shifting the embeddings to incorporate specific geometric properties.

3.2.1. Embedding Layer Training

For building word embeddings in the neural network, we use a hidden layer with initial weights set to pretrained embeddings. In our tests, the architectures with fewer layers performed best, likely due to the small gradient in the initial layers. As the number of layers increased, the initial layers were updated less effectively during backpropagation, causing the embedding layer to change too slowly to produce strongly semantically related word embeddings. As a result, we used a smaller network for the final experiments.

The algorithm for creating semantic embeddings using a neural network layer is as follows:

1. Load the pretrained embeddings.
2. Preprocess the input data.
3. Separate a portion of the training data to create a validation set. In our approach, the validation set was 20% of the training data.
   Next, the initial word vectors were combined with the pretrained embedding data. To achieve this, an embedding matrix was created, where the row at index *i* corresponds to the pretrained embedding of the word at index *i* in the vectorizer.
4. Load the embedding matrix into the embedding layer of the neural network, initializing it with pretrained word embeddings as weights.
5. Create a neural network-based embedding layer.
   In our experiments, we tested several architectures. The one that worked best was a CNN with the following configuration.

```
inputs = keras.Input(shape=(None,), dtype="int64")
x = word_embeddings_layer(int_sequences_input)
x = Conv1D(128, 5, activation="relu")(x)
x = MaxPooling1D(5)(x)
x = Conv1D(128, 5, activation="relu")(x)
x = GlobalMaxPooling1D()(x)
x = Dense(128, activation="relu")(x)
x = Dropout(0.5)(x)
outputs = layers.Dense(len(class_names), activation="softmax")(x)

model = Model(inputs=inputs, outputs=output)
```

6. Train the model using the prepared data.
7. Map the data into words in vocabulary embeddings created in hidden layer weights.

```
def get_word_embeddings_from_weights(model):
  weights = model.get_weights()[0]
  # Empty dict to store embeddings
  # The key is a unique word and the value is embedding
  embedding_dict = {}
  # Iterate through vocabulary
  for word in vocabulary:
  # Get index by word, then get proper embedding by index
      embedding_dict.update({
```

```
            word: weights[word_index.get(word)]
            })
    return embedding_dict
```

8. Save the newly created embeddings in a pickle format.

After testing several network architectures, we found that the models with fewer layers performed best. This is likely due to the small gradient in the initial layers. As the number of layers increases, the initial layers are less effectively updated during backpropagation, causing the embedding layer to change too slowly to produce strongly semantic word embeddings. For this reason, a smaller network was used for the final experiments, with the following configuration.

```
inputs = keras.Input(shape=(None,), dtype="int64")
x = word_embeddings_layer(int_sequences_input)
x = Conv1D(128, 5, activation="relu")(x)
x = GlobalMaxPooling1D()(x)
x = Dropout(0.5)(x)
outputs = layers.Dense(len(class_names), activation="softmax")(x)
model = Model(inputs=inputs, outputs=output)
```

3.2.2. Fine-Tuning of Pretrained Word Embeddings

This step uses word embeddings created in a self-supervised manner by fine-tuning pretrained embeddings.

The entire process for training embeddings is encapsulated in the following steps:

1. Load the pretrained embeddings. The algorithm uses GloVe embeddings as input, so the embeddings need to be converted into a dictionary.
2. Preprocess the input data (described in Section 3.1).
3. Prepare a co-occurrence matrix of words using the GloVe mechanism.
4. Train the model using the function provided by Mittens for fine-tuning GloVe embeddings [33].
5. Save the newly created embeddings in a pickle format.

3.2.3. Embedding Shifts

The goal of this step is to adjust word vectors in the representation space, enabling geometric operations on pretrained embeddings that incorporate semantics from the WordNet network. To achieve this, we used the implementation by Beringer et al., available on GitHub [37]. However, to adapt the code to the planned experiments, the following modifications were introduced:

- Modification of the pretrained embeddings: Since Beringer et al. originally used embeddings from the SpaCy library as a base [38], even though they also used the GloVe method for training, the data they used differed. Therefore, it was necessary to modify how the vectors were loaded. While SpaCy provides a dedicated function for this task, it had to be replaced with an alternative.
- Manual splitting of the training set into two smaller sets, one for training (60%) and one for validation (40%), aimed at proper hyperparameter selection.
- Adapting other features: Beringer et al. created embeddings by combining the embedding of a word and the embedding of the meaning of that word, for example, the word *tree* and its meaning *forest*, or *tree (structure)*. They called them *keyword embeddings*. In the case of the experiment conducted in this thesis, the embeddings of the word and the category were combined, like *cat* and *animal*, or *airplane (technology)*. An embedding created in this way is called a keyword–category embedding. The term *keyword embedding* is used to describe any word contained in categories, such as *cat*, *dog*, etc. The term *category embedding* describes the embedding of a specific category word, namely *animal*, *meal*, *technology*, or *vehicle*.

- Adding hyperparameter selection optimization:

Finally, for the sake of clarity, the algorithm design was as follows:

1. Upload or recreate non-optimized keyword–category embeddings. For each keyword in the dataset and its corresponding category, we created an embedding, as shown in Equation (1), which included both sample objects and category items:

$$kc_{fish(animal)} = e(fish, animal) = \frac{e(fish) + e(animal)}{2} \tag{1}$$

2. Load and preprocess the texts used to train the model.
3. For each epoch:
    3.1. For each sentence in the training set:
        3.1.1. Calculate the context embedding. The context itself is the set of words surrounding the keyword, as well as the keyword itself. We create the context embedding by averaging the embeddings of all the words that make up that context.
        3.1.2. Measure the cosine distance between the computed context embedding and all category embeddings. In this way, we select the closest category embedding and check whether it is a correct or incorrect match.
        3.1.3. Update keyword–category embeddings. The algorithm moves the keyword–category embedding closer to the context embedding if the selected category embedding turns out to be correct. If not, the algorithm moves the keyword–category embedding away from the context embedding. The alpha (α) and beta (β) coefficients determine how much we manipulate the vectors.
4. Save the newly modified keyword–category embeddings in the *pickle* format. The *pickle* module is a method for serializing and deserializing objects, developed in the Python programming language [39].

The top-k accuracy on the validation set was used as a measure of the quality of the embeddings to be used in selecting appropriate hyperparameters and selecting the best embeddings. This is a metric that determines whether the correct category is in the top k categories, where the input data are context embeddings from the validation set. Due to the small number of categories (4 in total), it was decided to evaluate the quality of the embeddings using the top-1 accuracy. Finally, a random search yielded the optimal hyperparameters.

3.3. Semantic Word Embedding Evaluation

We performed the evaluation of our method for building the word embeddings using two applications. We tested how they influence the quality of the classification, and we measured their distribution and plotted them using PCA projections. We also evaluated how the introduction of the embeddings influences the unsupervised processing of the data.

3.3.1. Text Classification

The first test for the newly created semantic word embeddings was based on a prediction of the category from the sentence. This can be carried out in many ways, for example, by using the embeddings as a layer in a deep neural network. Initially, however, it would be better to test the capabilities of the embeddings themselves without further training on the new architecture. Thus, we used a method based on measuring the distance between embeddings. In this case, our approach consisted of the following steps:

1. For each sentence in the training set:
    1.1. Preprocess the sentence.
    1.2. Convert the words from the sentence into word embeddings.
    1.3. Calculate the mean embedding from all the embeddings in the sentence.

1.4. Calculate the cosine distance between the mean sentence embedding and all category embeddings based on Equation (2):

$$distance(\bar{e}, c) = \cos(\bar{e}, c) = \frac{\bar{e} \cdot c}{||\bar{e}||\,||c||} \quad (2)$$

where $\bar{e}$ is the mean sentence embedding and c is the category embedding.

1.5. Select the assigned category by taking the category embedding with the smallest distance from the sentence embedding.

2. Calculate the accuracy score by taking the predictions for all sentences based on Equation (3):

$$Accuracy = \frac{Number\ of\ correct\ predictions}{Total\ number\ of\ predictions} \quad (3)$$

In addition, in modern research, weighted embeddings are typically used to improve the quality of the prediction [40,41]. Latent semantic analysis uses a term–document matrix that takes into account the frequency of a term's occurrence. This is accomplished by using the TF-IDF weighting schema, which is used to measure the importance of a term by its frequency of occurrence in a document.

To fully see the potential of word embeddings in classification, it is beneficial to combine them with a separate classification model, distinct from the DNN used to create the embeddings. We chose a KNN and Random Forest [42], which are non-neural classifiers, to avoid introducing any bias into the results. It should be noted that for the evaluation of the embeddings, any other classifier can be used.

3.3.2. Separability and Spatial Distribution

To measure the quality of the semantic word embeddings, we checked their spatial distribution by performing two tests. The first was an empirical test aimed at reducing the dimensionality of the embeddings and visualizing them in two-dimensional space using the Principal Component Analysis (PCA) method, which transforms multivariate data into a smaller number of representative variables, preserving as much accuracy as possible. In other words, the goal is to achieve a trade-off between the simplicity of the representation and accuracy [43].

Still considering the quality of the semantic word embeddings, we computed the relevant metrics, which prove the separability of the classes, as well as the correct distribution of the vectors in space. In view of this, we used the mean distance of all the keywords to their respective category words. The purpose of checking these distances is to test whether the average embedding of keywords in a given category yields a vector that is close to the category vector, evaluated based on Equation (4), with sample objects from the same semantic category.

$$\frac{e(computer) + e(CPU) + e(keyboard) + e(monitor) + e(TV) + e(phone)}{6} \approx e(technology) \quad (4)$$

The formula for calculating the average distance of keyword embeddings from their category embeddings is based on Equation (5).

$$meanDistanceToCategory(c, C) = \frac{1}{|C|}\sum_{k \in C} \cos(c, k) = \frac{1}{|C|}\sum_{k \in K} \frac{c \cdot k}{||c||\,||k||} \quad (5)$$

where

- c is a category embedding;
- C is a list of all the keyword embeddings in the category.

The local density of classes was another metric used. This is the mean distance of all the keywords to the mean vector of all the keywords in the category, calculated based on Equation (6).

$$categoryDensity(C) = \frac{1}{|C|} \sum_{k \in C} \cos(k, \frac{1}{|C|} \sum_{l \in C} l) \quad (6)$$

where

- c is a category embedding;
- C is a list of all the keyword embeddings in the category.

The Silhouette Coefficient was another metric used to examine class separability [44]. It is built with two values:

- a: the mean distance between a data sample and other data points in the same class (cluster).
- b: the mean distance between a data sample and other data points in the next nearest class.

Considering a single sample, the Silhouette Coefficient is calculated based on Equation (7):

$$s = \frac{b - a}{max(a, b)} \quad (7)$$

It is important to note that the Silhouette Coefficient is computed for only one sample, whereas if more than one sample were considered, it would be sufficient to take the average of the individual values. In order to test the separability of all classes, we calculated the average Silhouette Coefficient for all keyword embeddings.

4. Results

In this section, we present a summary of the results for creating semantic embeddings using the methods discussed above. Each section refers to a specific method for testing the embeddings and provides the results in both tabular and graphical form. To maintain methodological clarity, we developed a naming scheme to denote specific embedding methods, namely, original embeddings (OE), the basic embeddings that served as the basis for the creation of others; neural embeddings (NE), embeddings created using the embedding layer in the DNN-based classification process; fine-tuned embeddings (FE), embeddings created with fine-tuning, that is, using the GloVe training method on new semantic data; and geometrical embeddings (GE), word vectors that were created by moving vectors in space.

4.1. Text Classification

First, we tested the quality of text classification using the semantic embeddings mentioned above. It is worth recalling that a weighted variant of the vectors using the IDF measure was also used for testing. So additionally, in classification experiments, we added the results from the usage of the re-weighed embedded representations: original weighted embeddings (OWE), fine-tuned weighted embeddings (FWE), neural weighted embeddings (NWE), and geometrical weighted embeddings (GWE).

First, text classification was checked by selecting classes for sentences by checking the shortest distance between the mean sentence embedding and the category embedding. Figure 1 shows the accuracy results for all the embedding methods, sorted in descending order for better readability. It turns out that geometric embeddings perform best (93.8%), with a huge lead of 6 percentage points over second place, which was taken by neural embeddings (87.78%).

This is a logical consequence of how these embeddings were trained since the geometrical embeddings underwent the most drastic change and thus were moved the most in space compared to the other methods. Moving them exactly toward the category embeddings at the training level led to better results. The other two methods, namely neural embeddings and fine-tuned embeddings, performed comparably well, both around 87%.

Each of the new methods outperformed the original embeddings (83.89%), so all of these methods improved the classification quality.

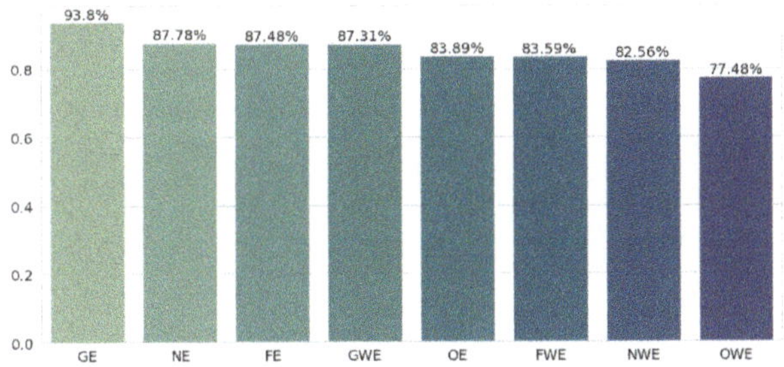

Figure 1. Classification results for the nearest neighbor method.

All the weighted embeddings perform significantly worse than their unweighted counterparts, with a difference of 5 or 6 percentage points, which was expected given that semantic embeddings occur in every sentence in the dataset. It turns out that only the geometrically weighted embeddings outperform the original embeddings.

Next, the classification quality was examined using the average sentence embeddings as features for the non-neural text classifiers. The results improved significantly over the previous classification method, and the ranking of the results also changed. This time (see Figure 2), fine-tuned embeddings and neural embeddings were the best, with the latter taking the advantage (97.95% and 96.58%, respectively). The geometric embedding method (92.39%) performed much worse, coming close to the original embeddings' result (91.97%).

Thus, it seems that despite the stronger shift in the vectors in space in the case of geometric embeddings, they lose some features and the classification suffers. In the case of the mean sentence distance to a category embedding, geometric embeddings showed their full potential, since both the method for creating embeddings and the method for testing them were based on the geometric aspects of vectors in space. In contrast, the Random Forest classification is based on individual features (vector dimensions), and each decision tree in the ensemble model consists of a limited number of features, so it is important that individual features convey as much value as possible, rather than the entire vector as a whole.

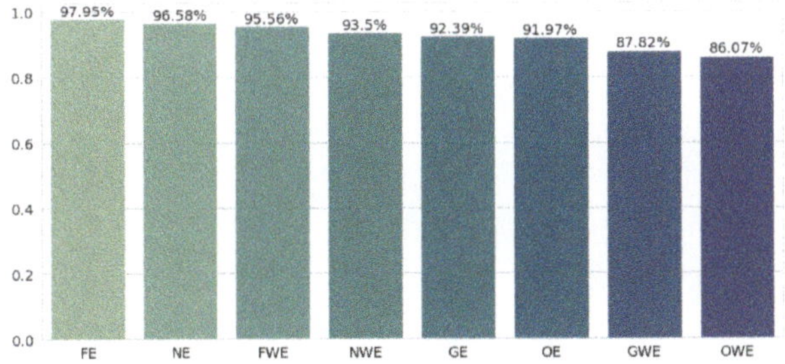

Figure 2. Classification results for the random forest classifier with mean sentence embedding.

Again, weighted embeddings achieve the worse results, although in this case, the difference between weighted embeddings and their unweighted counterparts is much smaller, similar to the difference in the source [40]. However, it turns out that even weighted embeddings, for the neural and fine-tuned methods (95.56% and 93.5%), perform better than the basic version of geometric embeddings (92.39%). Original weighted embeddings perform the worst, with an accuracy drop of almost 12 percentage points (97.95% to 86.07%).

Table 1 shows a summary of the results for Random Forest classification and mean sentence embedding distance.

Table 1. Classification results for weighted and unweighted semantic embeddings. (Underline—best results).

Embeddings	Classification with Random Forest	Classification with Mean Sentence Embedding Distance
Original	0.920	0.839
Original Weighted	0.861	0.775
Neural	0.966	0.875
Neural Weighted	0.935	0.836
Fine-tuned	<u>0.979</u>	0.878
Fine-tuned Weighted	0.956	0.826
Geometrical	0.924	<u>0.938</u>
Geometrical Weighted	0.878	0.873

4.2. Separability and Spatial Distribution

In the next test, the separability of the classes in space and the overall distribution of the vectors were tested. First, the distribution of keyword embeddings in space is shown using the PCA method to reduce dimensionality to two dimensions [43]. Figure 3 shows the distribution of the original embeddings.

As can be seen, even here, the classes seem to be relatively well separated. The classes *technology* and *vehicle* seem to be far apart in space, while *meal* and *animal* are closer together. Note that the keyword *fish* is much further away from the other keywords in the cluster, heading towards the category *meal*. Also, the word *kebab* seems to be far away from the correct cluster, even though it refers to a meal and does not have a meaning correlated with animals.

Figure 4 shows a visualization of the PCA for neural embeddings. In this case, the proximity of vectors within classes is visible, and each class becomes more separable. One can see a significant closeness to the corresponding classes of words that were previously more distant, such as *fish* and *kebab*, although the latter is somewhere roughly between the centers of the two clusters, so clustering with the K-means algorithm later in this section may still be somewhat problematic.

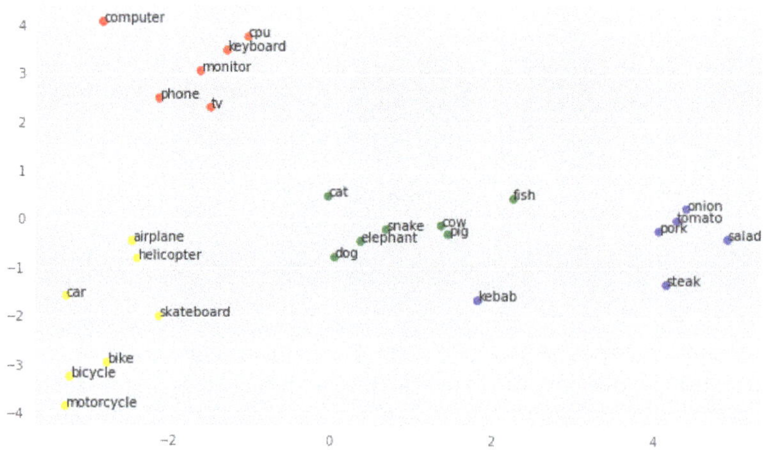

Figure 3. PCA visualization of original embeddings.

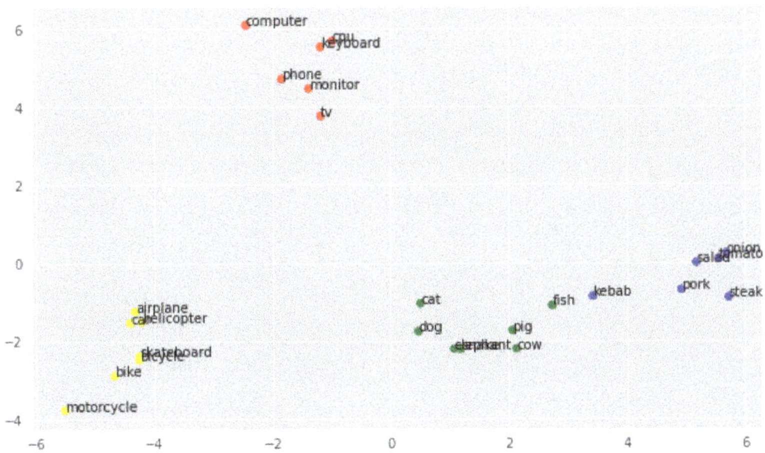

Figure 4. PCA visualization of neural embeddings.

Figure 5 shows how vectors decompose in two-dimensional space for fine-tuned embeddings, and this method seems to have the worst separability of all the methods so far. Once again, the words *fish* and *kebab*, which are separated from their classes while being very close to each other, proved to be problematic. Also, the classes *technology* and *vehicle* seem to be more similar to each other, which will be checked with the metrics shown later.

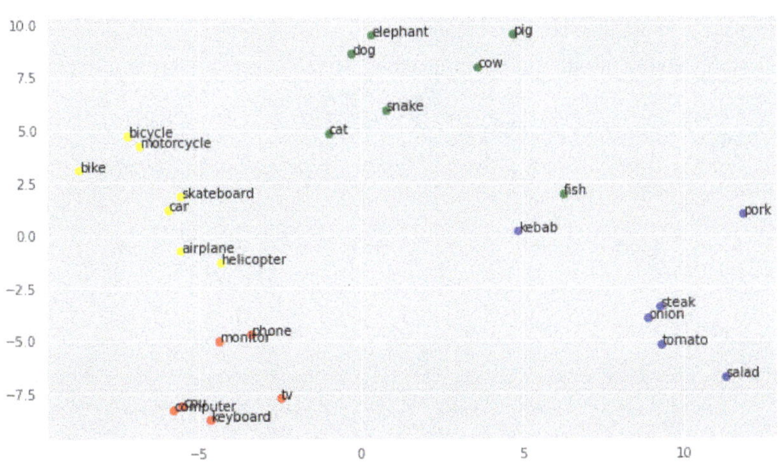

Figure 5. PCA visualization of fine-tuned embeddings.

The last of the visualizations of vectors in two-dimensional space using the PCA method concerns geometrical embeddings, as illustrated in Figure 6. In this case, the classes are brilliantly separable, the words within a single cluster are very close to each other, and the problematic cases are resolved. This was to be expected given the nature of the method used to create geometric embeddings, i.e., strong interference with the positions of vectors in space using the $alpha$ and $beta$ parameters.

Next, the mean distance from the semantic embeddings to the category embeddings was checked. The results are shown in Figure 7. It turns out that the geometric embeddings are by far the best here (0.17), improving the result of the original embeddings (0.6) by more than three times. The characteristics of the embedding displacement in the space method indicated that it would perform best for this particular test, but the results for the neural embeddings (0.617) and the fine-tuned embeddings (0.681) are surprising, as they were worse compared to the original embeddings.

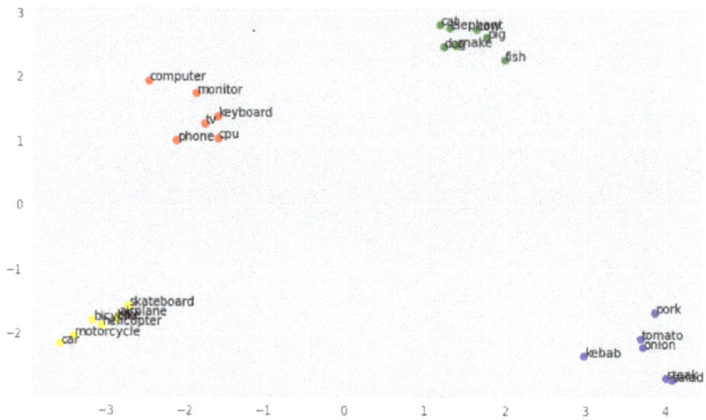

Figure 6. PCA visualization of geometrical embeddings.

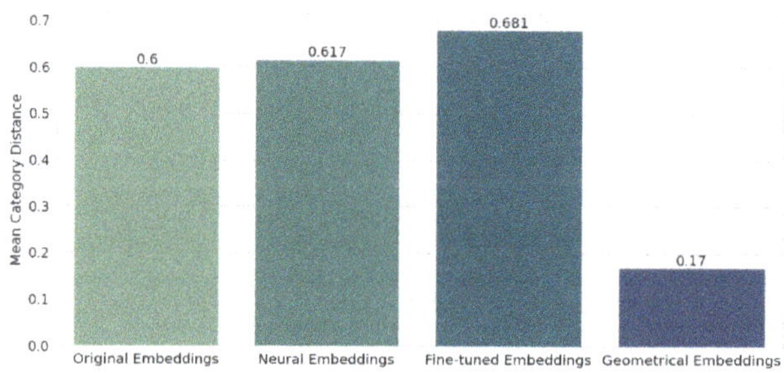

Figure 7. Results for usage of mean category distance.

In order to look deeper into the data, it was decided to also present the mean category distance relative to each category separately. As can be seen in Figure 8, the category that allowed the keyword embeddings to come closest to the category embeddings was the *animal* category. By far, the worst category in this respect was *technology*, although the geometric embeddings also did well here (0.201). The only category that managed to improve its performance (excluding geometric embeddings) was the *food* category, where the neural embeddings (0.614) performed slightly better than the original embeddings (0.631).

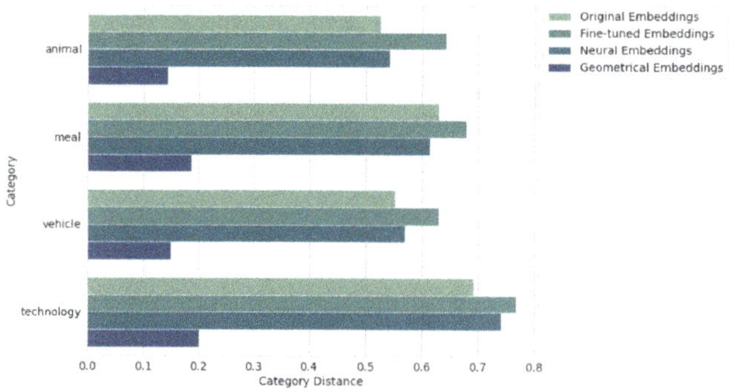

Figure 8. Results for mean category distance by category.

Next, we examined the category density metric, the results of which are shown in Figure 9. Here, the situation looks a bit better, as the results for two methods—neural (0.278) and, of course, geometric (0.105), improved over the original embeddings (0.321). This confirms what was observed when visualizing the PCA for the different types of embeddings, namely that geometric embeddings have excellent separability, and the keywords within classes are very close to each other, while fine-tuned embeddings blur the spatial differences between clusters, and individual words move away from their respective clusters.

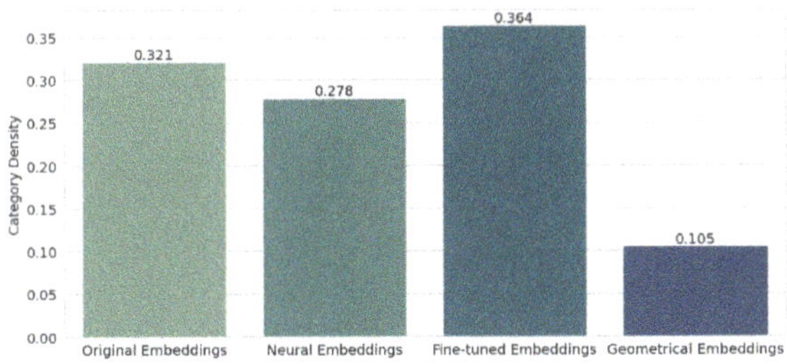

Figure 9. Results for Category Density.

Figure 10 shows the category density results for each category separately. Our suspicion about the problematic *meal* and *animal* classes for fine-tuned embeddings, observed in the PCA visualization in Figure 5, was confirmed. In fact, the values for these categories are very high (0.349 and 0.386, respectively), and the *technology* category also has a low density (0.384). Meanwhile, the least problematic category for most methods is the *vehicle* class.

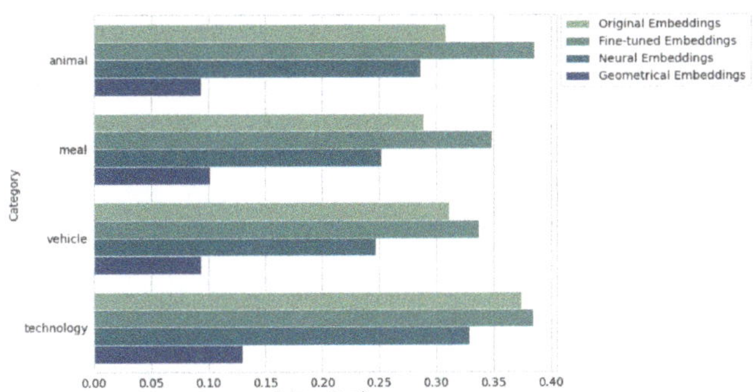

Figure 10. Results for category density by category.

The results for the previous two metrics are also reflected in the results for the Silhouette Score. The clusters for the geometric embeddings (0.678) are dense and well separated. The neural embeddings (0.37) still have better cluster quality than the original embeddings (0.24), while the fine-tuned embeddings (0.144) are the worst. Note that none of the embeddings have a silhouette value below zero, so there is little overlap between the categories.

The final test in this section is to see how the K-means method handles class separation. Table 2 shows the results for all metrics tested on the clusters generated by the K-means algorithm, comparing these clusters to the ground truth. It turns out that even in the case of the original embeddings, all these metrics reach the maximum value so that the clustering is performed flawlessly. The same is true for neural and geometric embeddings.

Table 2. Results for K-means clustering.

Metrics	Original	Neural	Fine-Tuned	Geometrical
Rand Index	1.000	1.000	0.887	1.000
Homogeneity	1.000	1.000	0.747	1.000
Completeness	1.000	1.000	0.762	1.000
V-measure	1.000	1.000	0.754	1.000

The only method with worse results is again the fine-tuning of pretrained word embeddings, which scores lower than the others for each metric. The misclassified keywords were *fish*, which was added to the cluster *meal* instead of *animal*; *phone*, which was added to the cluster *vehicle* instead of *technology*; and *tv*, which was added to the cluster *animal*, instead of *technology*.

The *technology* category was the most difficult to cluster correctly. This was not obvious from the PCA visualization (see Figure 5), which only shows the limitations of this method, while it is almost impossible to reduce 300-dimensional data to 2 dimensions without losing relevant information.

Table 3 shows a summary of the results for all metrics. For separability and spatial distribution, the method used to create the geometric embeddings performs by far the best, and the neural embeddings also slightly improve the results compared to the original embeddings. The worst embeddings are the fine-tuned embeddings, which seem to become blurred and degraded during the learning process, which was quite unexpected given that these embeddings performed best for the text classification problem using Random Forest.

Table 3. Summarized results for separability and spatial distribution.

Metrics	Original	Neural	Fine-Tuned	Geometrical
Mean Category Distance	0.600	0.617	0.681	0.170
Category Density	0.321	0.278	0.364	0.105
Silhouette Score	0.240	0.370	0.144	0.678
Rand Index	1.000	1.000	0.887	1.000
Homogeneity	1.000	1.000	0.747	1.000
Completeness	1.000	1.000	0.762	1.000
V-measure	1.000	1.000	0.754	1.000

5. Discussion

The contribution of our study is the introduction of a supervised method for aligning word embeddings by using the hidden layer of a neural network, their fine-tuning, and vector shifting in the embedding space. For the text classification task, a simpler method that examines the distance between the mean sentence embeddings and category embeddings showed that geometrical embeddings performed best. However, it is important to note that these methods rely heavily on the distance between embeddings, which may not always be the most relevant measure. When using embeddings as input data for the Random Forest classifier, geometrical embeddings performed significantly worse than the other two methods, with fine-tuned embeddings achieving higher accuracy by more than 5 percentage points (97.95% compared to 92.39%).

The implication of this is that when creating embeddings using the vector-shifting method, some information that could be useful for more complex tasks is lost. Additionally, excessively shifting the vectors in space can isolate these embeddings, causing the remaining untrained ones, which intuitively should be close by, to drift apart with each iteration of training. This suggests that for small, isolated sets of semantic word embeddings, geometrical embeddings will perform quite well. However, for open-ended problems with a larger vocabulary and more blurred category boundaries, other methods are likely to be more effective.

While fine-tuned embeddings performed well in the text classification task, they also have their drawbacks. Examining the distribution of vectors in space reveals that embeddings within a class tend to move farther apart, which weakens the semantic relationships between them. This becomes evident in test tasks, such as clustering, where the fine-tuned embeddings performed even worse than the original embeddings. On the other hand, enriching embeddings with domain knowledge improves their performance. Embeddings trained with the GloVe method on specialized texts acquire more semantic depth, as seen in the Random Forest classification results, where they achieved the second-best performance for category queries.

The most well-balanced method appears to be vector training using the trainable embedding layer. Neural embeddings yield very good results for text classification, both with the distance-based method and when using Random Forest. While the embeddings within classes may not be very dense, the classes themselves are highly separable.

6. Conclusions and Future Works

In this paper, we propose a method for aligning word embeddings using a supervised approach that employs the hidden layer of a neural network and shifts the embeddings into the specific categories they correspond to. We evaluate our approach from multiple perspectives: both in an application for supervised and non-supervised tasks and through the analysis of the vector distribution in the representation space. The tests confirm the methods' usability and provide a deeper understanding of the characteristics of each proposed method.

By comparing our results with state-of-the-art approaches and achieving better accuracy, we confirm the effectiveness of the proposed method. A deep analysis of the vector distributions reveals that there is no one-size-fits-all solution for generating semantic word embeddings; the choice of method should depend on specific design goals and expected outcomes. However, our approach extends traditional methods for creating word vectors.

The results of this research are promising, although some important issues remain unresolved, indicating that further development is needed. The potential applications of semantic word embeddings are vast, and developing new test environments could lead to interesting findings. One rapidly growing area of AI is recommendation systems. McKinsey reports that up to 35% of products purchased on Amazon and 75% of content watched on Netflix come from embedded recommendation systems [45]. Word embedding-based solutions have already been explored and identified as a promising alternative to traditional methods [46].

It is therefore worthwhile to explore whether semantic word embeddings can improve performance. In the context of test environments, embedding weights can be adjusted to enhance prediction accuracy. While IDF weights were used to limit the prediction model, future work could experiment with weights that further emphasize the semantics of the word vectors.

An interesting experiment could involve adding words to the dataset that are strongly associated with multiple categories. For instance, the word jaguar is primarily linked to an animal but also frequently appears in the context of a car brand. It would be valuable to see how different embedding creation methods handle such ambiguous data.

In terms of data modification, leveraging well-established datasets from areas like text classification or information retrieval could be beneficial. This would allow semantic embeddings to be tested on larger datasets and compared against state-of-the-art solutions. The approaches described in this paper have been proven to enhance widely used and well-researched word embedding methods. The *intersect_word2vec_format* method from the Gensim library enables merging the input hidden weight matrix from an existing model with a newly created vocabulary. By setting the *lockf* parameter to 1.0, the merged vectors can be updated during training.

Another direction worth exploring to extend our research is testing different architectures for embedding creation, particularly with modified cost functions to incorporate semantics. Future research could also expand this to include additional languages, broadening the scope and applicability of these methods.

Author Contributions: Conceptualization, J.S.; methodology, J.S.; software, M.O.; validation, M.O. and J.S.; data curation, M.O.; writing—original draft preparation, J.S. and M.O.; writing—review and editing, J.S. and P.W.; supervision, J.S.; funding acquisition, P.W.; project administration, J.S. All authors have read and agreed to the published version of this manuscript.

Funding: The research was supported partially by the project "Cloud Artificial Intelligence Service Engineering (CAISE) platform to create universal and smart services for various application areas", No. KPOD.05.10-IW.10-0005/24, as part of the European IPCEI-CIS program, financed by NRRP (National Recovery and Resilience Plan) and Gdansk University of Technology IDUB funds under grant no 3/1/2024/IDUB/II.1b/Am.

Institutional Review Board Statement: Not applicable.

Informed Consent Statement: Not applicable.

Data Availability Statement: The data presented in this study are available on request from the corresponding author. The data are not publicly available due to privacy restrictions

Conflicts of Interest: The authors declare no conflicts of interest.

References

1. Aizawa, A. An information-theoretic perspective of tf–idf measures. *Inf. Process. Manag.* **2003**, *39*, 45–65. [CrossRef]
2. Miller, G.A. WordNet: A lexical database for English. *Commun. Acm* **1995**, *38*, 39–41. [CrossRef]
3. Poli, R.; Healy, M.; Kameas, A. *Theory and Applications of Ontology: Computer Applications*; Springer: Berlin/Heidelberg, Germany, 2010.
4. Szymański, J.; Duch, W. Information retrieval with semantic memory model. *Cogn. Syst. Res.* **2012**, *14*, 84–100. [CrossRef]
5. Mikolov, T.; Chen, K.; Corrado, G.; Dean, J. Efficient Estimation of Word Representations in Vector Space. *arXiv* **2013**, arXiv:1301.3781.
6. Pennington, J.; Socher, R.; Manning, C. Glove: Global Vectors for Word Representation. In Proceedings of the 2014 Conference on Empirical Methods in Natural Language Processing (EMNLP), Doha, Qatar, 25–29 October 2014; Voume 14, pp. 1532–1543.
7. Dharma, E.M.; Gaol, F.L.; Warnars, H.; Soewito, B. The accuracy comparison among word2vec, glove, and fasttext towards convolution neural network (cnn) text classification. *J. Theor. Appl. Inf. Technol.* **2022**, *100*, 31.
8. Hernes, M.; Maleszka, M.; Nguyen, N.T.; Bytniewski, A. The automatic summarization of text documents in the Cognitive Integrated Management Information System. In Proceedings of the 2015 Federated Conference on Computer Science and Information Systems (FedCSIS), Lodz, Poland, 13–16 September 2015; IEEE: Piscataway, NJ, USA, 2015; pp. 1387–1396.
9. Rothe, S.; Schütze, H. Word embedding calculus in meaningful ultradense subspaces. In Proceedings of the 54th Annual Meeting of the Association for Computational Linguistics, Berlin, Germany, 7–12 August 2016; Volume 2: Short Papers, pp. 512–517.
10. Beringer, G.; Jablonski, M.; Januszewski, P.; Sobecki, A.; Szymański, J. Towards semantic-rich word embeddings. In Proceedings of the 2019 Federated Conference on Computer Science and Information Systems (FedCSIS), Leipzig, Germany, 1–4 September 2019; pp. 273–276.
11. Bengio, Y.; Ducharme, R.; Vincent, P. A Neural Probabilistic Language Model. *J. Mach. Learn. Res.* **2000**, *3*, 932–938.
12. Collobert, R.; Weston, J.; Bottou, L.; Karlen, M.; Kavukcuoglu, K.; Kuksa, P. Natural Language Processing (almost) from Scratch. *J. Mach. Learn. Res.* **2011**, *12*, 2493–2537.
13. Turian, J.; Ratinov, L.A.; Bengio, Y. Word Representations: A Simple and General Method for Semi-Supervised Learning. In Proceedings of the 48th Annual Meeting of the Association for Computational Linguistics, Uppsala, Sweden, 11–16 July 2010; Volume 2010, pp. 384–394.
14. Wang, Z.; Cui, X.; Gao, L.; Yin, Q.; Ke, L.; Zhang, S. A hybrid model of sentimental entity recognition on mobile social media. *Eurasip J. Wirel. Commun. Netw.* **2016**, *2016*, 1–12. [CrossRef]
15. Fonseca, E.R.; Rosa, J.L.G. An architecture for semantic role labeling on portuguese. In *Proceedings of the International Conference on Computational Processing of the Portuguese Language*; Springer: Berlin/Heidelberg, Germany, 2012; pp. 204–209.
16. Zhang, B.; Zhang, J.; Wan, F. The Common RNN Based Models with Sequence Tagging. In Proceedings of the 2020 2nd International Conference on Artificial Intelligence and Advanced Manufacture (AIAM), Manchester, UK, 15–17 October 2020; IEEE: Piscataway, NJ, USA, 2020; pp. 129–132.

17. Barnickel, T.; Weston, J.; Collobert, R.; Mewes, H.W.; Stümpflen, V. Large scale application of neural network based semantic role labeling for automated relation extraction from biomedical texts. *PLoS ONE* **2009**, *4*, e6393. [CrossRef]
18. Fonseca, E.R.; Rosa, J.L.G. A two-step convolutional neural network approach for semantic role labeling. In Proceedings of the 2013 International Joint Conference on Neural Networks (IJCNN), Dallas, TX, USA, 4–9 August 2013, IEEE: Piscataway, NJ, USA, 2013; pp. 1–7.
19. Available online: https://ronan.collobert.com/senna/ (accessed on 10 June 2024).
20. Joulin, A.; Grave, E.; Bojanowski, P.; Mikolov, T. Bag of tricks for efficient text classification. *arXiv* **2016**, arXiv:1607.01759.
21. Bojanowski, P.; Grave, E.; Joulin, A.; Mikolov, T. Enriching word vectors with subword information. *Trans. Assoc. Comput. Linguist.* **2017**, *5*, 135–146. [CrossRef]
22. Hossain, M.; Hoque, M.; Sarker, I. Text Classification Using Convolution Neural Networks with FastText Embedding. In *International Conference on Hybrid Intelligent Systems*; Springer International: Cham, Switzerland, 2021; pp. 103–113.
23. Joulin, A.; Grave, E.; Bojanowski, P.; Douze, M.; Jégou, H.; Mikolov, T. FastText.zip: Compressing text classification models. *arXiv* **2016**, arXiv:1612.03651.
24. Peters, M.E.; Neumann, M.; Iyyer, M.; Gardner, M.; Clark, C.; Lee, K.; Zettlemoyer, L. Deep Contextualized Word Representations. In *Proceedings of the 2018 Conference of the North American Chapter of the Association for Computational Linguistics: Human Language Technologies, Volume 1 (Long Papers)*; Walker, M., Ji, H., Stent, A., Eds.; Association for Computational Linguistics: New Orleans, LA, USA, 2018; pp. 2227–2237.
25. Liu, W.; Wen, B.; Gao, S.; Zheng, J.; Zheng, Y. A multi-label text classification model based on ELMo and attention. *MATEC Web Conf.* **2020**, *309*, 03015. [CrossRef]
26. Sherstinsky, A. Fundamentals of Recurrent Neural Network (RNN) and Long Short-Term Memory (LSTM) network. *Phys. D Nonlinear Phenom.* **2020**, *404*, 132306. [CrossRef]
27. Devlin, J.; Chang, M.W.; Lee, K.; Toutanova, K. BERT: Pre-training of Deep Bidirectional Transformers for Language Understanding. *arXiv* **2018**, arXiv:1810.04805.
28. Vaswani, A.; Shazeer, N.; Parmar, N.; Uszkoreit, J.; Jones, L.; Gomez, A.N.; Kaiser, L.; Polosukhin, I. Attention is All You Need. *arXiv* **2017**, arXiv:1706.03762.
29. Navigli, R. Word Sense Disambiguation: A Survey. *ACM Comput. Surv.* **2009**, *41*, 1–69. [CrossRef]
30. word2vec—Tool for Computing Continuous Distributed Representations of Words. 2013. Available online: https://code.google.com/archive/p/word2vec/ (accessed on 10 June 2024).
31. Pennington, J.; Socher, R.; Manning, C.D. GloVe. Available online: https://nlp.stanford.edu/projects/glove/ (accessed on 11 November 2024).
32. Al-Khatib, A.; El-Beltagy, S.R. A Simple and Effective Approach for Fine Tuning Pre-trained Word Embeddings for Improved Text Classification. *arXiv* **2019**, arXiv:1908.02579.
33. Dingwall, N.; Potts, C. Mittens: an Extension of GloVe for Learning Domain-Specialized Representations. In *Proceedings of the 2018 Conference of the North American Chapter of the Association for Computational Linguistics: Human Language Technologies, Volume 2 (Short Papers)*; Association for Computational Linguistics: New Orleans, LA, USA, 2018; pp. 212–217.
34. Kuang, S.; Davison, B.D. Learning word embeddings with chi-square weights for healthcare tweet classification. *Appl. Sci.* **2017**, *7*, 846. [CrossRef]
35. Yucalar, F. Developing an advanced software requirements classification model using bert: An empirical evaluation study on newly generated turkish data. *Appl. Sci.* **2023**, *13*, 11127. [CrossRef]
36. Krzeszewska, U.; Poniszewska-Marańda, A.; Ochelska-Mierzejewska, J. Systematic comparison of vectorization methods in classification context. *Appl. Sci.* **2022**, *12*, 5119. [CrossRef]
37. Beringer, G.; Jablonski, M.; Januszewski, P.; Sobecki, A.; Szymański, J. Optimization of semantic word embeddings. In Poceedings of the 2019 Federated Conference on Computer Science and Information Systems, Leipzig, Germany, 1–4 September 2019.
38. spaCy. Available Trained Pipelines for English. Available online: https://spacy.io/models/en (accessed on 3 November 2023).
39. Python Software Foundation. Pickle—Python Object Serialization. Available online: https://docs.python.org/3/library/pickle.html (accessed on 20 May 2023).
40. DS Lore. Text Classification with Word2Vec. Available online: http://nadbordrozd.github.io/blog/2016/05/20/text-classification-with-word2vec/ (accessed on 22 May 2023).
41. De Boom, C.; Van Canneyt, S.; Demeester, T.; Dhoedt, B. Representation learning for very short texts using weighted word embedding aggregation. *Pattern Recognit. Lett.* **2016**, *80*, 150–156. [CrossRef]
42. Vora, P.; Khara, M.; Kelkar, K. Classification of Tweets based on Emotions using Word Embedding and Random Forest Classifiers. *Int. J. Comput. Appl.* **2017**, *178*, 1–7. [CrossRef]
43. Jolliffe, I.T.; Jackson, J.E. A User's Guide to Principal Components. *Statistician* **1991**, *42*, 76–77. [CrossRef]
44. Rousseeuw, P.J. Silhouettes: A graphical aid to the interpretation and validation of cluster analysis. *J. Comput. Appl. Math.* **1987**, *20*, 53–65. [CrossRef]

45. Boudet, J.; Gregg, B.; Rathje, K.; Stein, E.; Vollhardt, K. The Future of Personalization—And How to Get Ready for It. 2019. Available online: https://static.sharedirecttech.com/clients/heronmartech6/programs/heronesites6/mckinsey-the-future-of-personalization-2019-06.pdf (accessed on 10 November 2024).
46. Musto, C.; Semeraro, G.; De Gemmis, M.; Lops, P. Word Embedding Techniques for Content-based Recommender Systems: An Empirical Evaluation. In *Recsys Posters*; CEUR-WS: Aachen, Germany, 2015.

applied sciences

MDPI

Article

Human Operator Mental Fatigue Assessment Based on Video: ML-Driven Approach and Its Application to HFAVD Dataset

Walaa Othman, Batol Hamoud, Nikolay Shilov * and Alexey Kashevnik

Saint Petersburg Federal Research Center of the Russian Academy of Sciences (SPC RAS), 199178 St. Petersburg, Russia; walaa_othman@itmo.ru (W.O.); bkhamud@itmo.ru (B.H.); alexey.kashevnik@iias.spb.su (A.K.)

* Correspondence: nick@iias.spb.su

Abstract: The detection of the human mental fatigue state holds immense significance due to its direct impact on work efficiency, specifically in system operation control. Numerous approaches have been proposed to address the challenge of fatigue detection, aiming to identify signs of fatigue and alert the individual. This paper introduces an approach to human mental fatigue assessment based on the application of machine learning techniques to the video of a working operator. For validation purposes, the approach was applied to a dataset, "Human Fatigue Assessment Based on Video Data" (HFAVD) integrating video data with features computed by using our computer vision deep learning models. The incorporated features encompass head movements represented by Euler angles (roll, pitch, and yaw), vital signs (blood pressure, heart rate, oxygen saturation, and respiratory rate), and eye and mouth states (blinking and yawning). The integration of these features eliminates the need for the manual calculation or detection of these parameters, and it obviates the requirement for sensors and external devices, which are commonly employed in existing datasets. The main objective of our work is to advance research in fatigue detection, particularly in work and academic settings. For this reason, we conducted a series of experiments by utilizing machine learning techniques to analyze the dataset and assess the fatigue state based on the features predicted by our models. The results reveal that the random forest technique consistently achieved the highest accuracy and F1-score across all experiments, predominantly exceeding 90%. These findings suggest that random forest is a highly promising technique for this task and prove the strong connection and association among the predicted features used to annotate the videos and the state of fatigue.

Keywords: operator fatigue detection; computer vision; physiological indicator; machine learning

1. Introduction

The issue of accurately detecting fatigue is of paramount importance, given the direct correlation between fatigue and endangering numerous lives, particularly those of complex system operators [1]. Operator fatigue is often the underlying cause of high-cost errors and mishaps when controlling complex systems, such as industrial robotic complexes, chemical reaction systems, or systems of multiple coordinated units [2].

Fatigue is typically induced when an individual does not receive an adequate amount of sleep or engages in extended periods of laborious activity. According to a comprehensive assessment of sleep patterns among the U.S. population, it was observed that 37% of the workforce received less than the recommended baseline of seven hours of sleep [3]. The principal challenge associated with fatigue is the difficulty in recognizing the onset of this state until it has escalated to a point where it significantly impairs work and system operation efficiency. Fatigued individuals often exhibit decreased decision-making speed and prolonged reaction times, which render them less effective in their professional roles and more susceptible to risks while working [4].

Numerous solutions have been proposed to tackle the issue of fatigue detection, with the primary aims of identifying signs of fatigue and subsequently alerting the concerned individual to this hazardous state. The majority of these methodologies employ machine learning classifiers or leverage deep learning models to construct a high-efficacy system for fatigue detection [5]. Most of the research on fatigue monitoring has concentrated on identifying fatigue during cognitive tasks, such as driving [6], and some researchers dedicated their endeavors to identifying signs of fatigue in natural-viewing situations, where individuals are not actively engaged in cognitive tasks [7]. Non-intrusive techniques, including those that utilize remote or webcam-based eye tracking, generally monitor variations in an individual's pupil response, blink rate, and eye movement patterns to deduce signs of fatigue during such cognitive tasks [8].

Within the fields of machine learning and deep learning, the selection of the training dataset is a critical factor that influences several aspects of the model, including its accuracy and generalizability. Consequently, numerous datasets have been curated, developed, and utilized for the advancement of fatigue or drowsiness detection techniques [5]. Some of these datasets rely on the use of sensors and devices to collect pertinent features associated with an individual's fatigue state and mental cognition. These include eye trackers [9], electroencephalogram (EEG) signals [10], and electrocardiography (ECG) signals [11], as well as other physiological indicators. However, the evolution of data collection techniques has led to the creation of more sophisticated datasets. These datasets comprise videos of drivers or operators engaged in a task or driving. The availability of such datasets opens up new avenues for experimentation, particularly with deep learning models. These models can process video data to recognize signs of fatigue with no need for any additional devices for the individual. Such non-intrusive methods enhance the applicability and usability of these approaches in safety and monitoring systems, thereby offering potential solutions for real-world fatigue detection challenges.

In this article, we present a selection of datasets that offer various advantages, particularly in terms of the wide range of the facial characteristics, behaviors, ethnic backgrounds, lighting conditions, recording settings, and facial poses they encompass. However, we also draw attention to a big limitation, that is, a noticeable trend of existing datasets of primarily focusing on simulating driving scenarios, which has created a lack of recordings that depict operator (computer user)-related tasks where fatigue exerts a substantial influence on both work performance and productivity. Therefore, our current research focuses on operator fatigue detection in work and academic settings. In this study, subjects experienced the causes of the two types of mental fatigue [12]. The first type is active fatigue, which arises from prolonged task-related perceptual–motor adjustments; for example, operators reading scientific articles develop cognitive strategies that enhance their ability to process complex information. This adaptation includes adjustments in reading speed and focused attention on key concepts. The second type is passive fatigue, which occurs due to extended periods of inactivity, such as sitting in front of a computer for a long time, leading to mental exhaustion. The originality of the proposal is in the usage of physiological indicators obtained by using computer vision for human fatigue state detection based on facial videos. The videos are taken from the public dataset OperatorEYEVP [13]. Therefore, we applied multiple advanced deep learning and computer vision models to analyze the dataset videos. In doing so, we successfully extracted physiological characteristics, which represents a novel and original contribution to the field. Unlike the usage of wearable or other sensors attached to the human body, computer vision techniques can be more useful in many situations, since they are non-intrusive and do not limit human mobility. Our contribution can be summarized as follows:

1. Innovative approach to non-intrusive fatigue detection: Diverging from conventional sensor-based research on mental fatigue, this study introduces a novel, non-intrusive method using deep learning, computer vision, and machine learning models. Unlike the traditional methods that leverage sensor-based data for mental fatigue detection, we extract features by using a simple camera and apply machine learning techniques

for fatigue prediction. This significant shift offers practical benefits due to its accessibility and user friendliness, demonstrating the potential of our approach as a valuable tool for fatigue research.

2. Estimation of physiological indicators: We utilize the OperatorEYEVP dataset [13], which contains a diverse collection of videos capturing the daily tasks of employees and students. Through the application of deep learning and computer vision models developed in our previous work, we estimate a set of physiological indicators that provide an evaluation of the subject's physiological condition in these videos.
3. Comprehensive dataset HFAVD: The resulting dataset includes physiological indicators besides other data (Landolt rings test results, VAS_F, and sensor data related to eye tracking and HRV). These indicators comprise head movement (as Euler angles: roll, pitch, and yaw), vital signs, and eye and mouth states (eye closure and yawning), which were automatically calculated and detected by using deep learning models, eliminating the need for manual intervention, which is often required by other existing datasets.
4. Applicability to work and academic settings: The resulting dataset holds particular significance for work and academic environments, where the impact of fatigue on work performance and efficiency is significant. Therefore, it has the potential to advance fatigue detection research in these contexts.

Our contribution provides a significant opportunity for researchers to advance their comprehension of the intricate relationship between mental fatigue and the diverse range of state changes displayed by subjects.

The paper is structured as follows: Section 2 describes several approaches and datasets for the task of fatigue and drowsiness detection. Section 3 begins by providing an overview of the original dataset we used for our research. Next, we explain how we made our own contribution to this dataset. Then, we describe the structure of models used with the details of the final tables that describe the dataset. Section 4 outlines the experiments we conducted to analyze the dataset with the results.

2. Related Work

In the context of monitoring drivers and operators, researchers often equate these terms due to the overlapping symptoms, notably including diminished concentration, delayed environmental processing, and reduced task performance, whether related to driving or computer-based work. This chapter provides an overview of current approaches and datasets utilized in the field of fatigue detection, encompassing both human drowsiness and fatigue resulting from work or study.

2.1. Existing Approaches to Human Fatigue Assessment

In this section, we detail the proposed approaches and methodologies employed for mental fatigue assessment for drivers and operators, since the biggest focus in the existing literature is on the mental fatigue caused by the driving situation.

The article [14] proposes a lightweight neural network model designed to address driver fatigue detection. The model comprises two primary components: object detection and fatigue assessment. Initially, a lightweight object detection network is developed to rapidly identify the opening and closing states of the driver's eyes and mouth in the video. Following this, the EYE-MOUTH (EM) driver fatigue detection model encodes these states and computes two critical metrics: the Percentage of Eyelid Closure over the Pupil (PERCLOS) and the Frequency of Open Mouth (FOM). Finally, a multi-feature fusion judgment algorithm is employed to assess the driver's fatigue state. The achieved accuracy was 98.30%.

Other researchers [15] introduced a system aimed at detecting drowsiness in drivers, featuring an architecture specifically designed to recognize signs of fatigue. The system incorporates four deep learning models, AlexNet, VGG-FaceNet, FlowImageNet, and ResNet, which analyze RGB video footage of drivers to facilitate drowsiness detection. The architecture focuses on four distinct feature categories: hand gestures, facial expressions, behavioral traits, and head movements. AlexNet is utilized to manage variations in environmental conditions, while VGG-FaceNet extracts facial features such as gender and ethnicity. Flow-ImageNet is responsible for assessing behavioral traits and head movements, and ResNet specializes in detecting hand gestures. The system classifies the detected features into four categories: alert, drowsy with eye blinking, yawning, and nodding. The outputs from these models are processed through an ensemble algorithm, culminating in a final classification via a SoftMax classifier. This system obtained an accuracy of 85%.

The authors of [16] utilized a pre-trained model employing Histogram of Oriented Gradients (HOG) [17] in conjunction with a linear Support Vector Machine (SVM) to extract the positions of the eyes, nose, and mouth, while also calculating the eye aspect ratio (EAR), mouth opening ratio (MOR), and nose length ratio (NLR). Once these three features are determined, drowsiness detection is performed on the extracted frames. Initially, adaptive thresholding is applied for the classification of blinking, yawning, and head tilting, followed by the implementation of various machine learning algorithms to further categorize the data into drowsy and non-drowsy. The SVM demonstrated the highest performance, achieving an accuracy of 96.4%.

Another proposed framework that facilitates drowsiness detection by examining the driver's facial features captured through a camera was presented in [18]. The initial step involves face detection, where the driver's face is identified within a video frame by utilizing Histogram of Oriented Gradients (HoG) features in conjunction with a Support Vector Machine (SVM). Subsequently, specific facial landmark positions are determined. These landmarks assist in estimating head orientation and identifying instances of blinking. By systematically monitoring blink duration, blink frequency, and PERCLOS (Percentage of Eyelid Closure over the Pupil over time), the level of driver drowsiness is assessed through a Fuzzy Inference System (FIS). The performance of the proposed system was tested against the subjective evaluation of the driver based on Karolinska Sleepiness Scale (KSS) [19], with an average normalized root mean square error (NRMSE) of 15.63%.

The study [20] introduces a real-time method for monitoring driver disturbances by utilizing Convolutional Neural Networks (CNNs). The detection process employs advanced CNN architectures, including InceptionV3, VGG16, and ResNet50. Consequently, the ResNet50 model outperforms the other deep learning architectures, attaining an accuracy of 93.69% with a loss of 0.6931.

The paper [21] introduces a cost-effective framework for predicting driver fatigue levels, which aims to detect fatigue in its initial stages. The framework assesses driver fatigue by analyzing behavioral cues related to the eyes, mouth, and head, utilizing an infrared (IR) camera. The features are detected by different architectures of CNNs, and the fatigue classification process is completed by using a logical model. To evaluate the effectiveness of the proposed fatigue prediction framework, experiments were conducted by using real datasets under both nighttime and daytime lighting conditions. The results indicate that the proposed method successfully predicts driver fatigue levels with an overall accuracy of 93.3%.

The authors of [22] introduced an algorithm for recognizing fatigue states. The methodology begins with the application of a Multitask Convolutional Neural Network (MTCNN) to detect human faces. Subsequently, DLIB (v.19.19.0), an open-source software library, is utilized to identify facial key points, enabling the extraction of fatigue feature vectors from each frame. These fatigue feature vectors from multiple frames are then concatenated into a temporal feature sequence, which is input into a Long Short-Term Memory (LSTM) network to derive a final fatigue feature value. The accuracy rates of fatigue detection of their method were 88% on the YawDD dataset [23] and 90% on their self-built dataset.

The article [24] employed an RGB-D camera to capture both RGB and infrared video of the face. The facial video is analyzed independently by using the Joint Approximate Diagonalization of Eigenmatrices (JADE) algorithm [25], Fourier transform, and Triangular Surface Patch (TSP) descriptor [26] to extract three key fatigue features: heart rate, eye opening level, and mouth opening level. Subsequently, the authors developed a novel Multimodal Fusion Recurrent Neural Network (MFRNN) that integrates these three features to enhance the accuracy of driver fatigue detection. Furthermore, a Recurrent Neural Network (RNN) layer is incorporated within the MFRNN to capture the temporal dynamics of the features. Fuzzy reasoning is combined with the RNN to extract the temporal information of the heart rate to address issues related to fuzziness and noise. The proposed method achieved fatigue detection accuracy rates of 86.75% and 91.67% across two distinct datasets.

The study [27] investigated the use of geometric measurements of facial features to assess mental fatigue in construction equipment operators. Seventeen male operators participated in excavation tasks, with data collected consistently over several days during the same morning hours under clear weather conditions. Mental fatigue was evaluated through both subjective NASA-TLX scores and objective electrodermal activity (EDA) recorded via the Empatica E4 sensor. The analysis focused on specific facial features, including the eyebrow, mouth contours, head movements, eye region, and overall facial area. The results revealed significant differences in these metrics between high and low fatigue states, particularly in the eye and facial area measurements, with mean differences of 45.88% and 26.9%, respectively. The findings indicate that geometric facial assessments can serve as an effective, non-invasive method for detecting mental fatigue in this occupational context.

The authors of the OperatorEYEVP dataset, which will be utilized in our experiments, developed a method for detecting mental fatigue based on human eye movements by using this dataset [28]. Their research was grounded in the analysis of eye-tracking data collected from human operators. Within the framework of the proposed methodology, they introduced a technique for identifying the most pertinent gaze characteristics associated with mental fatigue detection, incorporating various machine learning techniques. The experimental findings indicated that the most significant characteristics included the average velocity within the fixation area, the average curvature of the gaze trajectory, the minimum curvature of the gaze trajectory, the minimum saccade length, the percentage of fixations lasting less than 150 ms, and the proportion of time spent in fixations shorter than 150 ms. The processing of eye movement data using this method is conducted in real time, achieving a maximum accuracy of 85% and an F1-score of 80%, with the random forest algorithm demonstrating the best performance.

The research study reported in [29] identified mental fatigue in operators of a construction site by using flexible headband-based sensors for the collection of raw EEG data, along with deep learning networks. The study involved fifteen operators performing a one-hour excavation task on a construction site. The NASA-TLX score served as the ground truth for assessing mental fatigue, while brain activity patterns were captured by using a wearable EEG sensor. The collected raw EEG data were utilized to develop classification models based on deep learning techniques. The performance of these models, specifically Long Short-Term Memory (LSTM), Bidirectional LSTM (Bi-LSTM), and one-dimensional Convolutional Networks, was evaluated. The results indicated that the Bi-LSTM model significantly surpassed the other models, achieving an accuracy of 99.941%. These results underscore the effectiveness of the Bi-LSTM model for wearable sensor-based recognition and classification of mental fatigue, thereby enhancing health and safety measures on construction sites.

The article [30] presents an innovative methodology that employs three machine learning models in conjunction with multimodal data fusion to classify mental fatigue states. Data were collected during an excavation operation through the integration of electroencephalography (EEG), electrodermal activity (EDA), and video signals. Specifically, a headband was utilized to acquire EEG data from construction equipment operators, an E4 watch was attached to their wrists to collect EDA data, and a video camera was installed on the interior of the excavator's front screen to capture facial feature information. Furthermore, a questionnaire was administered to gather subjective assessments of mental fatigue. The decision tree model, utilizing multimodal sensor data fusion, demonstrated superior performance compared with other models, achieving an accuracy of 96.2%. The findings suggest that multimodal sensor data fusion has the potential to facilitate the development of a real-time system for classifying mental fatigue.

The paper [31] presents the development of a non-invasive smart cushion designed to assess the mental fatigue of construction equipment operators through the real-time continuous monitoring of heart rate and respiration signals. A laboratory experiment was conducted with twelve participants who performed simulated excavator operations for data collection. Subsequently, relevant time-domain and frequency-domain features of the heart rate and respiration signals were extracted to create a random forest classification model aimed at exploring the relationship between vital signs and self-reported mental fatigue, as measured by the NASA-TLX scale. The results indicated that the random forest model achieved a classification accuracy of 92%. Furthermore, the combination of heart rate and respiration features proved to be more effective than using a single feature for identifying and classifying mental fatigue among construction equipment operators.

Table 1 summarizes the various approaches discussed previously, highlighting the features utilized and the techniques implemented for fatigue detection, along with the accuracy achieved in each case. While numerous studies report high performance and favorable outcomes, certain limitations are evident. Notably, many methods rely heavily on features derived solely from the eyes and mouth to ascertain fatigue levels. Although these features are known to correlate significantly with fatigue, there are circumstances—such as the presence of sunglasses, reflections from lighting on eyewear, or mouth coverings due to medical reasons—where accurate detection may be compromised. Furthermore, some studies have employed fuzzy logic systems or logical models for fatigue classification. These systems are often hindered by their lack of learning capabilities and their dependence on expert knowledge, as their effectiveness is contingent upon the quality of the rules established by specialists. Conversely, other research has focused on deep learning methodologies, including CNNs and RNNs, which, while powerful, are computationally intensive and necessitate substantial datasets for effective training. Moreover, the metrics used to define the fatigue states were mostly subjective, such as NASA-TLX or data collected from sensors (EDA). Nevertheless, numerous studies investigating operators' mental fatigue have relied on sensors to gather the features utilized in their research. This reliance on sensors may be viewed as a limitation, as these devices can introduce discomfort for the participants involved in the studies. Additionally, a noticeable trend is the exclusive focus on detecting driver fatigue, which overlooks other contexts in which individuals, such as operators and students, may also experience fatigue. This narrow scope may limit the applicability of the findings across broader domains.

Table 1. Summary of approaches employed for fatigue detection using facial videos.

Article	Used Features	Technique Used for Fatigue Detection	Accuracy of Fatigue Detection Models
Cui et al. [14]	Eye and mouth features	A multi-feature fusion judgment algorithm	98.30%
Dua et al. [15]	Hand gestures, facial expressions, behavioral features, and head movements	SoftMax classifier	85%
Dey et al. [16]	Eye, mouth, and nose features	SVM	96.4%
Salzillo et al. [18]	Eye features	Fuzzy logic-based system	NRMSE of 15.63%
Minhas et al. [20]	Features detected by CNN layers	CNN (ResNet50)	93.69%
Kassem et al. [21]	Eye, mouth, and head features	Logical model	93.3%
Chen et al. [22]	Eyes and mouth features	LSTM	88% and 90%
Du et al. [24]	Heart rate, and eye and mouth features	MFRNN	86.75% and 91.67%

Table 1. *Cont.*

Article	Used Features	Technique Used for Fatigue Detection	Accuracy of Fatigue Detection Models
Mehmood et al. [27]	Eyebrows, mouth contours, head movements, eye region, and facial area	Static tests	Significant differences in these metrics between high and low fatigue states
Kashevnik et al. [28]	Eye-tracking data	Machine learning techniques	85%
Mehmood et al. [29]	EEG data collected from sensor	Deep learning techniques	99.941%
Mehmood et al. [30]	EEG and EDA data collected from sensors and facial features obtained from videos	Machine learning techniques	96.2%
Wang et al. [31]	Heart rate and respiration signals	Machine learning techniques	92%

In an effort to address the aforementioned limitations, our study focuses on detecting mental fatigue associated with cognitively taxing tasks. This form of fatigue parallels the exhaustion experienced by students and workers after extended periods of studying or working. Importantly, our approach extends beyond the exclusive reliance on eye and mouth features, incorporating multiple features to enhance fatigue recognition. It should be emphasized that all features employed in this study are derived from pre-trained deep learning models, eliminating the need for additional sensor use or retraining, thus reducing costs. Furthermore, we utilize cost-effective machine learning techniques, such as random forest, for fatigue detection based on these estimated features. This strategy makes our approach cost-effective, straightforward, and user-friendly.

2.2. Existing Datasets

This section outlines several widely utilized datasets designed specifically for the detection of fatigue and drowsiness. Our research primarily concentrates on video datasets, which do not contain data derived from physical sensors, and the videos are captured by using ordinary cameras.

The UTA-RLDD dataset [32] comprises approximately 30 h of RGB videos featuring 60 healthy participants from diverse ethnic backgrounds, with varying ages ranging from 20 to 59 years old. Each participant contributed 1 video for each of three distinct states, i.e., alertness, low vigilance, and drowsiness, totaling 180 videos. Recordings were captured from diverse angles in real-life environments against different backgrounds. A notable aspect of this dataset is the authenticity of drowsiness samples, which contrast with staged portrayals found in other datasets. Moreover, data collection involved the use of different cameras, with each participant recording themselves in a real-life indoor setting of their choosing by using either a cell phone or a web camera.

The YawDD dataset [23] comprises two publicly available sub-datasets featuring RGB videos. Participants in these videos engage in simulated driving scenarios, while the vehicle remains stationary. In the first sub-dataset, a camera is positioned beneath the front mirror of the car, capturing three to four videos per participant. This sub-dataset offers a total of 322 videos depicting drivers (57 males and 50 females), with and without glasses/sunglasses, representing diverse ethnic backgrounds. In the second sub-dataset, which includes 29 videos, the camera is mounted on the car's dashboard in front of the driver. Each video showcases various mouth conditions, including normal, talking/singing, and yawning.

However, some individuals in the YawDD dataset videos simulate yawning rather than naturally exhibiting the behavior. Hence, the authors of [22] decided to collect their own dataset to capture more authentic fatigue. Ten volunteers each participated in the creation of two videos: one depicting normal behavior and the other featuring signs of fatigue, including moments of eye closure, conversation, laughter, and yawning. The dataset aims to mitigate discrepancies in experimental outcomes arising from variations in skin color.

The NTHU-DDD dataset [33] consists of 360 (infrared (IR) and RGB) videos of 36 subjects showcasing a range of simulated driving situations, including regular driving, yawning, slow blinking, nodding off, and laughing. These recordings capture subjects under both daytime and nighttime lighting conditions. They were directed by an experimenter to perform a sequence of eight actions across five different scenarios: BareFace, Glasses, Sunglasses, Night-BareFace, and Night-Glasses.

The authors of [34] created the Licensed Crane Operators dataset, which features videos obtained through interviews conducted by the authors with five skilled crane operators. The videos encompasses three distinct behaviors: alertness, low vigilance, and fatigue. The recordings were captured from different camera angles and conducted across various scenarios, including working at computer stations, simulated or actual driving environments, and simulated crane operations. The participants exhibited diverse facial characteristics, behaviors, and ethnic backgrounds.

The authors of [35] introduced a multimodal database called DROZY. This database was constructed by using multimodal data obtained from 14 participants, who underwent three consecutive sessions of the authors' version of psychomotor vigilance tests (PVTs) proposed by [36] under conditions of escalating sleep deprivation. This test produces an objective measure of vigilance and, indirectly, of drowsiness. For each participant and each PVT session, the database includes time-synchronized raw data, encompassing Karolinska Sleepiness Scale (KSS) measurements [19], polysomnography (PSG) signals, PVT data (including reaction times), and near-infrared (NIR)-intensity and -range images of the face.

The creators of the FatigueView dataset [37] introduced a comprehensive, large-scale multi-camera dataset tailored for investigating drowsiness detection in real-world driving scenarios. The dataset includes recordings from both RGB and infrared (IR) cameras positioned at five different angles, capturing genuine instances of drowsy driving along with various visual indicators of drowsiness, ranging from subtle to overt. It encompasses 17,403 distinct instances of yawning across more than 124 million frames sourced from 95 participants.

The SUST-DDD dataset [38] comprises 2074 videos captured by participants by using their mobile phone cameras while experiencing fatigue and normal states during authentic driving scenarios. Notably, participants were not directed to exhibit specific behaviors while driving, thereby ensuring the authenticity and safety of the driving experience. Nineteen participants, aged between 24 and 42, were tasked with recording videos by using their personal phones positioned in front of the driver's seat when they felt drowsy or normal. The videos were recorded in periods of reduced daylight, such as driving to or from work.

In Table 2, we present the highlighted features of the above-mentioned datasets. It is evident that these datasets offer a considerable number of videos showcasing diverse facial characteristics, behaviors, ethnic backgrounds, lighting conditions, recording settings, and facial poses. Nonetheless, an observable trend unfolds, that is, the majority of existing datasets prioritize the simulation of driving scenarios. This emphasis consequently leads to an insufficiency of recordings that portray tasks associated with computer operation. This gap is particularly pertinent, as such tasks constitute the routine activities of numerous employees and students who may also experience fatigue and drowsiness. Moreover, upon analysis of datasets capturing tasks associated with office work or studying, it becomes evident that they lack a substantial number of informative features related to fatigue, such as vital signs, head movement, and the states of the eyes and mouth. Hence, the detection of these features is mandatory for the implementation of a fatigue recognition system using these dataset, which could be both computationally intensive and time-consuming. To address this limitation, we introduce our dataset, designed to rectify this deficiency by offering a more comprehensive collection of recordings relevant to the daily tasks of employees and students annotated with fatigue related features estimated by deep learning and computer vision methods. Such a dataset holds promise for facilitating fatigue detection research, given its direct relevance to work and academic settings, where fatigue significantly influences work performance and efficiency.

Table 2. Summary of datasets.

Dataset	Number of Subjects	Age Range	Number of Videos	Illumination	Camera Type	Human State
UTA-RLDD [32]	60	20–59	180	Daytime and nighttime	RGB	Alertness, low vigilance, and drowsiness
YawDD [23]	107	-	322 and 29	Daytime	RGB	Talking, yawning, and singing
Chen et al. [22]	10	-	20	Daytime	RGB	Talking, yawning, and laughing
NTHU-DDD [33]	36	-	360	Daytime and nighttime	RGB and IR	Nodding, regular driving, yawning, slow blinking, and laughing
Licensed Crane Operators [34]	5	30–50	-	Daytime	RGB	Alertness, low vigilance, and fatigue
DROZY [35]	14	22.7 ± 2.3	-	1st PVT: 8:30 a.m. 2nd PVT: 3:30 a.m. 3rd PVT: 12:00 p.m.	IR	Alertness, low vigilance, and drowsiness
FatigueView [37]	95	Over 18	124 million frames	Daytime and nighttime	RGB and IR	Normal, alertness, signs of sleepness, sleepy, and sleepness
SUST-DDD [38]	19	24–42	2074	Reduced daylight	RGB	Normal and drowsiness

3. Dataset Description

This article introduces a dataset comprising video data integrated with numerical metrics, offering a comprehensive assessment of the subject's physiological condition as depicted in the videos. The dataset provides parameters such as head movement in terms of Euler angles (roll, pitch, and yaw), vital signs, and eye and mouth states (eye closure and yawning), eliminating the necessity for manual calculation or detection of these parameters, which is commonly required by other existing datasets. In the following subsections, we describe the original dataset we worked with, explain our main contribution to the dataset, delineate the structure of the models used with a detailed explanation of the final tables.

3.1. OperatorEYEVP Dataset

For this purpose, we utilized the OperatorEYEVP dataset as introduced by [13]. This dataset features video recordings of ten unique individuals participating in diverse activities. These activities were documented at three distinct moments throughout the day (morning at 9:00 a.m., afternoon at 13:00 p.m., and evening at 18:00 p.m.), over a period spanning eight to ten days. The recordings occurred on weekdays from December 2022 to June 2023. The recorded data not only offer a frontal view of the participants' facial expressions, but they also include a set of additional information. This encompasses data pertaining to pupil and head movement (including specific coordinates), scene visuals, pulse rate for each interval, and Choice Reaction Time measured twice. Furthermore, the dataset incorporates responses to various questionnaires and scales, notably the VAS-F scale. The VAS-F scale, an 18-question measure, addresses the subjective experiences of fatigue among participants. It is worth noting that these questions were administered prior to the beginning of the experimental session. The timeline of a session is shown in Figure 1.

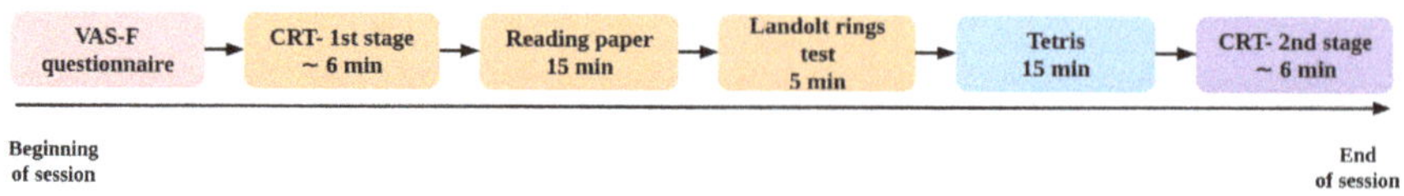

Figure 1. Timeline of each session.

The experimental procedure included a daily sleep quality survey conducted prior to the morning session. This was succeeded by the VAS-F questionnaire, a Choice Reaction Time (CRT) task, reading a text in a scientific style, performing the "Landolt rings" correction test, playing the "Tetris" game, and a second CRT. The latter was incorporated at the discretion of the authors, cognizant of the potential variability in operator fatigue levels between the inception and termination of the recording session. The aggregate duration of such sessions, on average, approximated one hour.

Throughout the Choice Reaction Time (CRT) assessment, a comprehensive group of parameters was recorded, such as the average reaction time, its standard deviation, and the number of errors made by participants during task performance. Participants engaged in the reading of scientific-style text, which can be a typical work-related activity. This task served as a static cognitive load task aimed at evaluating cognitive performance. The Landolt rings correction test is a well-established method for evaluating visual acuity. Based on [13], the following parameters were recorded during the Landolt rings test:

1. Time spent (t).
2. The total number of characters viewed up to the last selected character (N).
3. The total number of lines viewed (C).
4. The total number of characters to be crossed out (n).
5. The total number of characters crossed out (M).
6. Correctly selected letters (S).
7. Letter characters missed (P).
8. Wrongly selected characters (O).

These parameters facilitate the computation of various metrics, including attention productivity (K), mental performance (Au), and stability of attention concentration (Ku), using the following equations:

$$Au = \frac{N}{t} * \frac{M - O + P}{n} \quad (1)$$

$$K = \frac{(M - O) * 100}{n} \quad (2)$$

$$Ku = C * \frac{C}{P + O + 1} \quad (3)$$

Finally, the Tetris game served as a dynamic active task to investigate hand–eye coordination. The recorded variables included the number of games played, the levels reached, the scores achieved, and the lines cleared.

This dataset's demographic is primarily composed of individuals with a mean age of 27.3 years, extending from a minimum age of 21 to a maximum of 50 years. It is evident that the sample is skewed towards the younger age spectrum, with the majority below the age of 33. Furthermore, there were no reported instances of significant health issues that could impact vital signs, with the exception of the first participant, who exhibited symptoms of depression. However, it is important to note that the participants predominantly have lighter skin tones. This demographic characteristic may present limitations for future applications of this approach, particularly in assessing fatigue among older individuals, those with serious health conditions, and individuals with darker skin tones. This observation highlights the future need to expand the dataset to include a more diverse range of ages, health conditions, and ethnicities, thereby enhancing the generalizability of the approach.

3.2. Annotation of Dataset with Physiological Indicators

Our primary contribution to the dataset entails the annotation, performed at one-minute intervals, of the videos encompassing four main physiological indicators: blood pressure, heart rate, oxygen saturation, and respiratory rate. We selected the one-minute interval because, as we discuss in more detail in Section 3.3, the models used for parameter estimation make predictions at the level of seconds, frames, or minutes. Therefore, a one-minute interval was deemed the most suitable choice for synchronizing the estimations. Moreover, supplementary metrics were computed, including estimation of head pose via Euler angles (roll, pitch, and yaw), the proportion of frames exhibiting Euler angles surpassing 30 degrees relative to the total frames within a one-minute span, ratios of eye closure and mouth opening, instances of yawning, numbers of prolonged eye closures exceeding two seconds, and characteristics pertaining to breathing patterns, encompassing rhythmicity and stability.

The computation of the aforementioned parameters was carried out by using computer vision and deep learning approaches developed by members of our laboratory, which are discussed in the subsequent subsection. These approaches were applied to each video within the OperatorEYEVP dataset, and annotations were made for the 15 parameters at each minute of the recordings. For each participant, a comprehensive CSV file was created to document all the details of the session. This file encompassed information related to the CRT and Tetris tasks, as well as eye parameters and HRV parameters provided by the dataset in addition to the parameters extracted by our models with additional features we calculated from the coordinates of the eye-tracking data. A more in-depth exploration of the CSV files, including their structure and contents, is provided later in the discussion.

3.3. HFAVD Dataset Structure

In this subsection, we present an overview of the models utilized for extracting physiological features used in the annotation of the videos. These models leverage deep learning, machine learning, and computer vision techniques to estimate specific physiological indicators based on facial video recordings of the subjects, as shown in Figure 2. Through these models, we were able to estimate 15 features at each minute of every video, capturing a comprehensive range of physiological information. The physiological features were extracted by using deep learning models with a sampling frequency of one second, as will be detailed in subsequent sections. However, many of these features exhibit minimal variation within a one-second interval, which can result in numerous identical or highly similar samples. This redundancy may compromise the reliability of the experimental outcomes. Consequently, we opted to compute the average of these indicators for each minute and label them with the corresponding mental performance value. This approach was selected to retain as many data as possible, ensuring they remained adequate for the machine learning models. Therefore, increasing the sampling interval from 1 min to 5, for example, could lead to insufficient data, adversely affecting performance, while decreasing it would result in duplicated samples, thereby rendering the experiments invalid and unreliable.

The estimated physiological features were compiled and organized into comprehensive CSV files. These CSV files enhance the distinctiveness of the dataset compared with existing ones. They include almost every detectable or extractable feature from the subject while the latter performs various tasks, without the need for sensors or external devices apart from the widely available camera. This distinguishing characteristic contributes to the dataset's versatility and accessibility for researchers. These models are explained as follows.

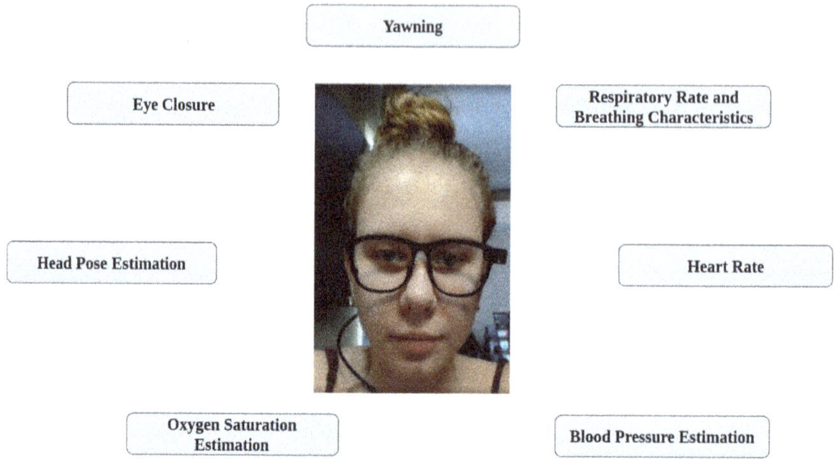

Figure 2. Models used to label the videos.

3.3.1. Respiratory Rate and Breathing Characteristics

To estimate the respiratory rate, the approach proposed by [39] was employed, encompassing the following steps: (a) Chest keypoint detection using OpenPose: The OpenPose framework was utilized to detect the keypoint corresponding to the chest region in the video frames. (b) Displacement detection using SelFlow: An Optical Flow-based neural network called SelFlow was employed to detect the displacement of the chest keypoint between consecutive frames. (c) Projection of displacement along the x- and y-axes: The displacement values obtained were projected onto the x- and y-axes, separating the movement into vertical (up/down) and horizontal (left/right) directions. (d) Signal processing techniques: Signal processing methods, including filtering and detrending, were applied to the displacement data to enhance the accuracy of respiratory rate estimation. (e) Calculation of true peak count and scaling: The true peak count was calculated based on the processed displacement data. This count was then scaled to estimate the number of breaths per minute. Furthermore, in addition to estimating the respiratory rate, various breathing characteristics, such as stability and rhythmicity, were assessed based on the amplitude and wavelength of the respiratory wave. The mean absolute error (MAE) for respiratory rate estimation was 1.5 breaths per minute (bpm). Regarding the limitations of this model, its effectiveness is restricted by its inability to handle body movement efficiently and its reliance on the subject maintaining a stationary position during recording. This requirement poses the risk of inaccurate estimations. Consequently, the model is unsuitable for use when the subject is walking, engaged in motion, or even driving on the road, due to the presence of vehicle vibrations.

3.3.2. Heart Rate

The heart rate estimation approach proposed by [40] was employed in this study. The methodology involved the following steps: (a) Extraction of the Region of Interest (Face): Landmarks obtained from the 3DDFA_V2 framework [41] were utilized to extract the Region of Interest, specifically the face, from the input data. (b) Processing through a Vision Transformer: The extracted face region was then processed through a Vision Transformer model, incorporating multi-skip connections. This process generated features from five different levels. (c) Block processing: The output from each level of the Vision Transformer was passed through a block structure. This block structure consisted of a Bidirectional Long Short-Term Memory (BiLSTM) layer, batch normalization, 1D convolution, and a fully connected layer. (d) Averaging of block outputs: The outputs from the five different blocks were averaged to obtain the minimum, maximum, and mean heart rates. (e) Weighted averaging: The minimum, maximum, and mean heart rates were then combined by using a weighted averaging scheme to estimate the final heart rate. This model demonstrated superior accuracy in heart rate estimation compared with previously published methods, as evidenced by its performance on the LGI-PPGI and V4V datasets, achieving mean absolute errors (MAEs) of 2.68 and 9.996 bpm, respectively. Nevertheless, it is important to note that the dataset utilized for training the model exhibited a lack of samples representing both very low and very high heart rates. Consequently, the model may produce inaccurate estimations when presented with subjects exhibiting specific medical conditions characterized by extreme heart rate values.

3.3.3. Blood Pressure

The estimation of blood pressure followed the approach proposed by [42]. The methodology consisted of the following steps: (a) Extraction of Regions of Interest: The left and right cheeks were identified as the Regions of Interest (ROIs) in each frame of every video. These sequential images were used as input for subsequent analysis. (b) Convolutional Neural Network (CNN) for spatial feature extraction: A CNN architecture was employed to capture spatial features from the ROIs. For systolic blood pressure estimation, EfficientNet_B3 was utilized for the left cheek, while EfficientNet_B5 was used for the right cheek. Conversely, for diastolic blood pressure estimation, an ensemble approach was adopted. This involved combining EfficientNet_B3 and ResNet50V2 for the left cheek and the right cheek, respectively. (c) Long Short-Term Memory (LSTM) for temporal feature extraction: The outputs obtained from the previous step were fed into an LSTM network. This enabled the extraction of temporal features within the sequence of images, considering the dynamic nature of blood pressure. (d) Fully connected layers for blood pressure estimation: Two fully connected layers were employed to derive the blood pressure values based on the extracted spatial and temporal features. Upon training and testing the proposed model, the mean absolute errors recorded were 11.8 and 10.7 mmHg for systolic and diastolic blood pressure, respectively. In parallel, the model's mean accuracies were found to be 89.5% for systolic blood pressure and 86.2% for diastolic blood pressure. The main challenge faced in training this model was the lack of diversity in the dataset, particularly in terms of the subjects' skin color. Most of the subjects had light-to-medium skin tones, which impacted the accuracy of the models in predicting blood pressure in individuals with dark skin tones. Another limitation was the dataset's imbalance, with a scarcity of unusual values for systolic and diastolic blood pressure.

3.3.4. Oxygen Saturation

The estimation of oxygen saturation in this study was conducted based on the method proposed by [43]. The procedure involved the following steps: (a) Extraction of face region: The face region was initially extracted by using the 3DDFA_V2 framework. This facilitated the isolation of the relevant area for subsequent analysis. (b) Input to VGG19: The extracted face region was then used as input for the VGG19 model, which had been pre-trained on the ImageNet dataset. This allowed for the extraction of relevant features from the face region. (c) Utilization of XGBoost: The output obtained from VGG19 was subsequently fed into an XGBoost algorithm. XGBoost is a machine learning framework that combines multiple decision trees to make predictions. In this case, it was employed to obtain the estimation of the oxygen saturation value. Following the training phase, the model was subjected to testing by using two distinct datasets. The derived mean absolute error (MAE) for the first test set was found to be 1.17%, while for the second test set, it was marginally lower, at 0.84%. The primary challenge encountered in training this model pertained to the limited number of samples depicting SpO2 levels below 85. This scarcity poses a significant obstacle for the models, particularly when examining subjects with specific health conditions that are associated with low SpO2 levels.

3.3.5. Head Pose

The determination of the head pose, characterized by Euler angles (roll, pitch, and yaw), was carried out by using the methodology described in [44]. The approach involved the following steps: (a) Face detection using YOLO Tiny: Initially, the YOLO Tiny framework was utilized to detect the face in the input data. This allowed for the identification of the Region of Interest. (b) Three-dimensional face reconstruction for landmark alignment: Subsequently, a 3D face reconstruction technique was applied to align the facial landmarks. This process ensured accurate landmark detection even for those not directly visible to the camera. (c) Landmark analysis for Euler angle calculation: Once the facial landmarks were detected, the transitions and rotations between landmarks across successive frames were analyzed. This analysis enabled the calculation of the Euler angles, providing information about the head pose. The limitation of the article is that the model proposed can only estimate head angles up to a maximum value of 70 degrees. This limitation restricts its applicability in scenarios where larger head angles need to be accurately detected. Additionally, the model's performance is slower compared with the Dlib detector, which itself loses accuracy when the head angle exceeds 35 degrees. This slower performance could hinder real-time applications that require fast and accurate head angle estimation.

3.3.6. Eye and Mouth State

The eye state was assessed through the utilization of a trained model. This model takes the detected face, identified via the FaceBoxes framework, as its input and generates an output that signifies whether the eyes are open or closed. Similarly, the yawning state is recognized by employing a modified version of MobileNet, as proposed by [45]. This model achieved an accuracy of 95.20%. However, a limitation of this model is that it relies on a private dataset for validation. This could potentially limit the generalizability of the findings, as the dataset is not representative of the broader population.

The videos in the OperatorEYEVP dataset exhibit a high degree of similarity to the conditions found in the datasets used for training the models. These similarities encompass factors such as lighting conditions, skin color, and the demographic characteristics of the subjects, who were healthy and young. Consequently, the likelihood of encountering extreme or abnormal values in vital signs, particularly oxygen saturation, is minimized. However, the insufficiency of sensor data in the OperatorEYEVP dataset prevents the calculation of estimation errors. Nevertheless, considering the aforementioned conditions, it is reasonable to assume that the models exhibit good accuracy. This assumption is further supported by the alignment of the obtained results with expectations and their consistency with the existing literature.

3.4. Data Example

As previously stated, inclusive CSV files were generated to meticulously record all the details of each video within the OperatorEYEVP dataset. In this subsection, we provide a thorough description of these files, elucidating the significance of each feature included therein, as well as their intended purpose. Additionally, we provide examples of how these files could potentially be utilized by other researchers.

For each participant, a CSV file encompassing 199 columns was generated to provide an extensive description of the session performed by the subject (see Table 3). The initial columns within the file include metadata associated with the session, including participant ID, session name, the type of activity undertaken, and the time of day during which the session occurred. Subsequently, information pertaining to the CRT sessions, specifically the first and second sessions, is listed. This information includes metrics such as the average reaction time, standard deviation of reaction time, and the number of errors made.

Table 3. The contents of the CSV files.

Type of Features	Features	Description
Info about the video	Four features included (participant ID, video name, time of day, and type of activity)	General info about the participant and the session
Questionnaire	One value presents VAS-F result	Subjective assessment of fatigue performed by the participants themselves
Task results	6 features describe CRT results	Metrics related to the performance during the CRT sessions
	19 features related to Landolt rings test results	Metrics related to the performance during the Landolt rings test session
Sensor features	60 eye movement strategy features	Eye characteristics regarding saccadic and fixation time and speed
	14 eye movement statistics features	Mean, standard deviation, minimum, maximum, 25th percentile, 50th percentile, and 75th percentile of each series on both coordinates
	79 HRV features	Features related to the HRV in multiple domains (time, frequency, and nonlinear)
	One value presents the actual heart rate	Real heart rate extracted from the PPI signals
Features extracted by our models	15 features, including head movement angles (roll, pitch, and yaw), vital signs (blood pressure, respiratory rate, heart rate, and oxygen saturation), and eye and mouth states (eye closure and yawning)	Detected features from each video at each minute using deep learning models proposed in our previous works [39,40,42–45]

Following that, we included the pertinent information related to the Landolt rings test, encompassing a variety of indicators and metrics. This includes many indicators about attention, stability of attention and work accuracy, the quantification of mental productivity through a mental productivity index, and the evaluation of mental performance. Additionally, we incorporated a concentration indicator, the qualitative characteristic of concentration, the capacity of visual information processing, and the processing rate. Furthermore, we listed additional information related to the test results, such as the total number of characters crossed out, letters correctly selected, letter characters missed, wrongly selected characters, total number of lines viewed, and other relevant variables.

Subsequent to this, we added a collection of features related to eye movement strategy (pupil movement) observed during each session. These features, totaling 60 in number, encompass measurements related to fixation and saccadic time and velocity, alongside additional parameters associated with acceleration. Additionally, an extensive set of 79 parameters were included, encompassing various domains such as time, frequency, and nonlinearity, all of which are related to heart rate variability. Notable examples within this set include VLF, LF, and HF band frequencies, as well as relative and absolute power measurements for these frequency bands, among numerous other relevant metrics. In addition, we extracted the absolute heart rate for each minute from the PPI signal provided. We thought that it would be useful to include both real and predicted heart rate (by our model (Section 3.3.2)), so that many experiments and comparisons may be carried out by researchers to evaluate the possibility of integrating and relying on computer science for such tasks. All the aforementioned features were provided by [13].

The next stage involved extracting the relevant features from the x- and y-coordinates from the eye movement data, which were represented as two separate time series for each activity. To accomplish this, we employed basic Python (v. 3.6.9) code in conjunction with the NumPy library to compute seven key features from each series, namely, mean, standard deviation, minimum, maximum, 25th percentile, 50th percentile, and 75th percentile.

Finally, we added the 15 features extracted by our models after processing the videos from the OperatorEYEVP dataset. These features included respiratory rate, rhythmicity and stability of breathing, eye closure ratio, mouth opening ratio, head Euler angles (roll, pitch, and yaw), the ratio of frames where any Euler angle exceeds 30 degrees relative to the total frames in one minute (angles > 30), the count of eye closures lasting more than 2 s (count of eye closure > 2 s), the count of yawns, heart rate, systolic and diastolic blood pressure, and blood oxygen saturation. The final column in the dataset represents the subjective assessment of fatigue, specifically measured by using the Visual Analogue Scale for Fatigue (VAS-F). The participants themselves provided this assessment, indicating their perceived level of fatigue.

We believe that our contribution to the OperatorEYEVP dataset offers a valuable opportunity for researchers to enhance their understanding of the relationship between mental fatigue and various responses exhibited by subjects. These responses encompass both observable cues, such as head movement and eye and mouth state, as well as underlying indicators, such as vital signs. Researchers can gain insights beyond fatigue alone by exploring the correlations between these responses and multiple features, including attention, work accuracy, concentration stability, and mental performance. We consider that our previous approaches and research have been employed in an optimal manner. The advantage of utilizing these approaches lies in their ability to provide access to all the aforementioned features without the need for sensors or external devices. This not only saves considerable time and effort in conducting studies but also eliminates the need to develop costly approaches to detecting specific features. Consequently, researchers can promptly apply and test their hypotheses and theories. Furthermore, the availability of annotating videos capturing various activities, not solely limited to fatigue, can provide a deeper understanding of the human mechanism.

4. The Proposed Approach and Its Application to the HFAVD Dataset

In this section, we delve into the experiments used to analyze the dataset, along with presenting the resulting outcomes. The dataset provided was utilized to detect fatigue levels based on the psychological traits extracted by our developed models, as explained in Section 3.2. The overall framework is illustrated in Figure 3.

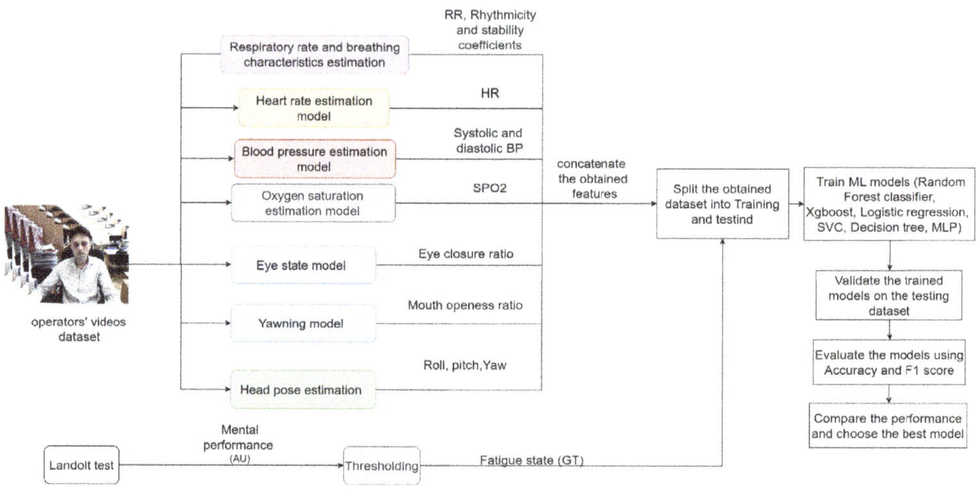

Figure 3. The overall scheme used for detecting the fatigue state.

As illustrated in the diagram, our approach involves extracting key physiological indicators from the videos of operators using a set of specialized models developed for this purpose. These indicators include heart rate, respiratory rate, blood pressure, oxygen saturation, eye closure ratio, head pose, and yawning frequency. Each of these physiological signals is calculated on a per-minute basis, ensuring a continuous, time-resolved input into the fatigue detection process. Once these indicators are extracted, they are used as input features for our fatigue detection model. This model is designed to assess the operator's fatigue state based on the mental performance (AU), which is derived from the results of the Landolt rings test—a well-known measure of visual acuity and attention. The Landolt test provides insights into the operator's cognitive state, specifically focusing on attention and concentration.

To determine the optimal threshold for AU, which separates the fatigue state from the non-fatigue state, we conducted a detailed analysis of the data. This analysis involved evaluating the mental performance across different AU values to identify the point at which fatigue became evident. In addition to AU, we also analyzed two other important cognitive measures: the attention level (K) and the stability of concentration (Ku). These metrics helped us gain a more comprehensive understanding of how attention and focus degrade over time, particularly as fatigue sets in.

We employed the results of the Landolt test, particularly mental performance, to assess the fatigue level due to its capacity to effectively capture the cognitive and attentional aspects associated with fatigue. Unlike subjective measures, such as the Visual Analog Scale for Fatigue (VAS-F), which rely on self-reported perceptions, which can be swayed by individual biases and emotional states, AU provides a more objective and standardized evaluation of fatigue. Table 4 shows the mental performance (Au), attention (K), and the stability of concentration of attention (Ku) for each participant.

Table 4. Mental performance (Au), attention (K), and the stability of concentration of attention (Ku) for each participant in the dataset.

Participant Id	Ku	K	Au
1	10–450	53.6–98.2	0.18–2.85
2	16–225	56.8–95.8	0.39–2.71
3	14–450	45.7–98.4	0.00–2.87
4	12–300	57.6–97.1	0.45–2.81
5	9–180	53.3–95.8	0.18–2.70
6	11–900	63.6–100.0	0.59–3.13
7	1–300	33.0–97.5	0.00–2.83
8	12–900	57.3–100.0	0.31–2.99
9	3–12	36.3–69.7	0.00–0.83
10	5–90	45.5–91.1	0.00–2.45

Given that attention decreases when a person is fatigued, we can draw meaningful insights from our data. For example, participant number 6 has the attention level above 60, which is considered good [46]. That indicates that the participant 6 is not fatigued. Therefore, we can conclude that the fatigue threshold for mental performance must be below 0.59, as individuals with mental performance scores above this threshold maintain high attention levels.

To predict the fatigue state, we based our study on the hypothesis that mental performance becomes lower with the increase in fatigue. Figure 4 illustrates the relationship between mental performance (as measured by the Landolt rings test) and the inverse of fatigue (1/Fatigue). We assume that there is a threshold value (a point or a small range of points) where fatigue level change causes a significant change in physiological indicators. We also assume the this change can be captured by a machine learning model.

We assessed each participant's mental performance, classifying them as fatigued if their performance fell below a predetermined threshold. To identify the optimal threshold, we conducted multiple experiments by using the random forest algorithm, testing thresholds from 0.2 (chosen empirically from our dataset) to 0.59 (discussed above) in increments of 0.015. This increment value was chosen experimentally, as performance differences were negligible with smaller increment values. We selected the random forest algorithm for evaluation due to its superior results demonstrated in our previous research [2]. Figure 5 shows the F1-score for different threshold values obtained through experiments.

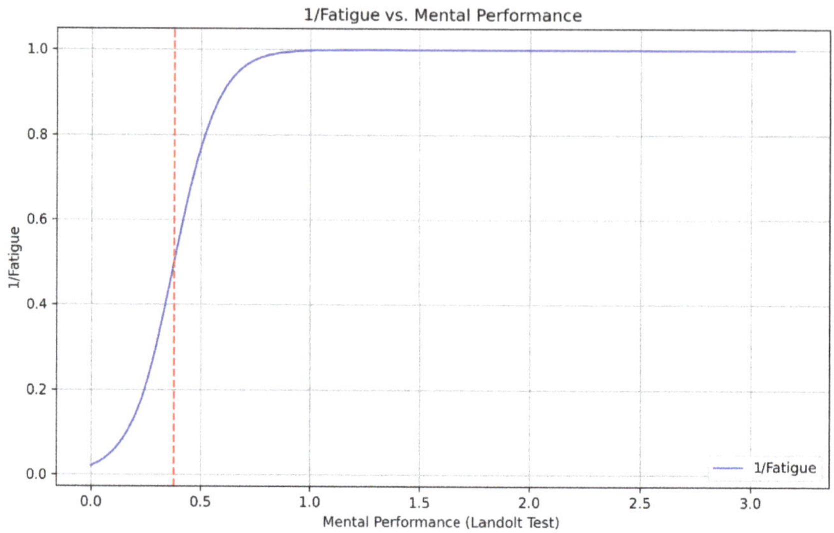

Figure 4. The relationship between the mental performance and the inverse of fatigue (red dotted line denotes an example threshold value separating fatigued and not fatigued states).

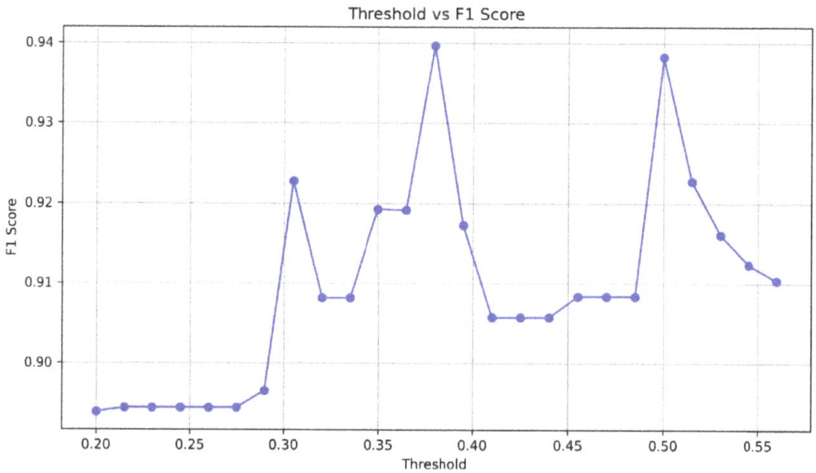

Figure 5. The relationship between the threshold and the F1-score.

Moreover, we recognized that vital signs vary significantly between individuals. To address this, we meticulously normalized these metrics before incorporating them into our predictive models. This normalization allowed our models to accurately account for individual differences, thereby providing more reliable predictions of fatigue. We used the min–max normalization function shown in Equation (4).

$$X' = \frac{X - X_{\min}}{X_{\max} - X_{\min}} \tag{4}$$

where the following apply:

- X is the original vital sign value.
- X' is the normalized vital sign value.
- $X_{\min}$ is the minimum value of the vital sign for the participant.

- X_{max} is the maximum value of the vital sign for the participant.

As we can see from the figure, there is a threshold value, 0.38, which achieves the best F1-score.

To predict the fatigue state, several methods were evaluated, including Support Vector Classifier (SVC), logistic regression, Multi-Layer Perceptron (MLP), decision tree, XGBoost, and random forest. Each method has distinct characteristics, which are shown in Table 5.

Table 5. The results of the models after normalizing the vital signs for each participant and considering fatigue states lower than 0.38.

Method	Distinct Characteristics	Accuracy	F1-Score
SVC	RBF kernel, C = 1.0, gamma = 'scale'	93.48%	0.4831
Logistic regression	L2 penalty, C = 0.001, solver = 'lbfgs'	96.27%	0.52147
MLP	Two hidden layers, ReLU activation, Adam solver, lr = 1×10^{-3}	96.27%	0.8719
Decision tree	Gini criterion, max_depth: 20, min_samples_leaf: 2, min_samples_split: 2	97.11%	0.8256
XGboost	lr = 0.1, max depth = 3, 100 estimators, gamma: 2, colsample_bytree: 0.8	98.10%	0.8658
Random forest	100 estimators, max_depth: 20, max_features: log2, min_samples_split: 2	98.1%	0.9465

The hyperparameters for each model were fine-tuned by using grid search to optimize performance. For SVC, we used a Radial Basis Function (RBF) kernel with a regularization parameter C = 1.0 and the kernel coefficient set to 'scale'. In logistic regression, we applied an L2 (Ridge) penalty with a regularization strength C = 0.001 and used the 'lbfgs' solver for optimization. The MLP classifier was configured with two hidden layers, with each using a linear activation function. The ReLU function was applied as the activation function for the neurons, Adam was selected as the solver, and the learning rate was set to 1×10^{-3}. For the decision tree, the Gini impurity was used as the splitting criterion, with a maximum tree depth of 20, a minimum of two samples required per leaf, and a minimum of two samples required to split a node. In the XGBoost model, the 'gbtree' booster was utilized. Key hyperparameters included a learning rate of 0.1, a maximum depth of 3, a minimum loss reduction (gamma) of 2, a feature subsample ratio (colsample_bytree) of 0.8, and 100 estimators (trees). The random forest classifier also employed 100 estimators with the Gini impurity criterion. It had a maximum tree depth of 20, used 'log2' as the maximum number of features to consider when splitting, and required a minimum of two samples to split a node. The environment used for these experiments was Python, with scikit-learn employed for SVC, logistic regression, MLP, decision tree, and random forest, while the XGBoost library was used for the XGBoost method. This setup ensures that the characteristics of each classification method are clearly presented and understood.

Table 5 shows the performance of the models using the threshold of 0.38 to detect the fatigue state based on the mental performance.

The random forest method addresses the issue of overfitting commonly associated with decision trees by combining the predictions of multiple trees to improve accuracy and robustness. While a single decision tree may capture noise in the data, random forests reduce this risk by averaging predictions across many trees, leading to better generalization. Random forest is an ensemble learning method that builds multiple decision trees by using bootstrap sampling (bagging) and random feature selection to reduce correlation between trees [47]. Each decision tree is constructed by using algorithms like ID3 (Iterative Dichotomiser 3), C4.5, or CART (Classification and Regression Trees). These algorithms split nodes based on criteria such as Gini impurity or entropy/information gain for classification problems, like detecting human fatigue. The final random forest prediction is made by aggregating results—majority voting for classification. It also uses out-of-bag (OOB) error for performance estimation and can calculate feature importance.

It is important to mention that the dataset, collected during working hours at various times of the day, reflects a greater number of non-fatigue state samples compared with fatigue state samples. For the testing dataset, this imbalance is evident, with the non-fatigue state (853 samples) samples being almost twice as numerous as the fatigue state samples (358 samples). This imbalance can make the accuracy metric less reliable, as it may not fully capture the performance of the models on the minority class. To address this issue, we also evaluated the F1-score, which is a more comprehensive measure by considering both precision and recall. As shown in Table 4, while all models exhibit high accuracy, there is significant variation in the F1-scores. Models such as SVC and logistic regression, despite their high accuracy, have relatively low F1-scores (0.4831 and 0.5215, respectively), indicating poorer performance in predicting the minority class. Conversely, models like MLP, decision tree, XGBoost, and random forest demonstrate both high accuracy and high F1-scores, with random forest achieving the highest F1-score, 0.9465. This indicates a better handling of the imbalanced data. The experiments revealed that the threshold of 0.38 yielded the best F1-score, with the random forest model achieving the F1-score of 0.9465 in fatigue prediction based on vital signs.

To analyze the effect of feature correlation on classification performance, it is essential to asses the relationship between the input variables. In this context, features exhibiting high correlation include num_eye_closure > 2 s and eye_closure_ratio, as well as rhythmicity_coeff and stability_coef, which were assessed by using Spearman rank correlation coefficient. However, other potential correlations, such as those between head Euler angles (roll, pitch, and yaw) with angles exceeding 30 degrees and the relationship between mouth_openness_ratio and num_yawns, were not captured by the Spearman method. This limitation may arise from the non-monotonic relationship between these variables. Notably, when training a random forest classifier without including the participant ID, the model achieved an F1-score of 0.7990 by using the original 15 features. Subsequent removing the redundant features, identified through correlation analysis, led to an improvement in the F1-score to 0.8384, highlighting the significance of feature selection in enhancing classification performance. This finding underscores the importance of addressing feature redundancy to optimize model performance in classification tasks.

Another crucial aspect is analyzing the most relevant characteristics for classification. Given the numerous variables used to characterize the problem, it is essential to determine their relevance and impact on classification performance. The relevance of each feature was analyzed, revealing that the most to least relevant features are BP Diastolic, participant ID, BP Systolic, eye_closure_ratio, heart_rate, oxygen_saturation, average pitch, average roll, mouth_openess_ratio, average yaw, average RR, rhythmicity_coeff, and stability_coef. The experiments shown in Table 6 revealed that using all 13 features resulted in an accuracy of 98.1% and an F1-score of 0.8658. Reducing the number of features to 11 increased the accuracy to 98.27% and the F1-score to 0.87481. A further reduction to nine features improved the accuracy to 98.43% and the F1-score to 0.8891. However, using only seven features slightly decreased the accuracy to 98.1% and dropped the F1-score to 0.8597. Leaving less than seven features produces significantly worse results. These results indicate that reducing the number of features can enhance classification performance up to a point, highlighting that not all features are equally relevant. Thus, focusing on the most significant features can enhance the model's efficiency and accuracy. Another notable observation is that the high importance of the participant ID feature means that the fatigue assessment models tend to be person-dependent. Table 7 shows the drop in the methods' performance after dropping the participant ID feature.

The participant ID's role in classification processes is underscored by a set of research observing differential reactions to changes in mental fatigue states among subjects, as evidenced in alterations in vital signs and eye movement. In one study, ref. [48] discovered that exposure to a fatigue-related task led to an increase in blood pressure in some participants, while others registered a decrease. This observation extended to respiratory rates, reinforcing the individual variance in physiological responses to fatigue. Moreover, ref. [49] found a correlation between mental fatigue and eye movement data (including blink frequency) in three of their subjects, as captured by an eye detection system, while the rest of the subjects did not show any significant relationship. These findings underscore the crucial nature of participant ID as a parameter. It enables machine learning models to account for person-specific dependencies, thereby enhancing their predictive accuracy for fatigue levels. Thus, the participant ID emerges as a significant factor in optimizing classification processes by accommodating individual variability.

Table 6. Performance of XGBoost model with varying number of features.

Number of Features	Used Features	Accuracy	F1-Score
All 13 features	BP Diastolic, participant ID, BP Systolic, eye_closure_ratio, heart_rate, oxygen_saturation, average pitch, average roll, mouth_openess_ratio, average yaw, average RR, rhythmicity_coeff, and stability_coef	98.1%	0.8658
Top 11 features	BP Diastolic, participant ID, BP Systolic, eye_closure_ratio, heart_rate, oxygen_saturation, average pitch, average roll, mouth_openess_ratio, average yaw, and average RR	98.27%	0.8748
Top 9 features	BP Diastolic, participant ID, BP Systolic, eye_closure_ratio, heart_rate, oxygen_saturation, average pitch, average roll, and mouth_openess_ratio	98.43%	0.8891
Top 7 features	BP Diastolic, participant ID, BP Systolic, eye_closure_ratio, heart_rate, oxygen_saturation, and average pitch	98.1%	0.8597

Table 7. Comparison of F1-scores for different classification methods with and without the participant ID feature.

Method	F1-Score with Participant ID	F1-Score Without Participant ID
Random forest	0.9465	0.8190
Logistic regression	0.52147	0.4788
XGBoost	0.8658	0.8128
Decision tree	0.8256	0.7237

Table 8 presents a comparative analysis of existing methodologies for detecting fatigue through images or videos without the use of sensors or specialized equipment. While our approach demonstrates good accuracy (98.43%), it is important to note that the dataset employed is relatively small compared with those utilized in other studies, such as [22]. The mentioned accuracy rates in the table were calculated based on different data, where fatigue was identified based on different metrics (subjective, objective, or based on monitoring some features); hence, the approaches listed cannot be compared directly. Nevertheless, the accuracy of our models is high enough for the proposed models to be used for fatigue detection. Additionally, this limitation stems from the scarcity of public datasets addressing mental fatigue among operators in work and academic contexts, highlighting a significant gap that our study aims to fill, encouraging further research efforts to address this. Nevertheless, the only approach that allows for a valid comparison is [28], as it utilized the same dataset. However, our approach demonstrates superior performance by achieving higher accuracy through the use of physiological features rather than relying on eye-tracking data. Additionally, it is noteworthy that many existing approaches primarily rely on features related to the eyes and mouth. However, factors such as the use of sunglasses, reflections from lighting on eyewear, or mouth coverings for medical purposes can affect the detection accuracy. In contrast, our method incorporates a diverse range of parameters that account for both the internal and external states of the subject, thereby enhancing the robustness of our approach in managing these challenges. Furthermore, the majority of the existing datasets rely on subjective indicators to evaluate fatigue levels among participants. These subjective measures can be influenced by personal perceptions and biases, which may compromise the validity of the findings. On the other hand, our study adopts a more robust approach by employing an objective metric to assess fatigue. This choice not only minimizes potential biases but also enhances the overall reliability and validity of our results.

Table 8. Comparison with existing approaches for fatigue detection.

Approach	Type of Fatigue	Number of Subject	Fatigue Assessment	Best Result Achieved
Cui et al. [14]	Driving fatigue	2000 facial images with fatigue state	Threshold based on eye and mouth features	98.30%
Dua et al. [15]	Driving fatigue	36 subjects (360 videos)	Facial landmark points and head posture data	85%
Dey et al. [16]	Driving fatigue	36 subjects	PERCLOS (Percentage of Eyelid Closure over the Pupil) and voice	96.4%
Salzillo et al. [18]	Driving fatigue	10 subjects	KSS (Karolinska Sleepiness Scale)	NRMSE of 15.63%
Minhas et al. [20]	Driving fatigue	Thousands of images of sleepy and active faces	Expert assessment	93.69%
Kassem et al. [21]	Driving fatigue	36 subjects (360 videos) + 2423 images of eyes	Facial landmark points and head posture data	93.3%
Chen et al. [22]	Driving fatigue	110 subjects (120 videos)	Yawns	90%
Du et al. [24]	Driving fatigue	20 subjects	KSS (Karolinska Sleepiness Scale)	91.67%
Kashevnik et al. [28]	Operator fatigue	10 subjects	Landolt rings test (mental performance indicator)	85%
Mehmood et al. [29]	Operator fatigue	15 subjects	NASA-TLX	99.941%
Mehmood et al. [30]	Operator fatigue	16 subjects	NASA-TLX	96.2%
Wang et al. [31]	Operator fatigue	12 subjects	NASA-TLX	92%
Our approach	Operator fatigue	10 subjects (817 videos)	Landolt rings test (mental performance indicator)	98.43%

5. Discussion

The results of the article shed light on the proposed association among mental performance, mental fatigue, and attention indicators. This is particularly noteworthy considering the existing literature that suggests strong correlations between mental fatigue and a range of physiological features, including head movement [50], vital signs (blood pressure, heart rate, oxygen saturation, and respiratory rate) [48,51], blinks [52], and yawns [53]. These findings provide empirical support for the idea that these physiological measures can serve as reliable indicators of mental fatigue. Moreover, the study's findings reveal that employing the identified mental performance threshold value of 0.38 leads to a high F1-score. These findings further indicate a robust relationship between mental performance and the aforementioned physiological features, which subsequently reflect the presence of mental fatigue.

In addition, the study findings not only reinforce the proposed connection among mental performance, mental fatigue, and attention indicators, but they also provide empirical evidence of the reliability and validity of utilizing specific physiological features and thresholds to assess mental fatigue. These insights contribute to a deeper understanding of the intricate relationship between mental performance and associated physiological manifestations, highlighting the potential for employing these measures in improving fatigue detection systems.

In our analysis, we believe that advanced deep learning techniques such as Transformers, CNNs, and RNNs are suited for more complex tasks, such as extracting deep layer features from video; hence, these type of models are used to extract vital signs from video data, a critical aspect of our research, after demonstrating their high capability due to their training on extensive datasets. Conversely, for the specific task at hand, traditional machine learning algorithms are more appropriate. This is primarily because they require significantly fewer data and are computationally efficient, allowing us to streamline our approach and avoid unnecessary complexity and time consumption.

Another reason that led us to exclude advanced deep learning techniques for the classification task is that the available data are insufficient for the effective implementation of deep learning algorithms. Additionally, it is noteworthy that most state-of-the-art (SOTA) fatigue assessment methodologies primarily address driving fatigue, often overlooking the fatigue associated with office work. Consequently, we consider our research to be novel and innovative, as it specifically targets this issue through objective measurement techniques, i.e., the mental performance assessment using the Landolt ring.

This study faced certain limitations due its relatively small size. However, this problem cannot be solved by merging our dataset with the other datasets mentioned in Section 2.2, because we collected different fatigue indicators in our dataset, such as VAS-F, Landolt rings test, and CRT. Based on our experiments described in [28], we found that Landolt rings test is the most reasonable in the considered case of fatigue detection. As a result, our models have been trained for the fatigue level estimated by the Landolt rings test, which is absent in the datasets mentioned in Section 2.2; hence, it is not possible to extend our dataset with the datasets introduced in Section 2.2. Additionally, another limitation was including group of participants with limited age diversity and ethnic background. As people age, heart rate and respiratory rate remain relatively stable, but maximal heart rate decreases, blood pressure—particularly systolic—tends to rise due to stiffening arteries, and oxygen saturation may decline, especially in those with chronic conditions. However, we have not studied how the data change across different age groups and ethnicities. These factors may hinder the generalizability of the proposed approach. One potential constraint that could impede the generalizability of the trained model is the fact that fatigue detection is human-dependent. This factor could pose challenges in applying the proposed model to novel subjects. Moreover, there remains ambiguity in determining the fatigue threshold, as the VAS-F result is subjective. To address this, we opted for the mental performance indicator as a replacement for fatigue. However, for greater reliability, it is recommended to establish an explicit value for the fatigue threshold.

Another challenge that opens up the discussion is the assessment of the validity of the data predicted by deep learning models. The majority of models employed for physiological indicator estimation have undergone testing on multiple datasets, distinct from the ones used for training. This is as detailed in the respective articles, which are cited in the reference section. The outcomes demonstrated high performance and low error margins, thereby validating their application to the current dataset. This dataset has a significant resemblance to the conditions present in the training datasets, including aspects such as lighting conditions and the skin tone and demographic characteristics of the subjects, who were predominantly healthy and young. As a result, the probability of encountering abnormal values in vital signs is considerably reduced. However, this limitation can be effectively addressed by expanding the training datasets to include a more diverse population. Specifically, incorporating individuals from various age ranges, ethnic backgrounds, and those who suffer from serious health conditions that may impact their vital signs will enhance the models' robustness. Doing so would better equip the models to be generalized across different demographic groups and health conditions, ultimately improving their accuracy and reliability in estimating physiological indicators for a broader population. This approach would not only mitigate existing biases but also ensure that the models can more accurately reflect the complexities of real-world health scenarios.

6. Conclusions

This research article emphasizes the significance of fatigue detection in system operation contexts and its direct impact on efficiency. The objective of this study is to assess the mental performance of operators who work long hours in front of their computers. Mental performance is considered the opposite of mental fatigue. Accordingly, we concentrate on office routines that do not involve physical work. This study starts with exploring various approaches and datasets that utilize videos, sensors, and devices to capture features associated with fatigue. This paper introduces an ML-based approach that integrates video data with features computed by using computer vision deep learning models. These features encompass head movement (roll, pitch, and yaw), vital signs (blood pressure, heart rate, oxygen saturation, and respiratory rate), and eye and mouth states (blinking and yawning).

By incorporating these features, the need for manual calculation or external sensors is eliminated, distinguishing it from existing datasets that rely on such methods. The primary objective of this research is to advance fatigue detection in work and academic settings.

The experiments were carried out by using the publicly available dataset OperatorEYEVP. The choice to apply the estimation process to this specific public dataset was motivated by the relevance of the videos provided, which depict the daily tasks of operators and students. As the main objective of this research is to investigate fatigue in this specific context, it was deemed appropriate to utilize this dataset. The specific task examined is reading scientific articles, as it necessitates a focused and alert mind to effectively analyze the information presented. Additionally, some individuals engage in gaming during their breaks, which is presented in the dataset used by the Tetris game, which also requires a significant level of concentration. Furthermore, we used an objective measurement of the mental performance by using the Landolt rings test to evaluate visual acuity. The physiologically indicative parameters, computed through computer vision techniques, were compiled into another publicly accessible dataset, HFAVD, presented in this paper. Future endeavors are directed towards applying this methodology to other video datasets and expanding the HFAVD dataset.

Through a series of experiments utilizing machine learning techniques, the dataset was analyzed to assess the fatigue state based on the predicted features by using deep learning models (Section 3.3). The mental performance indicator obtained from the Landolt rings test was utilized as a measure of mental fatigue. The primary objective was to investigate the relationship between this indicator and the attention metric in order to determine the optimal starting threshold for dividing our ground truth into fatigued and non-fatigued classes. Furthermore, a series of experiments were conducted to identify the most effective final value for identifying sessions characterized by mental fatigue in participants.

The research findings indicate that the threshold of 0.38 exhibits the highest potential for effectively discriminating between subjects experiencing mental fatigue and those who are not fatigued for achieving superior performance in the classification process. We employed a range of machine learning techniques, including Support Vector Classifier (SVC), logistic regression, Multi-Layer Perceptron (MLP), decision tree, XGBoost, and random forest. To optimize model performance, we utilized grid search to identify the best hyperparameters that yield the highest accuracy and F1-score. Our results indicate that most of these techniques achieved impressive accuracy levels, all exceeding 90%. However, we prioritized the F1-score as a critical performance metric, given that our test set was imbalanced. By attaining a high F1-score, we ensured that the models performed well across both classes, thereby mitigating potential bias issues. The outcomes consistently showed that random forest achieved the highest F1-score, 94%. These results highlight random forest as a highly promising technique for fatigue detection. XGBoost, decision tree, and MLP reached good F1-scores of 86%, 82%, and 87%, respectively.

It was also revealed that participant ID emerged as a significant factor in optimizing classification, showing that accounting for person-specific dependencies enhances the predictive accuracy and F1-score of machine learning models for the fatigue state. This observation was made by comparing the classification metrics when incorporating participant ID as a feature in the model versus when it was excluded. We found that the F1-score declined from 94% to 81%, highlighting the significance of individual differences in the context of mental fatigue classification.

Additionally, the concept of feature importance motivated us to investigate which features contribute most significantly to enhancing classification performance. Consequently, we conducted an experiment in which we sequentially analyzed the top n features based on their relevance. The results indicate that using 9 out of the 13 features achieved the highest accuracy, 98.43%. This finding suggests that we can streamline our approach by excluding certain less impactful features, such as those related to respiratory rate, thereby reducing the time required for implementation.

In summary, this study contributes to the topic of fatigue detection by presenting an ML-based approach and its application to a comprehensive dataset, demonstrating the effectiveness of machine learning in accurately assessing fatigue levels. The integration of video data and the use of deep learning techniques offer a more efficient and practical approach to identifying signs of fatigue, potentially enhancing safety in various contexts. Further research and development in this area can lead to the implementation of effective fatigue detection systems that mitigate risks and improve overall well-being.

Author Contributions: Conceptualization, N.S.; methodology, N.S. and W.O.; software, W.O. and B.H.; validation, N.S. and W.O.; formal analysis, N.S. and W.O.; investigation, N.S., W.O. and B.H.; resources, N.S.; data curation, A.K., W.O. and B.H.; writing—original draft preparation, A.K., W.O. and B.H.; writing—review and editing, N.S. and A.K.; visualization, W.O.; supervision, N.S.; project administration, N.S.; funding acquisition, N.S. All authors have read and agreed to the published version of the manuscript.

Funding: This research study was funded by the Russian Science Foundation (project 24-21-00300).

Institutional Review Board Statement: Not applicable.

Informed Consent Statement: Not applicable.

Data Availability Statement: The dataset is available at https://disk.yandex.ru/d/A49WtGSODl7eCw (accessed on 11 November 2024).

Conflicts of Interest: The authors declare no conflicts of interest.

Abbreviations

Bi-LSTM	Bidirectional Long Short-Term Memory
CNN	Convolutional Neural Networks
CRT	Choice Reaction Time
CSV	Comma-Separated Values
EAR	eye aspect ratio
ECG	electrocardiography signals
EEG	electroencephalogram signals
EDA	electrodermal activity
EM	EYE-MOUTH
FIS	Fuzzy Inference System
FOM	Frequency of Open Mouth
HF	high frequency
HFAVD	Human Fatigue Assessment Based on Video Data
HOG	Histogram of Oriented Gradients
HRV	heart rate variability
JADE	Joint Approximate Diagonalization of Eigenmatrices
KSS	Karolinska Sleepiness Scale
LF	low frequency
LSTM	Long Short-Term Memory
MAE	Mean Absolute Error
MFRNN	Multimodal Fusion Recurrent Neural Network
MLP	Multi-Layer Perceptron
MOR	mouth opening ratio
MTCNN	Multitask Convolutional Neural Network
NLR	nose length ratio
NRMSE	normalized root mean square error
PERCLOS	Percentage of Eyelid Closure over the Pupil
PSG	polysomnography

PVT	Psychomotor Vigilance Test
RBF	Radial Basis Function
RNN	Recurrent Neural Network
ROI	Region of Interest
TSP	Triangular Surface Patch
SVC	Support Vector Classifier
SVM	Support Vector Machine
VAS-F	Visual Analogue Scale to Evaluate Fatigue Severity
VLF	very low frequency

References

1. Zhu, T.; Zhang, C.; Wu, T.; Ouyang, Z.; Li, H.; Na, X.; Liang, J.; Li, W. Research on a Real-Time Driver Fatigue Detection Algorithm Based on Facial Video Sequences. *Appl. Sci.* **2022**, *12*, 2224. [CrossRef]
2. Shilov, N.; Othman, W.; Hamoud, B. Operator Fatigue Detection via Analysis of Physiological Indicators Estimated Using Computer Vision. In Proceedings of the 26th International Conference on Enterprise Information Systems—Volume 2: ICEIS, Angers, France, 28–30 April 2024 ; SciTePress: Setúbal, Portugal, 2024; pp. 422–432. [CrossRef]
3. Li, X.; Lian, X.; Liu, F. Rear-End Road Crash Characteristics Analysis Based on Chinese In-Depth Crash Study Data. In Proceedings of the 16th COTA International Conference of Transportation Professionals (CICTP 2016), Shanghai, China, 6–9 July 2016; pp. 1536–1545. [CrossRef]
4. Zhang, G.; Yau, K.K.; Zhang, X.; Li, Y. Traffic accidents involving fatigue driving and their extent of casualties. *Accid. Anal. Prev.* **2016**, *87*, 34–42. [CrossRef] [PubMed]
5. Alharbey, R.; Dessouky, M.M.; Sedik, A.; Siam, A.I.; Elaskily, M.A. Fatigue State Detection for Tired Persons in Presence of Driving Periods. *IEEE Access* **2022**, *10*, 79403–79418. [CrossRef]
6. Di Stasi, L.L.; Renner, R.; Catena, A.; Cañas, J.J.; Velichkovsky, B.M.; Pannasch, S. Towards a driver fatigue test based on the saccadic main sequence: A partial validation by subjective report data. *Transp. Res. Part C Emerg. Technol.* **2012**, *21*, 122–133. [CrossRef]
7. Yamada, Y.; Kobayashi, M. Fatigue Detection Model for Older Adults Using Eye-Tracking Data Gathered While Watching Video: Evaluation Against Diverse Fatiguing Tasks. In Proceedings of the 2017 IEEE International Conference on Healthcare Informatics (ICHI), Park City, UT, USA, 23–26 August 2017; pp. 275–284. [CrossRef]
8. Dawson, D.; Searle, A.K.; Paterson, J.L. Look before you (s)leep: Evaluating the use of fatigue detection technologies within a fatigue risk management system for the road transport industry. *Sleep Med. Rev.* **2014**, *18*, 141–152. [CrossRef]
9. Gomer, J.; Walker, A.; Gilles, F.; Duchowski, A. Eye-Tracking in a Dual-Task Design: Investigating Eye-Movements, Mental Workload, and Performance. *Proc. Hum. Factors Ergon. Soc. Annu. Meet.* **2008**, *52*, 1589–1593. [CrossRef]
10. Hu, X.; Lodewijks, G. Detecting fatigue in car drivers and aircraft pilots by using non-invasive measures: The value of differentiation of sleepiness and mental fatigue. *J. Saf. Res.* **2020**, *72*, 173–187. [CrossRef]
11. Matuz, A.; van der Linden, D.; Kisander, Z.; Hernádi, I.; Kázmér, K.; Csathó, Á. Enhanced cardiac vagal tone in mental fatigue: Analysis of heart rate variability in Time-on-Task, recovery, and reactivity. *PLoS ONE* **2021**, *16*, e0238670. [CrossRef]
12. Desmond, P.A.; Hancock, P.A. Active and passive fatigue states. In *Stress, Workload, and Fatigue*; CRC Press: Boca Raton, FL, USA, 2000; pp. 455–465.
13. Kovalenko, S.; Mamonov, A.; Kuznetsov, V.; Bulygin, A.; Shoshina, I.; Brak, I.; Kashevnik, A. OperatorEYEVP: Operator Dataset for Fatigue Detection Based on Eye Movements, Heart Rate Data, and Video Information. *Sensors* **2023**, *23*, 6197. [CrossRef]
14. Cui, Z.; Sun, H.M.; Yin, R.N.; Gao, L.; Sun, H.B.; Jia, R.S. Real-time detection method of driver fatigue state based on deep learning of face video. *Multimed. Tools Appl.* **2021**, *80*, 25495–25515. [CrossRef]
15. Dua, M.; Shakshi.; Singla, R.; Raj, S.; Jangra, A. Deep CNN models-based ensemble approach to driver drowsiness detection. *Neural Comput. Appl.* **2021**, *33*, 3155–3168. [CrossRef]
16. Dey, S.; Chowdhury, S.A.; Sultana, S.; Hossain, M.A.; Dey, M.; Das, S.K. Real Time Driver Fatigue Detection Based on Facial Behaviour along with Machine Learning Approaches. In Proceedings of the 2019 IEEE International Conference on Signal Processing, Information, Communication & Systems (SPICSCON), Dhaka, Bangladesh, 28–30 November 2019; pp. 135–140. [CrossRef]
17. Dalal, N.; Triggs, B. Histograms of oriented gradients for human detection. In Proceedings of the 2005 IEEE Computer Society Conference on Computer Vision and Pattern Recognition (CVPR'05), San Diego, CA, USA, 20–25 June 2005; Volume 1, pp. 886–893. [CrossRef]
18. Salzillo, G.; Natale, C.; Fioccola, G.B.; Landolfi, E. Evaluation of Driver Drowsiness based on Real-Time Face Analysis. In Proceedings of the 2020 IEEE International Conference on Systems, Man, and Cybernetics (SMC), Toronto, ON, Canada, 11–14 October 2020; pp. 328–335. [CrossRef]
19. Shahid, A.; Wilkinson, K.; Marcu, S.; Shapiro, C.M., Karolinska Sleepiness Scale (KSS). In *STOP, THAT and One Hundred Other Sleep Scales*; Shahid, A., Wilkinson, K., Marcu, S., Shapiro, C.M., Eds.; Springer: New York, NY, USA, 2012; pp. 209–210. [CrossRef]
20. Minhas, A.A.; Jabbar, S.; Farhan, M.; ul Islam, M.N. A smart analysis of driver fatigue and drowsiness detection using convolutional neural networks. *Multimed. Tools Appl.* **2022**, *81*, 26969–26986. [CrossRef]

21. Kassem, H.A.; Chowdhury, M.; Abawajy, J.H. Drivers Fatigue Level Prediction Using Facial, and Head Behavior Information. *IEEE Access* **2021**, *9*, 121686–121697. [CrossRef]
22. Chen, L.; Xin, G.; Liu, Y.; Huang, J. Driver Fatigue Detection Based on Facial Key Points and LSTM. *Secur. Commun. Networks* **2021**, *2021*, 5383573. [CrossRef]
23. Abtahi, S.; Omidyeganeh, M.; Shirmohammadi, S.; Hariri, B. YawDD: A yawning detection dataset. In Proceedings of the 5th ACM Multimedia Systems Conference, Singapore, 19 March 2014; pp. 24–28. [CrossRef]
24. Du, G.; Li, T.; Li, C.; Liu, P.X.; Li, D. Vision-Based Fatigue Driving Recognition Method Integrating Heart Rate and Facial Features. *IEEE Trans. Intell. Transp. Syst.* **2021**, *22*, 3089–3100. [CrossRef]
25. Cardoso, J.; Souloumiac, A. Blind beamforming for non-gaussian signals. *IEE Proc. F (Radar Signal Process.)* **1993**, *140*, 362–370. [CrossRef]
26. Papazov, C.; Marks, T.K.; Jones, M. Real-time 3D head pose and facial landmark estimation from depth images using triangular surface patch features. In Proceedings of the 2015 IEEE Conference on Computer Vision and Pattern Recognition (CVPR), Boston, MA, USA, 7–12 June 2015; pp. 4722–4730. [CrossRef]
27. Mehmood, I.; Li, H.; Umer, W.; Ma, J.; Saad Shakeel, M.; Anwer, S.; Fordjour Antwi-Afari, M.; Tariq, S.; Wu, H. Non-invasive detection of mental fatigue in construction equipment operators through geometric measurements of facial features. *J. Saf. Res.* **2024**, *89*, 234–250. [CrossRef]
28. Kashevnik, A.; Kovalenko, S.; Mamonov, A.; Hamoud, B.; Bulygin, A.; Kuznetsov, V.; Shoshina, I.; Brak, I.; Kiselev, G. Intelligent Human Operator Mental Fatigue Assessment Method Based on Gaze Movement Monitoring. *Sensors* **2024**, *24*, 6805. [CrossRef]
29. Mehmood, I.; Li, H.; Qarout, Y.; Umer, W.; Anwer, S.; Wu, H.; Hussain, M.; Fordjour Antwi-Afari, M. Deep learning-based construction equipment operators' mental fatigue classification using wearable EEG sensor data. *Adv. Eng. Inform.* **2023**, *56*, 101978. [CrossRef]
30. Mehmood, I.; Li, H.; Umer, W.; Arsalan, A.; Anwer, S.; Mirza, M.A.; Ma, J.; Antwi-Afari, M.F. Multimodal integration for data-driven classification of mental fatigue during construction equipment operations: Incorporating electroencephalography, electrodermal activity, and video signals. *Dev. Built Environ.* **2023**, *15*, 100198. [CrossRef]
31. Wang, L.; Li, H.; Yao, Y.; Han, D.; Yu, C.; Lyu, W.; Wu, H. Smart cushion-based non-invasive mental fatigue assessment of construction equipment operators: A feasible study. *Adv. Eng. Inform.* **2023**, *58*, 102134. [CrossRef]
32. Ghoddoosian, R.; Galib, M.; Athitsos, V. A Realistic Dataset and Baseline Temporal Model for Early Drowsiness Detection. In Proceedings of the 2019 IEEE/CVF Conference on Computer Vision and Pattern Recognition Workshops (CVPRW), Long Beach, CA, USA, 16–20 June 2019; pp. 178–187.
33. Weng, C.H.; Lai, Y.H.; Lai, S.H. Driver Drowsiness Detection via a Hierarchical Temporal Deep Belief Network. In *Computer Vision—ACCV 2016 Workshops*; Chen, C.S., Lu, J., Ma, K.K., Eds.; Springer: Cham, Switzerland, 2017; pp. 117–133.
34. Liu, P.; Chi, H.L.; Li, X.; Guo, J. Effects of dataset characteristics on the performance of fatigue detection for crane operators using hybrid deep neural networks. *Autom. Constr.* **2021**, *132*, 103901. [CrossRef]
35. Massoz, Q.; Langohr, T.; François, C.; Verly, J.G. The ULg multimodality drowsiness database (called DROZY) and examples of use. In Proceedings of the 2016 IEEE Winter Conference on Applications of Computer Vision (WACV), Lake Placid, NY, USA, 7–10 March 2016; pp. 1–7. [CrossRef]
36. Basner, M.; Dinges, D.F. Maximizing Sensitivity of the Psychomotor Vigilance Test (PVT) to Sleep Loss. *Sleep* **2011**, *34*, 581–591. [CrossRef] [PubMed]
37. Yang, C.; Yang, Z.; Li, W.; See, J. FatigueView: A Multi-Camera Video Dataset for Vision-Based Drowsiness Detection. *IEEE Trans. Intell. Transp. Syst.* **2023**, *24*, 233–246. [CrossRef]
38. Kavalci Yilmaz, E.; Akcayol, M.A. SUST-DDD: A Real-Drive Dataset for Driver Drowsiness Detection. In Proceedings of the 31st IEEE Conference of Open Innovations Association FRUCT, Helsinki, Finland, 27–29 April 2022. [CrossRef]
39. Othman, W.; Kashevnik, A.; Ryabchikov, I.; Shilov, N. Contactless Camera-Based Approach for Driver Respiratory Rate Estimation in Vehicle Cabin. *Lect. Notes Netw. Syst.* **2022**, *5431*, 429–442.
40. Othman, W.; Kashevnik, A.; Ali, A.; Shilov, N.; Ryumin, D. Remote Heart Rate Estimation Based on Transformer with Multi-Skip Connection Decoder: Method and Evaluation in the Wild. *Sensors* **2024**, *24*, 775. [CrossRef]
41. Guo, J.; Zhu, X.; Yang, Y.; Yang, F.; Lei, Z.; Li, S.Z. Towards Fast, Accurate and Stable 3D Dense Face Alignment. *arXiv* **2021**, arXiv:2009.09960.
42. Hamoud, B.; Kashevnik, A.; Othman, W.; Shilov, N. Neural Network Model Combination for Video-Based Blood Pressure Estimation: New Approach and Evaluation. *Sensors* **2023**, *23*, 1753. [CrossRef]
43. Hamoud, B.; Othman, W.; Shilov, N.; Kashevnik, A. Contactless Oxygen Saturation Detection Based on Face Analysis: An Approach and Case Study. In Proceedings of the 2023 33rd Conference of Open Innovations Association (FRUCT), Zilina, Slovakia, 24–26 May 2023; pp. 54–62. [CrossRef]
44. Kashevnik, A.; Ali, A.; Lashkov, I.; Zubok, D. Human Head Angle Detection Based on Image Analysis. In *Proceedings of the Future Technologies Conference (FTC) 2020, Volume 1*; Arai, K., Kapoor, S., Bhatia, R., Eds.; Springer: Cham, Switzerland, 2021; pp. 233–242.
45. Hasan, F.; Kashevnik, A. State-of-the-Art Analysis of Modern Drowsiness Detection Algorithms Based on Computer Vision. In Proceedings of the 2021 29th Conference of Open Innovations Association (FRUCT), Tampere, Finland, 12–14 May 2021; pp. 141–149. [CrossRef]

46. Bourdone Test—Landolt Rings Online. Available online: https://metodorf.com/tests/bourdon/bourdonlandolt.php (accessed on 11 November 2024).
47. Hastie, T.; Tibshirani, R.; Friedman, J. *The Elements of Statistical Learning: Data Mining, Inference, and Prediction*, 2nd ed.; Springer: New York, NY, USA, 2009.
48. Meng, J.; Zhao, B.; Ma, Y.; Yiyu, J.; Nie, B. Effects of fatigue on the physiological parameters of labor employees. *Nat. Hazards* **2014**, *74*, 1127–1140. [CrossRef]
49. Sampei, K.; Ogawa, M.; Torres, C.C.C.; Sato, M.; Miki, N. Mental Fatigue Monitoring Using a Wearable Transparent Eye Detection System. *Micromachines* **2016**, *7*, 20. [CrossRef]
50. Ye, M.; Zhang, W.; Cao, P.; Liu, K. Driver Fatigue Detection Based on Residual Channel Attention Network and Head Pose Estimation. *Appl. Sci.* **2021**, *11*, 9195. [CrossRef]
51. Luo, H.; Lee, P.A.; Clay, I.; Jaggi, M.; De Luca, V. Assessment of Fatigue Using Wearable Sensors: A Pilot Study. *Digit. Biomarkers* **2020**, *4*, 59–72. [CrossRef] [PubMed]
52. Zhao, Q.; Nie, B.; Bian, T.; Ma, X.; Sha, L.; Wang, K.; Meng, J. Experimental study on eye movement characteristics of fatigue of selected college students. *Res. Sq.* 2023, *preprint*. [CrossRef]
53. Gupta, N.K.; Bari, A.K.; Kumar, S.; Garg, D.; Gupta, K. Review Paper on Yawning Detection Prediction System for Driver Drowsiness. In Proceedings of the 2021 5th International Conference on Trends in Electronics and Informatics (ICOEI), Tirunelveli, India, 3–5 June 2021; pp. 1–6.

applied sciences

MDPI

Article

Brainwaves in the Cloud: Cognitive Workload Monitoring Using Deep Gated Neural Network and Industrial Internet of Things

Muhammad Abrar Afzal [1,*], Zhenyu Gu [1,*], Syed Umer Bukhari [2] and Bilal Afzal [3]

1 School of Design, Minhang Campus, Shanghai Jiao Tong University, Shanghai 200040, China
2 Department of Computer Science & Engineering, Chalmers University of Technology, 41756 Gothenburg, Sweden; syedu@chalmers.se
3 School of Management & Economics, Qingshuihe Campus, University of Electronic Science and Technology of China, Chengdu 611731, China; bilal.afzal@outlook.com
* Correspondence: abrar456@sjtu.edu.cn (M.A.A.); zygu@sjtu.edu.cn (Z.G.)

Abstract: Monitoring and classifying cognitive workload in real time is vital for optimizing human–machine interactions and enhancing performance while ensuring safety, particularly in industrial scenarios. Considering this significance, the authors aim to formulate a cognitive workload monitoring system (CWMS) by leveraging the deep gated neural network (DGNN), a hybrid model integrating bi-directional long short-term memory (Bi-LSTM) and gated recurrent unit (GRU) networks. In our experimental setup, each of the four virtual users is equipped with a Raspberry Pi Zero W module to ensure efficient data transmission, thereby enhancing the reliability and efficacy of the monitoring process. This seamless monitoring framework utilizes the constrained application protocol (CoAP) and the Things Board platform to evaluate cognitive workload in real time. The most popular EEG benchmark dataset, the STEW is utilized for workload classification in this study. We employ the short-time Fourier transformation (STFT) to extract frequency bands corresponding to users in both high and low cognitive workload modes. The proposed DGNN models achieve a perfect accuracy of 99.45%, outperforming every previous state-of-the-art model. We meticulously monitored critical parameters, including latency, classification processing time, and cognitive workload levels. This research demonstrates the importance of continuous monitoring for increasing productivity and safety in industries by introducing a novel method of real-time cognitive workload monitoring. The implementation codes for each experiment are documented and made available for reproducibility.

Keywords: human–machine interface; mental workload; cognitive workload monitoring system; deep gated neural network; electroencephalogram; brain–computer interface applications

1. Introduction

Cognitive workload is a global issue, alongside the ongoing advancements of the industrial revolution, where new technologies are continually emerging [1]. Industry 4.0 marked a transition from traditional manufacturing methods to digital manufacturing [2], driven by innovations like advanced robot arms [3], industrial Internet of Things (IIoT) [4], artificial intelligence (AI) [5], and cloud computing [6]. These innovations generate a significant amount of data [7] and contribute to a disruptive workplace process, which has become a new source of stress [8]. Industry 5.0, in contrast to its predecessor Industry 4.0, heralds a new era of smooth integration between cutting-edge technology and human–machine interactions [9]. This integration aims to enhance production and efficiency, emphasizing the significance of improving human–machine interactions to boost overall performance [10,11].

In this context, implementing a new cognitive workload monitoring system (CWMS) incorporating deep gated neural network (DGNN) and industrial Internet of Things (IIoT) technologies is essential. This system tracks and classifies cognitive workloads in real time, thereby improving the performance and safety of human–machine interactions. The CWMS holds significance in multiple industrial contexts where human-robot collaboration is pervasive [12–15]. Cognitive workload, encompassing mental exertion [16] across various functions such as memory [17], decision-making [18], attention [19], perception [20], problem-solving [21], and information processing [22], serves as a foundational element. Continuous monitoring and real-time analysis of workload levels are essential for improving efficiency [23], reducing errors [24], and addressing safety concerns [25], especially in critical industrial environments.

A spectrum of cutting-edge technologies contributes to the assessment of cognitive workload. Electrocardiography (ECG) monitors the heart's electrical activity [26], providing valuable data on physiological stress levels. Eye-tracking captures gaze patterns, offering insights into visual attention and cognitive processing [27]. Facial expression analysis deciphers real-time emotional states [28], and task performance analysis evaluates metrics related to specific activities [29]. Functional near-infrared spectroscopy (fNIRS), measures changes in blood oxygen levels [30], revealing neural activity associated with cognitive workload. Among all these methods, electroencephalography (EEG) stands out for its effectiveness [31], providing a detailed understanding of cognitive processes with real-time monitoring, non-invasiveness, and portability. Also, these methods leverage machine learning [32], and deep learning [33] algorithms for autonomous identification and classification of cognitive states.

A critical aspect that requires attention is the underexplored intersection of cognitive workload assessment, IIoT, and advanced neural network methodologies for real-time workload monitoring and analysis. There is a lack of comprehensive studies harnessing the combined potential of EEG, IIoT, and refined neural network methods. This creates a gap in developing efficient and accurate cognitive workload monitoring systems that are capable of handling the challenges posed by industrial interactions [1]. In our current study, we aim to bridge this identified gap by focusing on EEG signals for real-time cognitive workload classification and monitoring. Our approach introduces the DGNN model, comprising a dual-layer architecture that incorporates both Bi-LSTM [34] and GRU [35] networks, providing a state-of-the-art framework for cognitive workload classification. The DGNN methodology harnesses the STEW dataset [36], including data from both low and high cognitive workload modes for training and evaluation. The model achieves a remarkable accuracy of 99.45%. Moreover, the article initiates the development of a real-time cognitive workload monitoring system though an experiment that integrates IIoT gateways, the constrained application protocol (CoAP) [37], an AI server, and the Things Board open-source IoT platform for live data monitoring [38]. In this experiment, four virtual users participate, each connected to the AI server, which employs the DGNN model for real-time cognitive workload analysis via IIoT gateways using CoAP. Additionally, the CWMS provides real-time data on parameters such as latency, classification processing time, and cognitive workload levels.

This paper offers the following contributions:

1. This study introduces a novel deep gated neural network (DGNN) designed for classifying workloads in real-world scenarios using EEG data.
2. The proposed approach achieves superior performance in classifying various workload levels compared to previous studies.
3. This study presents a distinctive cognitive workload monitoring system (CWMS) that combines IIoT gateways, CoAP, and an AI server for real-time monitoring and analysis of workload parameters.

The subsequent sections of this article are organized as follows. Section 2 presents a detailed analysis of current studies on the categorization of cognitive workload. Section 3 provides an explanation of the materials and methods used in the study. This includes the preprocessing of EEG data, the design of the DGNN model, and the implementation of the real-time CWMS. In Section 4, the results of both the proposed method and CWMS are presented, along with a thorough analysis. Section 5 provides the conclusion, while Section 6 pertains to the future works.

2. Literature Review

Electroencephalography (EEG) is recognized as a physiological index that is capable of continuously measuring cognitive load [39]. However, EEG signals exhibit characteristics such as weakness, noise, and non-stationarity among subjects, making the identification of robust features a persistent challenge [40]. Traditional methods for cognitive workload classification from EEG data encompass various analytical techniques and feature extraction methods. These include statistical tests, validating differences among features by examining changes in power within specific frequency bands, and time–domain features, such as amplitude, latency, and waveform morphology, providing insights into the temporal characteristics of EEG signals [41–43]. Additionally, power spectral density analysis reveals the distribution of power across frequency bands [44], aiding in the inference of workload variations. Event-related potentials (ERPs) reflect the brain's neural responses to stimuli or events, with the analysis of ERP amplitude and latency elucidating cognitive processing and workload demands [45]. EEG signals are also segmented into different frequency bands, including delta, theta, alpha, beta, and gamma [46], with the activity within these bands providing information about cognitive workload levels, particularly in the gamma band, known for its role in higher cognitive functions [47]. Furthermore, traditional classification methods such as support vector machines (SVM) [48], k-nearest neighbors (kNN) [49], and naive Bayes classifiers [50] have been widely employed for cognitive workload classification based on EEG data.

Significant advancements have been made in understanding cognitive workload using EEG, using machine learning approaches to make valuable contributions towards its assessment. Machine learning algorithms apply features learned from a training dataset to perform tasks such as classification or regression [51]. On the other hand, deep learning is a specific subclass of machine learning that constructs models based on neural networks, such as convolutional networks or long short-term memory networks. Unlike traditional machine learning, deep learning models have the capability to learn complex patterns and representations directly from raw data, eliminating the need for manual feature extraction [52,53]. With automated feature learning, deep learning models can achieve a high recognition accuracy by training directly on raw datasets, allowing for more efficient and accurate cognitive workload classification [53].

In exploring the landscape of cognitive workload detection, several studies have contributed significant insights leveraging machine learning approaches. Momeni et al. (2019) [54] utilized extreme gradient boosting (XGB) algorithms to detect cognitive workload, showing promising results in drone control simulation experiments. Ramírez-Moreno et al. (2021) [55] emphasized the feasibility of short-calibration methods for mental fatigue modeling through biometric signals. Moreno et al. (2021) focused on mental fatigue and human–machine interactions, while Zanetti et al. (2021) [56] addressed job execution support using wearable EEG devices. Cao et al. (2022) [57] introduced a novel framework using hybrid EEG-fNIRS and bivariate functional brain connectivity (FBC) features for workload classification. Additionally, Liu et al. (2023) [58] developed a real-time system for identifying pilots' cognitive workload levels using a wireless EEG headset, contributing to EEG-based methodologies for real-time assessments in aviation. Despite achieving a lower accuracy rate compared to previous studies, Liu et al.'s research provides valuable insights into cognitive workload assessments in aviation contexts.

Studies using deep learning approaches include Sharma et al. (2021) [59], who developed a cascade one-dimensional convolution neural network (1DCNN) and bidirectional long short-term memory (BLSTM) model for binary and ternary classification of mental workload (MWL). Dolmans et al. (2021) [60] proposed a deep neural network (DNN) that flexibly makes use of multiple modalities, including galvanic skin response, functional near-infrared spectrograms, and eye movements, to classify perceived mental workload

(PMWL) accurately on a seven-level workload scale. Afzal et al. (2023) [61] proposed a bi-directional gated network (BDGN) for cognitive workload classification in Industry 5.0 settings, integrating LSTM and GRU and demonstrating promising results in a simulated environment. The reviewed recent studies are summarized in Table 1.

Table 1. Summary of the performance of previous studies reviewed.

Year	Dataset	Proposed Methodology	Metric	Metric Value	Reference
2019	Lab experiment	Extreme gradient boosting	Accuracy	86%	[54]
2021	Lab experiment	Multiple linear regression	Accuracy	88%	[55]
2022	STEW	Random forest model	Accuracy	83.6%	[56]
2022	Technical University Berlin	Support vector machine	Accuracy	83%	[57]
2023	Lab experiment	K-nearest neighbor	Accuracy	87.57%	[58]
2021	STEW	One-dimensional convolution neural network	Accuracy	95.36% and 96.77%	[59]
2021	Lab experiment	Deep neural network	Accuracy	98%	[60]
2023	STEW	Bi-directional gated network	Accuracy	98%	[61]

While existing studies have made strides in cognitive workload detection using EEG and traditional methods, they often face challenges with lower classification accuracies and a lack of real-time monitoring capabilities. These limitations hinder their suitability for advanced industrial settings. Furthermore, there is a notable gap in the literature regarding the development of a unique real-time cognitive workload monitoring system tailored specifically for such environments. This study aims to address these shortcomings by introducing a novel approach that achieves a higher accuracy and enables real-time monitoring using advanced technologies like the Things Board platform and IIoT.

3. Materials and Methods

This section introduce a cognitive workload monitoring system that is designed specifically for industrial applications utilizing EEG signals. The focal point of our approach is the proposed deep gated neural network (DGNN), an innovative recurrent neural network (RNN) architecture. The simultaneous task EEG workload (STEW) dataset, sourced from the IEEE database [62], is selected as the source of raw EEG data. This dataset underwent preprocessing to ensure data integrity and optimal conditions for achieving a high classification accuracy. Furthermore, we performed correlation and frequency domain analyses. Our system's robustness is augmented by the integration of industrial Internet of Things (IIoT) gateways, the constrained application protocol (CoAP), and the Things Board platform. Figure 1 illustrates the proposed cognitive workload monitoring framework, showcasing the system's workflow, interconnected components, and data flow that underpin the system's functionality.

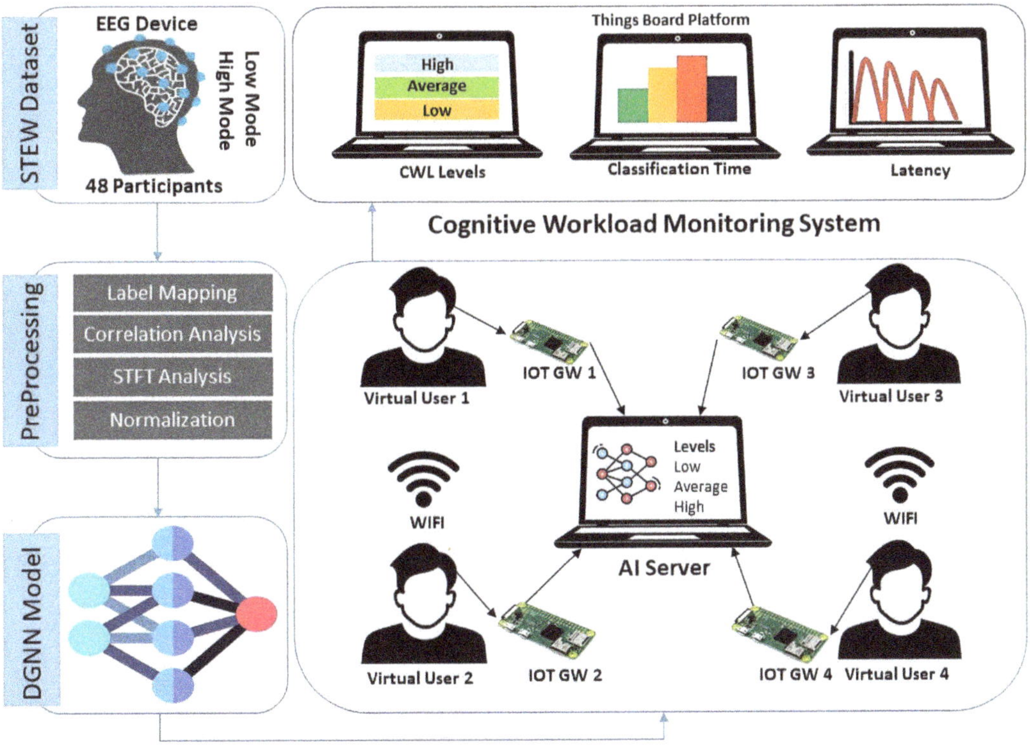

Figure 1. The proposed cognitive workload monitoring framework.

3.1. The STEW Dataset

Wei Lun Lim et al. [62] curated this freely accessible dataset, which comprises raw EEG data collected from 48 subjects engaged in a multitasking workload experiment utilizing the SIMKAP multitasking test at Nanyang Technological University. The brain activity of the subjects during rest was recorded prior to the commencement of the test and is also included in the dataset. The STEW dataset offers insights into brain activity during both rest and task execution states. Each subject contributed 19,200 samples for each state, captured using the Emotiv EPOC device, featuring 14 channels: AF3, F7, F3, FC5, T7, P7, O1, O2, P8, T8, FC6, F4, F8, and AF4. The device operates at a sampling frequency of 128 Hz.

During the experiment, each task-specific session lasted for 3 min per subject. The participants engaged in activities such as comparing the numbers in two windows, crossing out matching numbers, and simultaneously answering accompanying questions. Following each stage, the participants rated their perceived cognitive workload on a scale of 1 to 9. Figure 2 provides an overview of the STEW dataset, detailing the experimental setup, data collection process, and the structure of the dataset.

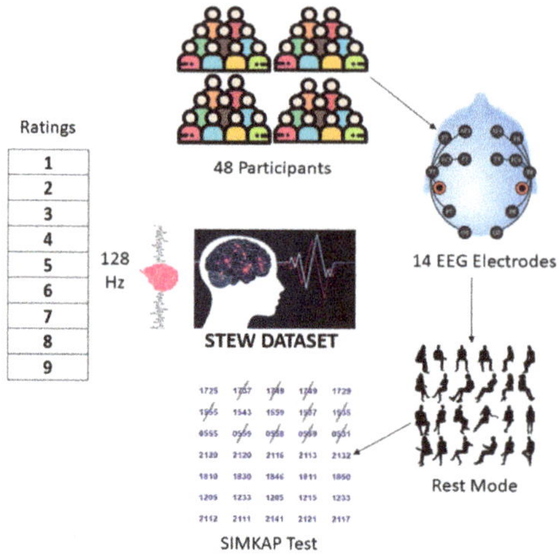

Figure 2. Overview of STEW dataset.

3.2. Data Preprocessing

EEG signals are inherently noisy due to various artifacts introduced during recording, such as eye movements, muscle activity, and head movements. Removing this noise, therefore, becomes a crucial step in the analysis. We have cleaned the raw data and applied a band-pass filter to retain frequencies within the 0.5 to 80 Hz range. In this analysis, the rest mode is considered as the low mode, and task execution is considered as the high mode. The initial data collection time was 3 min, but the first 15 s and the last 15 s were removed by the STEW data team to mitigate inter-task effects. Ratings for subjects 5, 24, and 42 were unavailable and were provided using an average method. Following segmentation, the EEG data were annotated with subjective workload ratings assigned by the participants, categorized into three levels: low (ratings 1–3), average (ratings 4–6), and high (ratings 7–9). For classification purposes, the EEG data from low- and high-workload subjects were merged into two files, each containing 921,600 rows and 15 columns. The first 14 columns represent Emotiv 14 channels EEG data, while the last column represents workload ratings considered as labels. Figure 3a,b showcases the workload ratings assigned by the participants in both low and high modes.

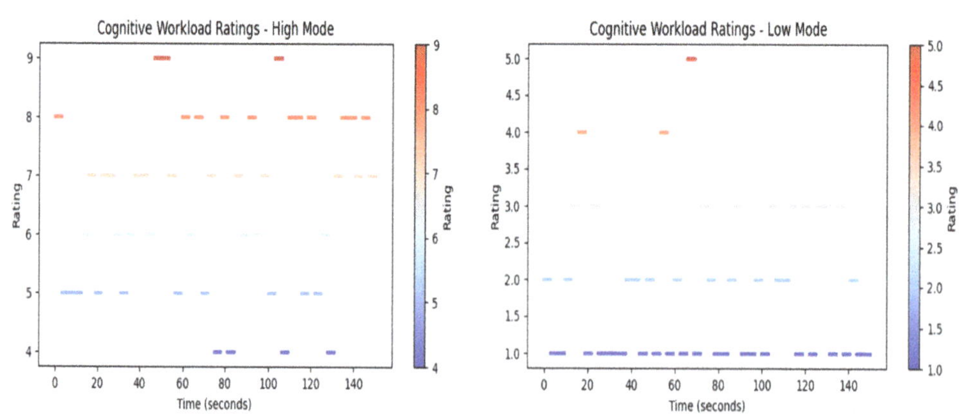

(a) (b)

Figure 3. (**a**,**b**) Participants' workload ratings in low and high modes.

3.3. EEG Channels Correlation Analysis

In this analysis, the correlation between EEG channels was computed separately for the low and high workload modes. This approach provides understanding into the interrelationship and synchronization of neural activity across different brain regions during different workload conditions [63]. By examining these correlation patterns, we can gain better insight into how cognitive workload affects brain dynamics and connectivity. Figure 4 shows the heat map of the correlation analysis of the channels transmitting low mode data. Strong positive correlations are evident between channels 6 and 7, channels 8 and 9, and channels 11 and 12, indicating robust synchronization during periods of high cognitive demand. Moderately positive correlations are observed among channels 2, 4, 8, 9, 10, 11, 12, and 13, suggesting substantial connectivity and interactions between these channels under similar conditions. Weaker positive correlations are found between channels 1, 2, 3, 7, 10, and 13, highlighting lesser but notable degrees of connectivity among these channels. Conversely, negative correlations are observed between Channels 3, 5, 6, 8, 11, and 12, indicating regions where activity patterns are divergent or inversely related during a low workload.

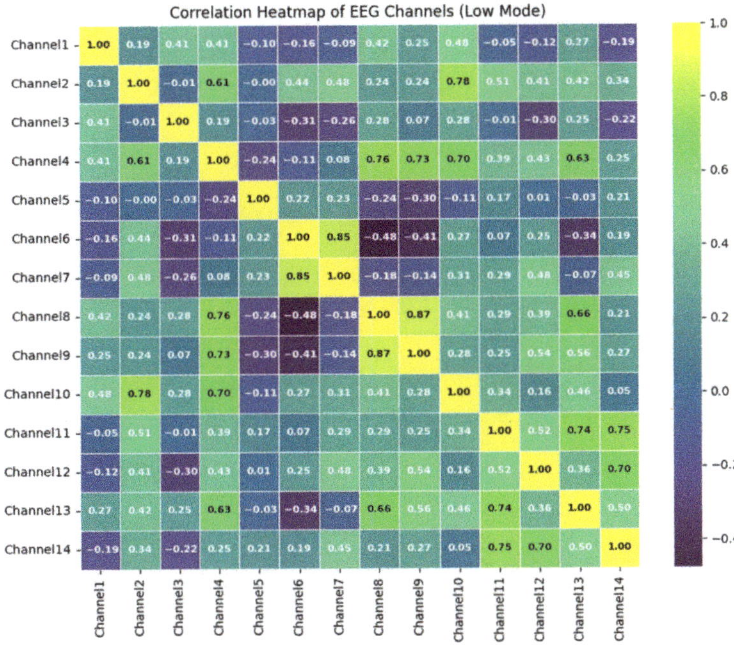

Figure 4. EEG channels correlation heat map (low mode).

Figure 5 represents the heat map of the correlation analysis of the channels transmitting high-mode data. Channels 7 and 6 demonstrate a robust synchronization, highlighting a strong interaction between these channels. Additionally, channels 8 and 9 exhibit a very strong correlation, indicating highly similar activity patterns. Moderate to high positive correlations are observed among channels 2, 4, 8, 9, 10, 11, 12, and 13, suggesting significant interconnected activity. Channels 11 and 13 show notable synchronization patterns. Conversely, some channels exhibit weak to moderate positive correlations, while others display negative correlations, indicating varying degrees of interaction and divergence in activity. These findings provide a comprehensive view of how different EEG channels interact under high-workload conditions, shedding light on both synchronized regions and areas with contrasting activity patterns.

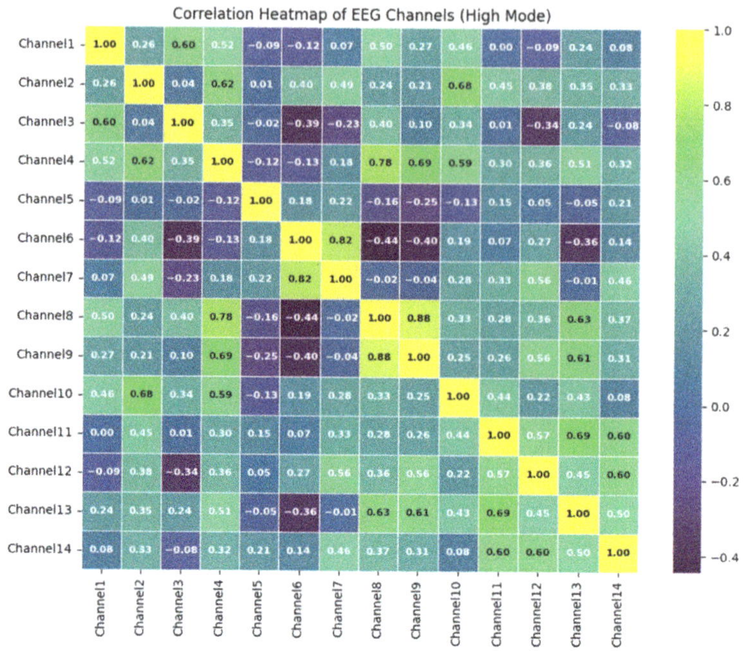

Figure 5. EEG channels correlation heat map (high mode).

3.4. Short-Time Fourier Transformation

We conducted a short-time Fourier transform (STFT) analysis to investigate the time-varying frequency characteristics and power distribution of EEG data. A notable advantage of utilizing STFT is its suitability for analyzing non-stationary EEG signals [64]. Unlike conventional Fourier analysis methods, which assume signal stationarity throughout, STFT breaks down the signal into smaller, overlapping windows. This approach is particularly beneficial for EEG data, as it accommodates the dynamic nature of brain activity, which can vary significantly over time due to shifts in cognitive states.

The equation for the short-time Fourier transform (STFT) is as follows:

$$x(\tau, f) = \int_{-\infty}^{\infty} x(t) w(t-\tau) e^{-i2\pi f t} dt \tag{1}$$

where $x(\tau, f)$ is the STFT of the signal $x(t)$ at time τ and frequency f.

$x(t)$ is the input signal.

$w(t)$ is the window function.

τ represents time.

f represents time.

Initially, the raw EEG data of subject 1 for both low and high workload modes were acquired, and all 14 channels were selected for subsequent analysis. With a sampling frequency of 128 Hz, the data were processed to capture frequency bands with the frequency ranges mentioned in Table 2.

The resulting visualizations depicted the power spectrum distribution across different frequency bands for both low mode and high mode conditions. Specifically, Figure 6a,b illustrate the frequency band visualization of subject 1 in the both low mode and high mode.

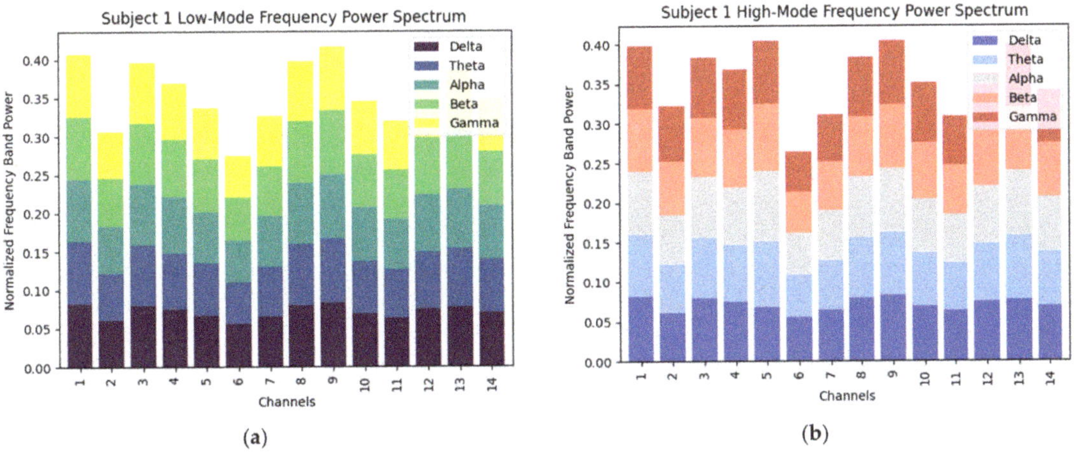

Figure 6. (**a**,**b**) Subject 1 frequency power spectrum (low mode vs. high mode).

Table 2. Captured frequency bands.

Band	Frequency Range
Delta	0.5–4 Hz
Theta	4–8 Hz
Alpha	8–13 Hz
Beta	13–30 Hz
Gamma	30–80 Hz

3.5. Data Preparation

The inherent structure of most natural data often proves inadequate for optimal model training due to its complexity and variability. To address this, our study represents EEG signals in the form of a matrix, which is considered a best practice for handling such multidimensional data. EEG data underwent preprocessing to remove noise and artifacts, ensuring that the signals were clean and reliable for further analysis. Following this, Z-score normalization was performed to standardize the data, bringing all the values into a common scale with a mean of zero and a standard deviation of one.

For the experiments, an 80/20 train–test split was employed, where 80% of the data were allocated for training the model and 20% were allocated for testing its performance. Additionally, 12.5% of the training data were set aside for validation purposes, allowing for an independent evaluation during the training phase. This resulted in a data split of 70% for training, 20% for testing, and 10% for validation. Considering all 48 participants in the analysis, the dimensions of the input data fed into the model were (921600, 15) for the low workload mode and (921600, 15) for the high workload mode. This extensive dataset ensures that the model has sufficient examples to learn from, ultimately enhancing its ability to generalize and perform well on unseen data.

3.6. Deep Gated Neural Network Model

The deep gated neural network (DGNN) is meticulously designed to achieve precise cognitive workload classification by effectively processing EEG data through a layered architecture. It begins with an input layer, followed by two bidirectional long short-term memory (Bi-LSTM) layers, each equipped with 64 and 32 units, respectively, ensuring comprehensive temporal understanding of the EEG data. Subsequently, the model integrates two gated recurrent unit (GRU) layers, each with 64 and 32 units, enhancing the network's capacity to capture both short and long-range dependencies within the data. The output layer employs a Softmax activation function to categorize predictions into three workload levels: low, average, and high. Figure 7 represents the architecture of the DGNN, detailing the sequence of layers and the number of neurons in each layer.

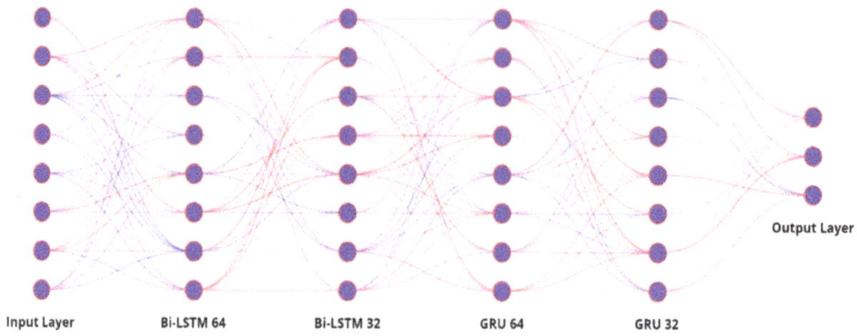

Figure 7. Deep gated neural network architecture.

The detailed formulation of the deep gated neural network (DGNN) is pivotal for accurately capturing and processing the complex temporal dependencies within EEG data, thereby enabling precise classification of cognitive workload levels.

1. First bi-directional LSTM layer:

$$\overrightarrow{h_t} = LSTM(x_t, \overrightarrow{h_{t-1}}) \quad (2)$$

Here, x_t represents the input EEG data at time step t, and $\overrightarrow{h_{t-1}}$ is the hidden state. The forward pass captures the temporal dependencies in the forward direction.

$$\overleftarrow{h_t} = LSTM(x_t, \overleftarrow{h_{t+1}}) \quad (3)$$

In the backward pass, $\overleftarrow{h_{t+1}}$ is the hidden state from the next time step. This captures the temporal dependencies in the reverse direction.

$$h_t = \left[\overrightarrow{h_t}, \overleftarrow{h_t}\right] \quad (4)$$

The hidden states from both the forward and backward passes are concatenated to form the final hidden state, h_t for the first Bi-LSTM layer. This provides a comprehensive representation of EEG data by considering information from both directions.

2. Second bi-directional LSTM layer

$$\overrightarrow{h_t}^{(2)} = LSTM(\overrightarrow{h_t}, \overrightarrow{h_{t-1}}^{(2)}) \quad (5)$$

Here, $\overrightarrow{h_t}$ is the output from the first Bi-LSTM layer at time step t, and $\overrightarrow{h_{t-1}}^{(2)}$ is the hidden state from the previous time step in the second Bi-LSTM layer's forward pass.

$$\overleftarrow{h_t}^{(2)} = LSTM(\overleftarrow{h_t}, \overleftarrow{h_{t+1}}^{(2)}) \quad (6)$$

Similarly, $\overleftarrow{h_t}$ is the output from the first Bi-LSTM layer, and $\overleftarrow{h_{t+1}}^{(2)}$ is the hidden state from the next time step in the second Bi-LSTM layer's backward pass.

$$h_t^{(2)} = \left[\overrightarrow{h_t}^{(2)}, \overleftarrow{h_t}^{(2)}\right] \tag{7}$$

The final hidden state $h_t^{(2)}$ for the second Bi-LSTM layer is obtained by concatenating the forward and backward hidden states. This ensures that the network has a robust understanding of the temporal patterns in the EEG data from both directions across two layers.

3. First GRU layer:

$$h_t^{(GRU1)} = GRU(h_t^{(2)}, h_{t-1}^{(GRU1)}) \tag{8}$$

Here, $h_t^{(2)}$ is the output from the second Bi-LSTM layer at time step t, and $h_{t-1}^{(GRU1)}$ is the hidden state from the previous time step in first GRU layer. The GRU layer captures dependencies in the data while being computationally efficient and addressing the vanishing gradient problem.

4. Second GRU layer:

$$h_t^{(GRU2)} = GRU(h_t^{(GRU1)}, h_{t-1}^{(GRU2)}) \tag{9}$$

Here, $h_t^{(GRU1)}$ is the output from the first GRU layer at time step t, and $h_{t-1}^{(GRU2)}$ is the hidden state from the previous time step in the second GRU layer. This layer further refines the temporal representations, capturing more complex patterns.

5. Output layer:

$$y = Softmax\ (W.\ h_t^{(GRU2)} + b) \tag{10}$$

In the output layer, the final hidden state $h_t^{(GRU2)}$ from the second GRU layer is passed through a dense layer with a Softmax activation function. Here:

W : weight matrix;

b : bias vector;

y : output probabilities for each class (low, average, high workload).

The Softmax activation function ensures that the model's output represents a probability distribution, enabling clear classification of cognitive workload levels.

The training process of the model is crafted to ensure not just efficiency but also robustness in its learning. The model is trained using the categorical cross-entropy loss function and optimized using the Adam optimizer. During the training process, the model undergoes 50 epochs, with a batch size of 32, effectively learning and reaching convergence. The Adam optimizer utilizes a default learning rate of 0.001, ensuring stable and efficient training dynamics. Additionally, dropout layers are incorporated to prevent overfitting, and 5-fold cross-validation is utilized to ensure robust model evaluation. Upon completion of training, the performance evaluation of the DGNN model is conducted using a set of metrics. The confusion matrix offers a clear overview by comparing the actual workload levels (low, average, high) with the model's predictions. In addition, we use fundamental evaluation metrics like accuracy [65], precision [66], recall [67], and F1 score [68] to thoroughly evaluate the model's ability to accurately classify workload levels.

3.7. Cognitive Workload Monitoring System

The DGNN model is deployed on an Ubuntu server (version 20.04) as part of the cognitive workload monitoring experiment (CWME). The experiment involves four virtual users, represented by laptops, whose EEG data is transmitted wirelessly to Raspberry Pi Zero W modules. These modules, selected for their affordability, compactness, and wireless capabilities, establish connections via Wi-Fi to receive real-time EEG data. Each virtual user sends their EEG data to their designated Raspberry Pi module, which functions as a gateway. Operating on a legacy 32-bit OS installed on a 32 GB memory card, these modules transmit live data to the Ubuntu server. While these modules can serve multiple virtual users simultaneously, in this scenario, each module is dedicated to serving a single virtual user. The DGNN model undertakes critical classification tasks, leveraging its advanced neural network architecture for efficient analysis.

The Ubuntu server takes responsibility for processing cognitive workload data and transmit it to Things Board for live monitoring. As a monitoring platform, Things Board presents various parameters such as latency, DGNN classification time, and workload type. To efficiently manage historical data, integration with a PostgreSQL database is employed due to its support for unstructured information. Figure 8 illustrates this four-layer process within a unified framework, showcasing the seamless flow of data from virtual users to the monitoring interface.

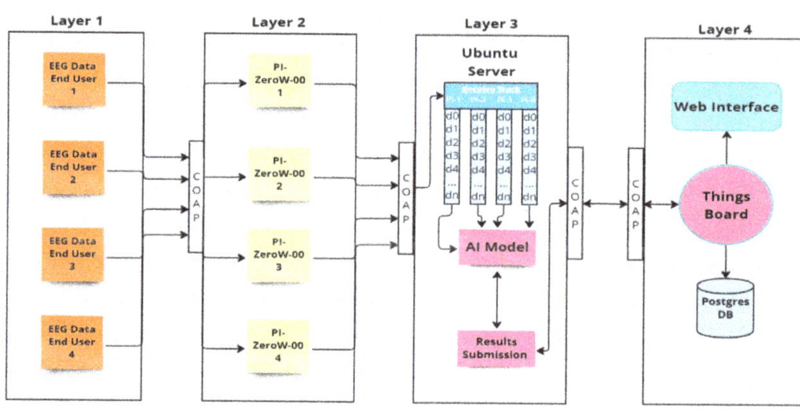

Figure 8. Multi-layer cognitive workload monitoring framework.

Communication across the layers is facilitated by the constrained application protocol (CoAP). It follows a pattern where data move from virtual users (Layer 1) to Raspberry Pi modules (Layer 2) in a one-to-one manner. Then, the data move from the Raspberry Pi modules to the Ubuntu server with the deployed DGNN model (Layer 3), which operates in a many-to-one fashion. Finally, the data are sent from the Ubuntu server to the Things Board module (Layer 4) in a one-to-one manner.

4. Results and Analysis

This section presents the results obtained from the proposed deep gated neural network (DGNN) and compares them with existing studies. Additionally, it includes the results from real-time monitoring in the cognitive workload monitoring system (CWMS). The Jupyter Lab tool was utilized for analysis due to its powerful interactive computing capabilities, and the proposed DGNN model was designed using the Python programming language.

4.1. The Results Obtained from the DGNN Model

The DGNN model integrates bidirectional long short-term memory (Bi-LSTM) and gated recurrent unit (GRU) architectures to analyze EEG data efficiently. The Bi-LSTM and GRU components within the DGNN framework enable comprehensive analysis of temporal dependencies and intricate patterns within EEG signals. Bi-LSTM excels in capturing long-term dependencies, while GRU enhances the model's ability to learn from sequential data efficiently. Our findings showcase a remarkable achievement in cognitive workload classification. The DGNN model demonstrates an impressive accuracy rate of 99.45%, as depicted in Figure 9a. The accuracy curve, spanning 50 training epochs, reveals a consistent improvement. In contrast, Figure 9b illustrates the curve of loss values, depicting a continual decrease as the model iteratively learns from the training data. These figures provide a clear visualization of the model's performance dynamics, underscoring its robustness and efficacy in cognitive workload classification. This suggests promising applications in practical industrial settings. It can enhance safety protocols, optimize task allocation, and improve overall productivity and worker well-being.

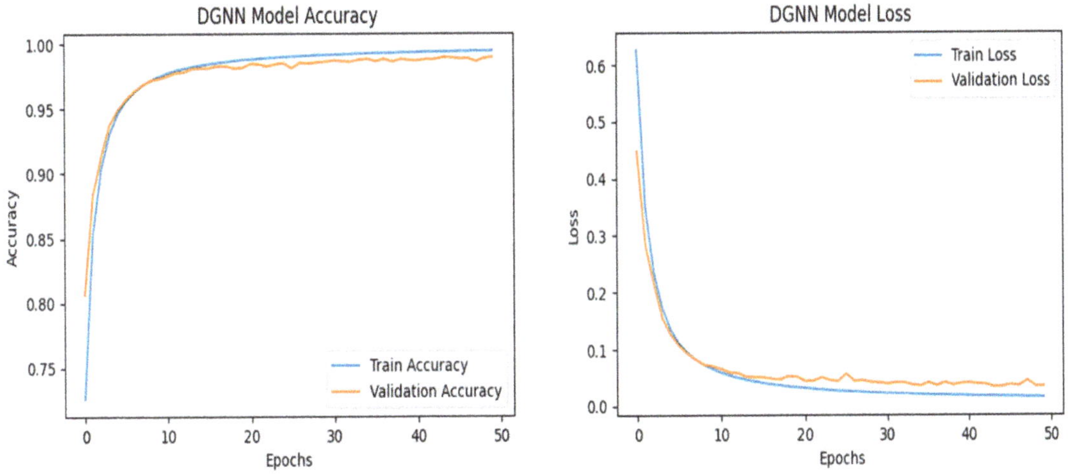

Figure 9. (**a**,**b**) Deep gated neural network model accuracy and loss.

The confusion matrix serves as a crucial tool for evaluating the accuracy and reliability of cognitive workload level predictions. In Figure 10, the confusion matrix of our DGNN model showcases significant achievements in workload classification. Specifically, the model accurately classified 93,807 instances as '0' for 'low workload levels'. Furthermore, it correctly identified 98,229 instances as '1' for 'medium workload levels', demonstrating its effectiveness in recognizing moderate cognitive workload levels. Moreover, the model achieved precise classifications in 172,444 instances as '2' for 'high workload levels', indicating its proficiency in accurately identifying instances of high cognitive workloads.

These findings underscore the importance of the confusion matrix in revealing how effectively the DGNN model handles different workload levels. By quantifying its classification outcomes, we gain valuable insights into the strengths and areas for improvement of the model, ultimately contributing to the enhancement of cognitive workload assessment techniques.

The classification metrics, as presented in Table 3, provide a detailed evaluation of the model's ability to accurately classify cognitive workload levels based on the STEW EEG dataset. These metrics includes precision, recall, and F1 score values for each workload category, including low, average, and high. With precision scores of 1.00 for low workloads and 0.98 for the average and high workload categories, the model demonstrates a high accuracy in identifying instances belonging to specific workload levels. Moreover, recall scores of 1.00 for low workloads and 0.98 for the average and high workload categories indicate the model's ability to capture a high proportion of the actual instances within each workload level. The F1 score values further confirm the model's balanced performance across all workload categories, with values of 1.00 for low workloads and 0.98 for the average and high workload categories.

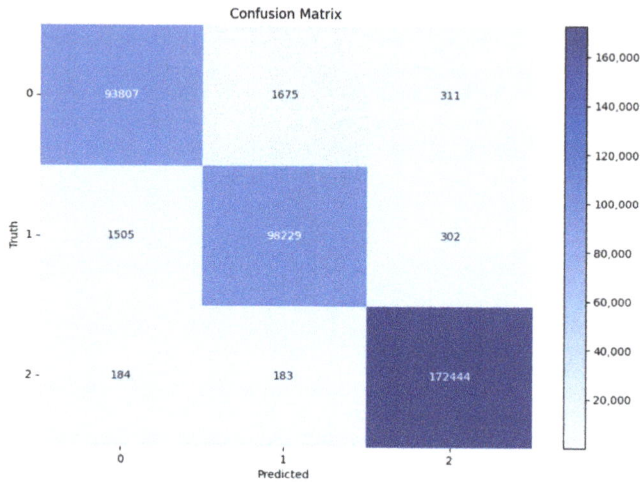

Figure 10. Confusion matrix of DGNN model.

Table 3. Classification metrics.

Labels	Precision	Recall	F1 Score
Low	1.00	1.00	1.00
Average	0.98	0.98	0.98
High	0.98	0.98	0.98

In our study, we conducted a comparative analysis of our proposed DGNN model with existing studies in cognitive workload classification, encompassing various methodologies including machine learning and deep learning approaches. This analysis involved evaluating the methodologies employed in these studies and comparing their achieved accuracy rates.

The existing methodologies in these studies included techniques such as XGBoost (XGB), multiple linear regression (MLR), random forests (RFs), support vector machines (SVMs), k-nearest neighbors (KNN), 1D convolutional neural networks (1D CNNs), deep neural networks (DNNs), and bidirectional gated networks (BDGNs). Previous research has demonstrated lower accuracy rates and methodologies that may not fully address the evolving needs of cognitive workload classification. However, the DGNN model stands out prominently in this analysis, showcasing the highest accuracy among the compared studies. Specifically, our model achieves an impressive accuracy rate of 99.45% in accurately classifying cognitive workload levels using EEG data. Figure 11 represents the plotted comparison of state-of-the-art results from Table 1, illustrating the superior performance of our DGNN model relative to these existing methodologies.

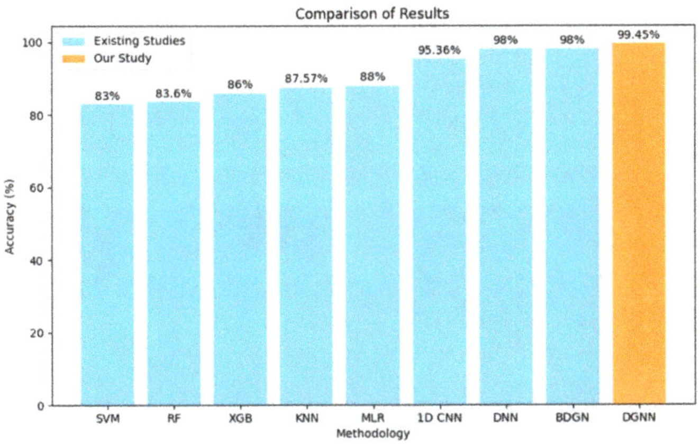

Figure 11. Plotted comparison of state-of-the-art results from Table 1.

4.2. Cognitive Workload Monitoring System

The real-time monitoring of the CWMS is evaluated using the Things Board platform, allowing us to track various parameters, including end-to-end latency, model classification processing time, and cognitive workload levels. The Raspberry Pi Zero W gateways efficiently manage data transmission between the EEG device and the AI server without significant delays, ensuring smooth operation. These modules are essential in this setup due to their cost-effectiveness, making them ideal for large-scale experiments involving multiple users and allowing for scalable deployments. Their small size ensures easy integration into experimental setups, particularly with wearable EEG devices, keeping the setup portable and unobtrusive. Additionally, the built-in Wi-Fi functionality enables seamless wireless communication between the virtual users and the Ubuntu server. The Things Board platform provides a user-friendly interface for visualizing and analyzing the results, with real-time tracking capabilities crucial for monitoring the system's performance.

Our experiment involved five laptops, with four representing virtual users and one for the live monitoring and classification results. Additionally, four Raspberry Pi Zero W gateways are used to facilitate data transmission between the EEG device and the AI server. To accommodate the connectivity and power needs of Raspberry Pi modules, a single Wi-Fi router and power cables are used. Figure 12 illustrates the experimental setup, depicting the interconnected components.

A set of monitoring data is presented, where each figure comprises three distinct graphs and a table representing the data from four gateways. In our scenario, each gateway serves only one virtual user; thus, the four gateways correspond to four virtual users. The first graph in each figure displays end-to-end latency, providing insights into the transmission delays. The second graph showcases AI model classification processing time, indicating the duration taken for cognitive workload assessment. Lastly, the third graph delineates workload levels, where 0 signifies a low workload level, 1 denotes an average workload level, and 2 indicates a high workload level. The corresponding table in each graph offers a direct representation of workload classification.

Figure 13 displays the information from gateway-001 on Things Board. We observe the latest latency recorded as 237 ms, indicating the time taken for data transmission. The classification processing time is noted as 4621.21 ms, reflecting the duration for cognitive workload assessment. Additionally, the workload levels are showcased separately, providing a comprehensive overview of the system's performance. This detailed visualization aids in identifying potential issues and optimizing the network's efficiency.

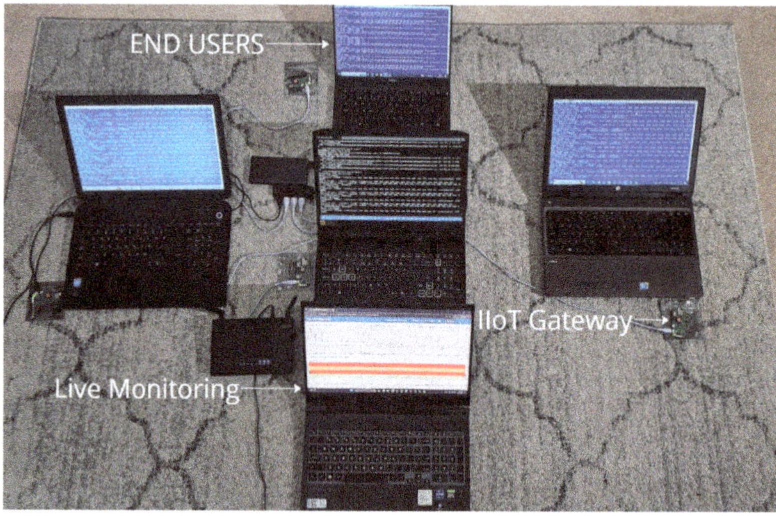

Figure 12. Cognitive workload classification monitoring experiment.

Figure 13. Gateway-001 cognitive workload monitoring data graphs.

Figure 14 presents the monitoring data from gateway-002 on Things Board, encompassing three critical graphs and a comprehensive table. The first graph indicates an end-to-end latency of 459 ms, a metric that measures the time taken for the data to travel from the virtual user to the server and back, reflecting the network's efficiency. Higher latency can adversely affect user experience, particularly in real-time applications. The second graph showcases the classification processing time, recorded at 6717.97 ms, which is significantly higher than that of gateway-001, suggesting potential variability in computational load or hardware performance across different gateways. The third graph delineates the workload levels for gateway-002, distinguishing between low (0), average (1), and high (2) workload levels.

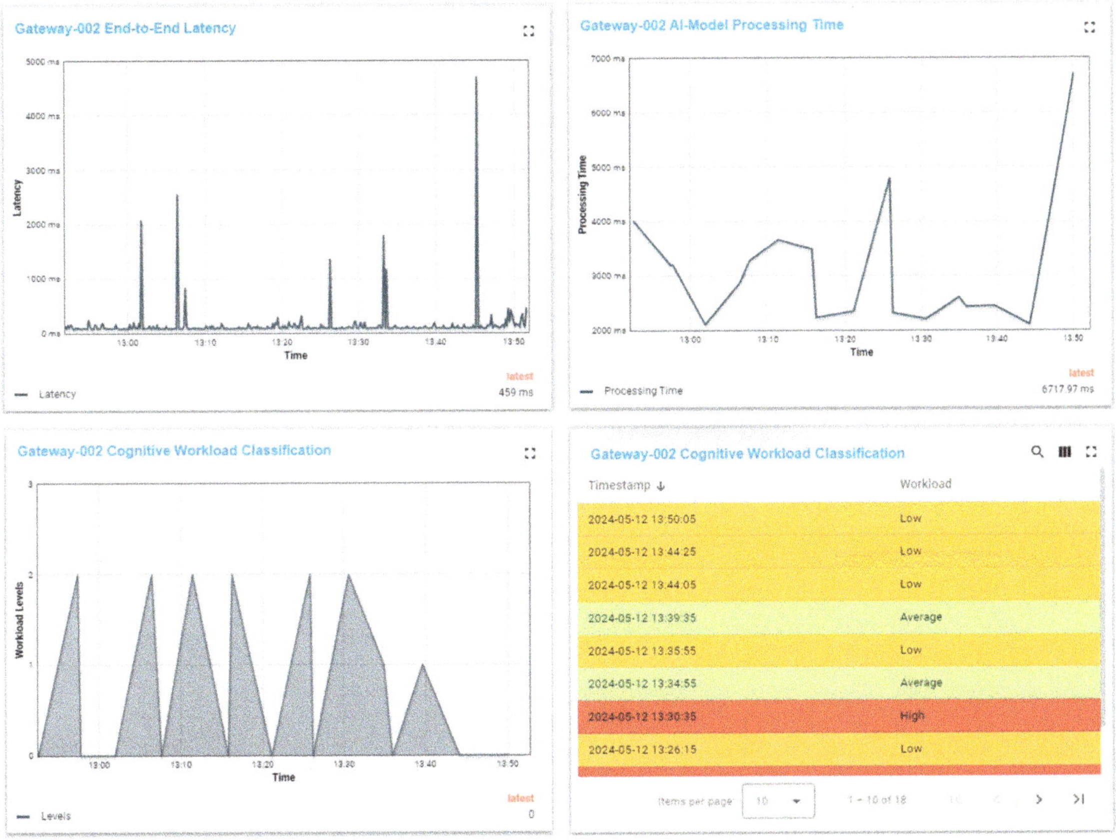

Figure 14. Gateway-002 cognitive workload monitoring data graphs.

Figure 15 displays the data from gateway-003 on Things Board, following the same structure as the previous figures but with distinct performance characteristics. The latest recorded latency for gateway-003 is 196 ms, which is notably lower than that of gateway-002, indicating a more efficient data transmission process and enhancing the responsiveness of applications relying on this gateway. The classification processing time for gateway-003 is measured at 3849.74 ms, which is substantially shorter compared to gateway-002. This suggests that gateway-003 may have more efficient processing capabilities or less computational load at the time of measurement. The workload levels graph for gateway-003 shows the distribution of cognitive workload levels, providing a visual representation of the virtual user's cognitive load.

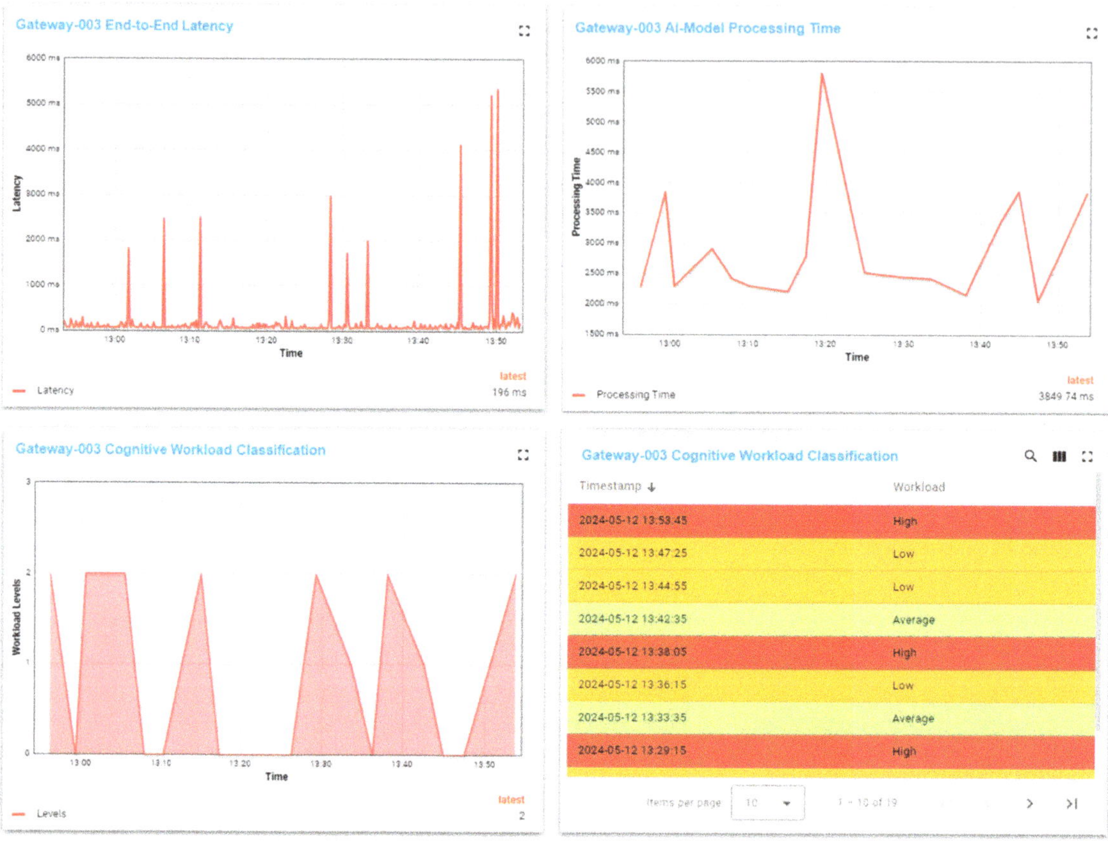

Figure 15. Gateway-003 cognitive workload monitoring data graphs.

Figure 16 represents the monitoring data from gateway-004 on Thing Board, showcasing its unique performance metrics. The latest latency recorded for gateway-004 is 132 ms: the lowest among all four gateways. This minimal latency suggests that gateway-004 provides the most efficient data transmission, which is critical for applications requiring real-time data processing.

The classification processing time for gateway-004 stands at 2866.82 ms, which is also the shortest among the four gateways. This efficiency in processing indicates a robust performance in cognitive workload assessment, potentially due to better hardware or lower concurrent computational demands. The third graph in Figure 16 illustrates the workload levels for gateway-004, continuing the pattern of distinguishing between low, average, and high workload levels. The table in this figure offers a clear and concise representation of the workload classifications, aiding in performance evaluation and decision-making.

We have proposed a system that enables continuous data transmission, real-time classification, and effective monitoring of results by integrating the DGNN model with Raspberry Pi, CoAP, and Things Board. While we have utilized the STEW dataset in this study, validating our model with real human participants is crucial. We plan to recruit healthy participants from the university, adhering to ethical guidelines and obtaining informed consent. EEG data will be collected during cognitive tasks, and the system's performance will be evaluated based on classification accuracy, sensitivity, specificity, and response time.

Figure 16. Gateway-004 cognitive workload monitoring data graphs.

5. Conclusions

This study introduces a novel approach for cognitive workload classification using EEG signals and advanced deep learning techniques. Leveraging the STEW dataset, comprehensive preprocessing and correlation analyses were conducted, followed by detailed short-time Fourier transformation analysis to extract relevant features. Our proposed deep gated neural network (DGNN) model, integrating Bi-LSTM and GRU layers, achieved an impressive accuracy of 99.45%. Precision, recall, and F1 score metrics further validated the efficacy of our model in effectively classifying cognitive workload across distinct categories. A comparative analysis with existing methodologies underscores the superior performance and robustness of the DGNN model in this domain.

To validate the proposed approach, this study conducted the cognitive workload monitoring experiment (CWME), deploying the DGNN model on a classification server. The deployment leveraged Raspberry Pi Zero W modules as gateways, employing the constrained application protocol (CoAP) for efficient data transfer to the Things Board monitoring platform. Things Board facilitated real-time visualization and analysis of critical parameters such as end-to-end latency, AI model classification processing time, and workload levels, providing actionable understandings into system performance. This setup enhances operational efficiency by enabling prompt cognitive workload assessments, supporting dynamic decision-making processes. The proposed research provides practical solutions for managing cognitive workload in industrial settings amidst ongoing digital transformation and automation [69].

The successful implementation and validation of the DGNN model in real-time cognitive workload monitoring hold significant implications for various fields, particularly in industrial settings. By accurately assessing cognitive workload levels using EEG signals, organizations can proactively manage workforce health and safety. This approach enables timely interventions to mitigate cognitive fatigue, thereby reducing the risk of errors and accidents in high-risk environments. Moreover, the ability to monitor workload levels in real-time fosters adaptive work environments, potentially enhancing overall productivity and employee well-being.

Despite the promising results achieved in this study, transitioning from virtual users to real human participants poses several significant challenges. Variability in EEG signal quality and ensuring participant comfort with wireless headsets are critical considerations that can impact data accuracy and participant engagement. Adhering to study protocols and ethical guidelines for EEG data collection is essential to maintain data integrity and participant trust. Additionally, the wide cost range of wireless EEG devices, spanning from 150 to 30,000 USD, and the variability in battery life (3 h to 10 h) present financial constraints and usability issues that must be carefully managed [70,71].

6. Future Works

While this study has laid a solid foundation for real-time cognitive workload monitoring using EEG signals and advanced deep learning techniques, several avenues for further research and development are essential to enhance the applicability and effectiveness of our approach in practical settings. A crucial next step involves validating our proposed deep gated neural network (DGNN) model with real human participants. While virtual users provide valuable insights, incorporating real-world factors such as individual EEG variations, stress levels, and environmental influences will enhance the model's robustness and applicability across diverse work environments. Longitudinal studies involving diverse participant groups will be essential to evaluate the model's performance under varying cognitive workload conditions over extended periods.

Improving the usability of wireless EEG headsets represents another significant avenue for future research. Current limitations, such as discomfort during prolonged wear and restricted mobility, need to be addressed through innovative ergonomic designs and lightweight materials. Enhancing headset ergonomics to minimize discomfort and fatigue, alongside optimizing sensor placement for better signal quality, will be crucial for facilitating long-term, real-world deployment. Future research should focus on scaling the DGNN model to accommodate multiple users concurrently and evaluating its performance under diverse operational conditions and noise levels. Assessing the system's robustness and reliability across different industrial contexts will be essential for broader adoption and practical implementation.

Additionally, integrating real-time alarm systems into our monitoring platform using tools like Things Board offers an opportunity to enhance responsiveness to cognitive workload fluctuations. Implementing proactive alert mechanisms based on workload classifications will empower timely interventions and adaptive decision-making, thereby improving operational efficiency and safety in dynamic settings. Addressing the challenges ahead will be crucial for fully realizing the potential of our approach in practical, real-world applications.

Author Contributions: Conceptualization, M.A.A. and S.U.B.; methodology, M.A.A. and S.U.B.; software, M.A.A.; data curation, M.A.A. and S.U.B.; writing—original draft preparation, M.A.A. and B.A.; writing—review and editing, M.A.A., Z.G. and B.A.; visualization, M.A.A. and S.U.B.; supervision, Z.G. All authors have read and agreed to the published version of the manuscript.

Funding: This research received no external funding.

Institutional Review Board Statement: Not applicable.

Informed Consent Statement: Not applicable.

Data Availability Statement: The data for this study are available at: https://ieee-dataport.org/open-access/stew-simultaneous-task-eeg-workload-dataset (accessed on 20 April 2024).

Conflicts of Interest: The authors declare no conflicts of interest.

Code Availability Statement: The implementation code for this article is available at: https://github.com/abrarafzal4567/Cognitive-Workload-Monitoring-System.git (accessed on 19th May 2024).

References

1. Goetz, C.; Bavaresco, R.; Kunst, R.; Barbosa, J. Industrial Intelligence in the Care of Workers' Mental Health: A Review of Status and Challenges. *Int. J. Ind. Ergon.* **2022**, *87*, 103234. [CrossRef]
2. Jiang, X. Incorporating Service Design for Industry 4.0: A Scientometric Review for Green and Digital Transformation Driven by Service Design. In Proceedings of the 2020 Management Science Informatization and Economic Innovation Development Conference (MSIEID), Guangzhou, China, 18–20 December 2020; pp. 296–299. [CrossRef]
3. Atmoko, R.A.; Yang, D. Online Monitoring & Controlling Industrial Arm Robot Using MQTT Protocol. In Proceedings of the 2018 IEEE International Conference on Robotics, Bio mimetics, and Intelligent Computational Systems (Robionetics), Bandung, Indonesia, 8–10 August 2018; pp. 12–16. [CrossRef]
4. Bhatt, V.; Bindal, A.K. Smart Hardware Development under Industrial IOT (IIOT) 4.0: A Survey Report. In Proceedings of the 2021 6th International Conference on Signal Processing, Computing and Control (ISPCC), Solan, India, 7–9 October 2021; pp. 262–265. [CrossRef]
5. Kumar, S.; Chandra, S.; Shukla, R.; Panigrahi, L. Industry 4.0 Based Machine Learning Models for Anomalous Product Detection and Classification. In Proceedings of the 2022 OPJU International Technology Conference on Emerging Technologies for Sustainable Development (OTCON), Raigarh, Chhattisgarh, India, 8–10 February 2023; pp. 1–6. [CrossRef]
6. Hussain, R.F.; Pakravan, A. Analyzing the Performance of Smart Industry 4.0 Applications on Cloud Computing Systems. In Proceedings of the 2020 IEEE 22nd International Conference on High Performance Computing and Communications; IEEE 18th International Conference on Smart City; IEEE 6th International Conference on Data Science and Systems (HPCC/SmartCity/DSS), Yanuca Island, Cuvu, Fiji, 14–16 December 2020; pp. 11–18. [CrossRef]
7. Xu, L.; Xu, E.; Li, L. Industry 4.0: State of the Art and Future Trends. *Int. J. Prod. Res.* **2018**, *56*, 2941–2962. [CrossRef]
8. Arpaia, P.; Moccaldi, N.; Prevete, R.; Sannino, I.; Tedesco, A. A Wearable EEG Instrument for Real Time Frontal Asymmetry Monitoring in Worker Stress Analysis. *IEEE Trans. Instrum. Meas.* **2020**, *69*, 8335–8343. [CrossRef]
9. Afzal, B.; Li, X.; AndezLara, A. The Innovation Journey and Crossroads of Sustainability, Resilience and Human centeredness: A Systematic Literature Review Innovation Journey. *Transform. Gov. People Process Policy* 2024. [CrossRef]
10. Prassida, G.F.; Asfari, U. A Conceptual Model for the Acceptance of Collaborative Robots in Industry 5.0. *Procedia Comput. Sci.* **2022**, *197*, 61–67. [CrossRef]
11. Lin, C.; Lukodono, R. Classification of Mental Workload in Human robot Collaboration Using Machine Learning Based on Physiological Feedback. *J. Manuf. Syst.* **2022**, *65*, 673–685. [CrossRef]
12. Sodhro, A.H.; Sennersten, C.; Ahmad, A. Towards Cognitive Authentication for Smart Healthcare Applications. *Sensors* **2022**, *22*, 2101. [CrossRef]
13. Panchetti, T.; Pietrantoni, L.; Puzzo, G.; Gualtieri, L.; Fraboni, F. Assessing the Relationship between Cognitive Workload, Workstation Design, User Acceptance and Trust in Collaborative Robots. *Appl. Sci.* **2023**, *13*, 1720. [CrossRef]
14. Loizaga, E.; Eyam, A.; Bastida, L.; Luis, J. A Comprehensive Study of Human Factors, Sensory Principles, and Commercial Solutions for Future Human Centered Working Operations in Industry 5.0. *IEEE Access* **2023**, *11*, 53806–53829. [CrossRef]
15. Villani, V.; Pini, F.; Leali, F.; Secchi, C. Survey on Human–Robot Collaboration in Industrial Settings: Safety, Intuitive Interfaces and Applications. *Mechatronics* **2018**, *55*, 248–266. [CrossRef]
16. Tao, D.; Tan, H.; Wang, H.; Zhang, X.; Qu, X.; Zhang, T. A Systematic Review of Physiological Measures of Mental Workload. *Int. J. Environ. Res. Public Health* **2019**, *16*, 2716. [CrossRef] [PubMed]
17. Dehais, F.; Lafont, A.; Roy, R.; Fairclough, S. A Neuro ergonomics Approach to Mental Workload, Engagement and Human Performance. *Front. Neurosci.* **2020**, *14*, 268. [CrossRef]
18. Acker, V.; Parmentier, D.; Vlerick, P.; Saldien, J. Understanding Mental Workload: From a Clarifying Concept Analysis toward an Implementable Framework. *Cogn. Technol. Work.* **2018**, *20*, 351–365. [CrossRef]
19. Longo, L.; Wickens, C.D.; Hancock, G.; Hancock, P.A. Human Mental Workload: A Survey and a Novel Inclusive Definition. *Front. Psychol.* **2022**, *13*, 883321. [CrossRef] [PubMed]
20. Young, M.S.; Brookhuis, K.A.; Wickens, C.D.; Hancock, P.A. State of Science: Mental Workload in Ergonomics. *Ergonomics* **2015**, *58*, 1–17. [CrossRef] [PubMed]
21. Charles, R.L.; Nixon, J. Measuring Mental Workload Using Physiological Measures: A Systematic Review. *Appl. Ergon.* **2019**, *74*, 221–232. [CrossRef] [PubMed]
22. Heard, J.; Harriott, C.E.; Adams, J.A. A Survey of Workload Assessment Algorithms. *IEEE Trans. Hum. Mach. Syst.* **2018**, *48*, 434–451. [CrossRef]

23. Shriram, R.; Sundhararajan, M.; Daimiwal, N. EEG Based Cognitive Workload Assessment for Maximum Efficiency. *IOSR J. Electron. Commun. Eng. (IOSRJECE)* **2012**, *7*, 34–38.
24. Bhavsar, P.; Srinivasan, B.; Srinivasan, R. Pupilometer Based Real Time Monitoring of Operator's Cognitive Workload to Prevent Human Error during Abnormal Situations. *Ind. Eng. Chem. Res.* **2015**, *55*, 3372–3382. [CrossRef]
25. Boehm, U.; Matzke, D.; Gretton, M.; Castro, S.; Cooper, J.; Skinner, M.; Strayer, D.; Heathcote, A. Real time Prediction of Shor timescale Fluctuations in Cognitive Workload. *Cogn. Res. Princ. Implic.* **2021**, *6*, 30. [CrossRef]
26. Nigusse, A.B.; Mengistie, D.A.; Malengier, B.; Tseghai, G.B.; Van Langenhove, L. Wearable Smart Textiles for LongTerm Electrocardiography Monitoring—A Review. *Sensors* **2021**, *21*, 4174. [CrossRef] [PubMed]
27. Skaramagkas, V.; Giannakakis, G.; Ktistakis, E.; Manousos, D.; Karatzanis, I.; Tachos, N.; Tripoliti, E.E.; Marias, K.; Fotiadis, D.I.; Tsiknakis, M. Review of Eye Tracking Metrics Involved in Emotional and Cognitive Processes. *IEEE Rev. Biomed. Eng.* **2023**, *16*, 260–277. [CrossRef]
28. Soni, K.; Kukade, J.; Jagtap, D.; Parkhi, D.P.; Barpha, V.S. Real Time Emotion Detection System: A Hybrid Approach of Computer Vision and Machine Learning Techniques. In *Proceedings of Data Analytics and Management*; Swaroop, A., Polkowski,, Z., Correia, S.D., Virdee, B., Eds.; Springer Nature Singapore: Singapore, 2024; pp. 417–430.
29. Shridhar, M.; Thomason, J.; Gordon, D.; Bisk, Y.; Han, W.; Mottaghi, R.; Zettlemoyer, L.; Fox, D. ALFRED: A Benchmark for Interpreting Grounded Instructions for Everyday Tasks. In Proceedings of the 2020 IEEE/CVF Conference on Computer Vision and Pattern Recognition (CVPR), Seattle, WA, USA, 13–19 June 2020; pp. 10737–10746. [CrossRef]
30. Peng, C.; Hou, X. Applications of Functional Near infrared Spectroscopy (FNIRS) in Neonates. *Neurosci. Res.* **2021**, *170*, 18–23. [CrossRef] [PubMed]
31. Benbadis, S.; Beniczky, S.; Bertram, E.; MacIver, S.; Moshé, S. The Role of EEG in Patients with Suspected Epilepsy. *Epileptic Disord. Int. Epilepsy J. Videotape* **2020**, *22*, 143–155. [CrossRef]
32. Zhao, G.; Liu, Y.J.; Shi, Y. Real Time Assessment of the Cross Task Mental Workload Using Physiological Measures during Anomaly Detection. *IEEE Trans. Hum. Mach. Syst.* **2018**, *48*, 149–160. [CrossRef]
33. Zhang, Q.; Yuan, Z.; Chen, H.; Li, X. Identifying Mental Workload Using EEG and Deep Learning. In Proceedings of the 2019 Chinese Automation Congress (CAC), Hangzhou, China, 22–24 November 2019; pp. 1138–1142. [CrossRef]
34. Saleh, M.; Paquelet, S.; Castel, P.; Hoarau, M.; Pallamin, N.; Lewkowicz, D. An Efficient Deep Learning Based Solution for the Recognition of Relative Changes in Mental Workload Using Wearable Sensors. In Proceedings of the 2023 IEEE SENSORS, Vienna, Austria, 29 October–1 November 2023; pp. 1–4. [CrossRef]
35. Gupta, S.S.; Taori, T.; Ladekar, M.; Manthalkar, R.; Gajre, S.; Joshi, Y. Classification of Cross Task Cognitive Workload Using Deep Recurrent Network with Modelling of Temporal Dynamics. *Biomed. Signal Process. Control* **2021**, *70*, 103070. [CrossRef]
36. Saini, M.; Satija, U.; Upadhayay, M.D. One dimensional Convolutional Neural Network Architecture for Classification of Mental Tasks from Electroencephalogram. *Biomed. Signal Process. Control* **2022**, *74*, 103494. [CrossRef]
37. Oliver, S.; Purusothaman, T. Lightweight and Secure Mutual Authentication Scheme for IoT Devices Using CoAP Protocol. *Comput. Syst. Sci. Eng.* **2022**, *41*, 767–780. [CrossRef]
38. Bestari, D.; Wibowo, A. IoT Based Real Time Weather Monitoring System Using Telegram Bot and Things board Platform. *Int. J. Interact. Mob. Technol. (Ijim)* **2023**, *17*, 4–19. [CrossRef]
39. Antonenko, P.; Paas, F.; Grabner, R.; Gog, T. Using Electroencephalography to Measure Cognitive Load. *Educ. Psychol. Rev.* **2010**, *22*, 425–438. [CrossRef]
40. Zhou, Y.; Huang, S.; Xu, Z.; Wang, P.; Wu, X.; Zhang, D. Cognitive Workload Recognition Using EEG Signals and Machine Learning: A Review. *IEEE Trans. Cogn. Dev. Syst.* **2021**, *14*, 799–818. [CrossRef]
41. Smith, M.; Gevins, A.; Brown, H.; Karnik, A.; Du, R. Monitoring Task Loading with Multivariate EEG Measures during Complex Forms of Human Computer Interaction. *Hum. Factors* **2001**, *43*, 366–380. [CrossRef] [PubMed]
42. Lemm, S.; Blankertz, B.; Dickhaus, T.; Müller, K. Introduction to Machine Learning for Brain Imaging. *Multivar. Decod. Brain Read.* **2011**, *56*, 387–399. [CrossRef] [PubMed]
43. Sciaraffa, N.; Aricò, P.; Borghini, G.; Di Flumeri, G.; Florio, A.; Babiloni, F. On the Use of Machine Learning for EEG Based Workload Assessment: Algorithms Comparison in a Realistic Task. In *Human Mental Workload: Models and Applications: Proceedings of the Third International Symposium, H-WORKLOAD 2019, Rome, Italy, 14–15 November 2019*; Springer: Cham, Switzerland, 2019; pp. 170–185. [CrossRef]
44. Wei, Q.; Xiao, M.; Lu, Z. A Comparative Study of Canonical Correlation Analysis and Power Spectral Density Analysis for SSVEP Detection. In Proceedings of the 2011 Third International Conference on Intelligent Human Machine Systems and Cybernetics, Hangzhou, China, 26–27 August 2011; Volume 2, pp. 7–10. [CrossRef]
45. Woodman, G.F. A Brief Introduction to the Use of Event related Potentials in Studies of Perception and Attention. *Atten. Percept. Psychophys.* **2010**, *72*, 2031–2046. [CrossRef] [PubMed]
46. Simfukwe, C.; Han, S.; Jeong, H.T.; Youn, Y.C. QEEG as Biomarker for Alzheimer's disease: Investigating Relative PSD Difference and Coherence Analysis. *Neuropsychiatr. Dis. Treat.* **2023**, *19*, 2423–2437. [CrossRef] [PubMed]
47. Shadpour, S.; Shafqat, A.; Toy, S.; Jing, Z.; Attwood, K.; Moussavi, Z.; Shafiei, S.B. Developing Cognitive Workload and Performance Evaluation Models Using Functional Brain Network Analysis. *Npj Aging* **2023**, 9. [CrossRef] [PubMed]

48. Chu, H.; Cao, Y.; Jiang, J.; Yang, J.; Huang, M.; Li, Q.; Jiang, C.; Jiao, X. Optimized EEG–FNIRS Based Mental Workload Detection Method for Practical Applications. 2021. Available online: https://www.researchsquare.com/article/rs-683529/v1 (accessed on 10 April 2024).
49. Zhao, X.; Yin, J.; Chen, Z.; He, S. Workload Classification Model for Specializing Virtual Machine Operating System. In Proceedings of the 2013 IEEE Sixth International Conference on Cloud Computing, Santa Clara, CA, USA, 28 June–3 July 2013; pp. 343–350. [CrossRef]
50. Wozniak, D.; Zahabi, M. Cognitive Workload Classification of Law Enforcement Officers Using Physiological Responses. *Appl. Ergon.* **2024**, *119*, 104305. [CrossRef] [PubMed]
51. Xin, Y.; Kong, L.; Liu, Z.; Chen, Y.; Li, Y.; Zhu, H.; Mingcheng, G.; Hou, H.; Wang, C. Machine Learning and Deep Learning Methods for Cybersecurity. *IEEE Access* **2018**, *6*, 35365–35381. [CrossRef]
52. Heaton, J. Ian Good fellow, Yoshua Bengio, and Aaron Courville: Deep Learning: The MIT Press, 2016, 800 Pp, ISBN: 0262035618. *Genet. Program. Evolvable Mach.* **2017**, *19*, 305–307. [CrossRef]
53. Janiesch, C.; Zschech, P.; Heinrich, K. Machine Learning and Deep Learning. *Electron Mark.* **2021**, *31*, 685–695. [CrossRef]
54. Momeni, N.; Dell'Agnola, F.; Valdés, A.; Atienza, D. Real Time Cognitive Workload Monitoring Based on Machine Learning Using Physiological Signals in Rescue Missions. In Proceedings of the 2019 41st Annual International Conference of the IEEE Engineering in Medicine and Biology Society (EMBC), Berlin, Germany, 23–27 July 2019; Volume 2019. [CrossRef]
55. Ramírez-Moreno, M.A.; Carrillo-Tijerina, P.; Candela-Leal, M.O.; Alanis-Espinosa, M.; Tudón-Martínez, J.C.; Roman-Flores, A.; Ramírez-Mendoza, R.A.; Lozoya-Santos, J.d.J. Evaluation of a Fast Test Based on Biometric Signals to Assess Mental Fatigue at the Workplace—A Pilot Study. *Int. J. Environ. Res. Public Health* **2021**, *18*, 11891. [CrossRef]
56. Zanetti, R.; Valdés, A.; Aminifar, A.; Atienza, D. Real Time EEG Based Cognitive Workload Monitoring on Wearable Devices. *IEEE Trans. Biomed. Eng.* **2021**, *69*, 265–277. [CrossRef]
57. Cao, J.; Garro, E.M.; Zhao, Y. EEG/FNIRS Based Workload Classification Using Functional Brain Connectivity and Machine Learning. *Sensors* **2022**, *22*, 7623. [CrossRef]
58. Liu, C.; Zhang, C.; Sun, L.; Liu, K.; Liu, H.; Zhu, W.; Jiang, C. Detection of Pilot's Mental Workload Using a Wireless EEG Headset in Airfield Traffic Pattern Tasks. *Entropy* **2023**, *25*, 1035. [CrossRef] [PubMed]
59. Sharma, V.; Ahirwal, M. Quantification of Mental Workload Using a Cascaded Deep One dimensional Convolution Neural Network and Bidirectional Long Short Term Memory Model. 2021. Available online: https://www.techrxiv.org/users/680833/articles/677310-quantification-of-mental-workload-using-a-cascaded-deep-one-dimensional-convolution-neural-network-and-bi-directional-long-short-term-memory-model (accessed on 10 April 2024).
60. Dolmans, T.C.; Poel, M.; van't Klooster, J.W.J.; Veldkamp, B.P. Perceived Mental Workload Classification Using Intermediate Fusion Multimodal Deep Learning. *Front. Hum. Neurosci.* **2021**, *14*, 609096. [CrossRef] [PubMed]
61. Afzal, M.A.; Gu, Z.; Afzal, B.; Bukhari, S.U. Cognitive Workload Classification in Industry 5.0 Applications: Electroencephalography Based Bi Directional Gated Network Approach. *Electronics* **2023**, *12*, 4008. [CrossRef]
62. Lim, W.L.; Sourina, O.; Wang, L.P. STEW: Simultaneous Task EEG Workload Data Set. *IEEE Trans. Neural Syst. Rehabil. Eng.* **2018**, *26*, 2106–2114. [CrossRef]
63. Dimitrakopoulos, G.; Kakkos, I.; Dai, Z.; Lim, J.; de Souza, J.; Bezerianos, A.; Sun, Y. Task Independent Mental Workload Classification Based upon Common Multiband EEG Cortical Connectivity. *IEEE Trans. Neural Syst. Rehabil. Eng.* **2017**, *25*, 1940–1949. [CrossRef] [PubMed]
64. Zhan, Y.; Halliday, D.; Jiang, P.; Xuguang, L.; Feng, J. Detecting Time dependent Coherence between Nonstationary Electrophysiological Signals A Combined Statistical and Time frequency Approach. *J. Neurosci. Methods* **2006**, *156*, 322–332. [CrossRef]
65. Caroline, C.; Wilson, M.; Scannella, S. Online ECG based Features for Cognitive Load Assessment. In Proceedings of the 2019 IEEE International Conference on Systems, Man and Cybernetics (SMC), Bari, Italy, 6–9 October 2019; pp. 3710–3717. [CrossRef]
66. Park, J.; Zahabi, M. Comparison of Cognitive Workload Assessment Techniques in EMG based Prosthetic Device Studies. In Proceedings of the 2020 IEEE International Conference on Systems, Man, and Cybernetics (SMC), Toronto, ON, Canada, 11–14 October 2020; pp. 1242–1248. [CrossRef]
67. Bozkir, E.; Geisler, D.; Kasneci, E. Person Independent, Privacy Preserving, and Real Time Assessment of Cognitive Load Using Eye Tracking in a Virtual Reality Setup. In Proceedings of the 2019 IEEE Conference on Virtual Reality and 3D User Interfaces (VR), Osaka, Japan, 23–27 March 2019; pp. 1834–1837. [CrossRef]
68. Knisely, B.; Joyner, J.; VaughnCooke, M. Cognitive Task Analysis and Workload Classification. *Methods X* **2021**, *8*, 101235. [CrossRef]
69. Ghobakhloo, M.; Mahdiraji, A.H.; Iranmanesh, M.; Sadeghi, J. From Industry 4.0 Digital Manufacturing to Industry 5.0 Digital Society: A Roadmap toward Human Centric, Sustainable, and Resilient Production. *Inf. Syst. Front.* **2024**, 1–33. [CrossRef]
70. Wireless EEG Headset [Product Webpage]. 2024. Available online: https://www.aliexpress.com/ (accessed on 10 April 2024).
71. He, C.; Chen, Y.; Phang, C.; Stevenson, C.; Chen, I.; Jung, T.; Ko, L. Diversity and Suitability of the State of the Art Wearable and Wireless EEG Systems Review. *IEEE J. Biomed. Health Inform.* **2023**, *27*, 3830–3843. [CrossRef] [PubMed]

applied sciences

MDPI

Article

Motion Sickness in Mixed-Reality Situational Awareness System

Rain Eric Haamer [1,*], Nika Mikhailava [1], Veronika Podliesnova [1], Raido Saremat [2], Tõnis Lusmägi [2], Ana Petrinec [2] and Gholamreza Anbarjafari [1,3,4]

1 iCV Research Lab, University of Tartu, 51009 Tartu, Estonia; shb@icv.tuit.ut.ee (G.A.)
2 Vegvisir, 11312 Tallinn, Estonia
3 PwC Advisory, 00180 Helsinki, Finland
4 Institute of Higher Education, Yildiz Technical University, Besiktas, Istanbul 34349, Turkey
* Correspondence: eric@icv.tuit.ut.ee

Abstract: This research focuses on enhancing the user experience within a Mixed-Reality Situational Awareness System (MRSAS). The study employed the Simulator Sickness Questionnaire (SSQ) in order to gauge and quantify the user experience and to compare the effects of changes to the system. As the results of SSQ are very dependant on inherent motion sickness susceptibility, the Motion Sickness Susceptibility Questionnaire (MSQ) was used to normalize the results. The experimental conditions were tested on a simulated setup which was also compared to its real-life counterpart. This simulated setup was adjusted to best match the conditions found in the real system by using post-processing effects. The test subjects in this research primarily consisted of 17–28 years old university students representing both male and female genders as well as a secondary set with a larger age range but predominantly male. In total, there were 41 unique test subjects in this study. The parameters that were analyzed in this study were the Field of View (FoV) of the headset, the effects of peripheral and general blurring, camera distortions, camera white balance and users adaptability to VR over time. All of the results are presented as the average of multiple user results and as scaled by user MSQ. The findings suggest that SSQ scores increase rapidly in the first 10–20 min of testing and level off at around 40–50 min. Repeated exposure to VR reduces MS buildup, and a FoV of 49–54 is ideal for a MRSAS setup. Additionally camera based effects like lens distortion and automatic white balance had negligible effests on MS. In this study a new MSQ based SSQ normalization technique was also developed and utilized for comparison. While the experiments in this research were primarily conducted with the goal of improving the physical Vegvisir system, the results themselves may be applicable for a broader array of VR/MR awareness systems and can help improve the UX of future applications.

Keywords: virtual reality; mixed reality; motion sickness; SSQ; MSQ; user experience; MRSAS

1. Introduction

Motion sickness is an uncomfortable phenomenon experienced by individuals in response to Virtual Reality (VR) environments or motion stimuli. This phenomenon necessitates reliable measurement methods to assess its impact. There are ways to accurately measure motion sickness in users. A very objective method is to measure body sway amplitude, blood pressure, body temperature, sweating, change of skin conductivity, heart rate, EGG, and EEG, as these can indicate motion sickness [1]. These measurement methods are great for obtaining very accurate Motion Sickness (MS) readings for a single participant but the setup is not very portable and may not provide accurate results on in-field setups. As an alternative method, subjective user-based questionnaire scores have widely been used and been proven to be reliable [2]. Two significant instruments employed in this context are the Simulator Sickness Questionnaire (SSQ) and the Motion Sickness Susceptibility Questionnaire (MSQ) [3].

A large contributor to MS is latency in VR, and how it breaks immersion [4]. There are multiple types of latencies that can be explored in a MRSAS system, two of the main ones being glass-to-glass latency and motion-to-picture latency. The former describes the latency between a change in the camera input being reflected on the VR Head-Mounted Display (HMD). The latter describes the latency between motion of the user and this being reflected in the HMD. Motion-to-picture latency and its effects on MS have been widely explored and is the primary consideration for most VR applications; as such, all additions made to a VR system in the hopes of reducing MS should be weighed against their computational cost and, by extension, their effects on latency [5].

1.1. Mixed-Reality Situational Awareness System

The term Mixed-Reality Situational Awareness System refers to a 360 degree real time camera pass-through system for a VR device. The functionality is similar to direct camera feed to HMD systems like FPV drone HMDs. The primary difference is the addition of head tracking on the HMD coupled with 360 degree camera feed projection, mimicking a traditional VR experience.

1.2. Simulator Sickness Questionnaire

The SSQ is a survey-based tool developed by Kennedy et al. [3] to quantify the severity of motion sickness symptoms experienced by participants in response to VR stimuli or simulators. Comprising 16 items, the questionnaire prompts respondents to rate their symptoms on a scale ranging from 0 to 3, reflecting the intensity of their discomfort. The items cover various aspects of motion sickness, including nausea, disorientation, and general discomfort. The SSQ effectively gauges the extent of an individual's motion sickness experience, allowing for a comprehensive assessment of their susceptibility to this phenomenon. During this study, only the overall SSQ score was used to evaluate different conditions.

The literature references a rapid evaluation technique for the Simulator Sickness Questionnaire SSQ, termed the verbal Fast Motion Sickness (FMS) score. Although evidence suggests a correlation between FMS and SSQ, the FMS method was not adopted in these particular tests due to concerns about the method's precision [6].

1.3. Motion Sickness Susceptibility Questionnaire

The MSQ was proposed by Frank et al. [7] and operates as a survey-based measure aimed at evaluating an individual's susceptibility to motion sickness. By employing questions and response scales tailored to assessing one's predisposition to motion sickness, the MSQ serves as a pre-screening tool. The questionnaire delves into factors such as age, gender, and prior VR experience, aiming to identify potential contributors to motion sickness susceptibility. A higher MSQ score typically suggests a higher likelihood of experiencing motion sickness when exposed to VR environments or stimuli.

In this study, MSQ was used as a way to counteract the effects of inherent motion sickness susceptibility on different users when evaluating SSQ scores. Both questionnaire have been included in the Supplementary Materials.

2. Related Work

The field of VR has undergone significant development, transitioning from its initial phase of exploration to a rapidly growing domain with diverse applications in areas such as entertainment, education, and medical rehabilitation. Despite the promising potential of VR, a significant obstacle to its widespread adoption is the occurrence of motion sickness caused by virtual environments [8–15]. The purpose of this Section is to provide a comprehensive analysis of the causes, measurement techniques, and potential solutions for motion sickness induced by VR. This will be achieved by reviewing and analyzing the current body of literature on the topic.

Full 360-degree VR videos and the time consistency of stitching the scene has been shown to cause MS [16]. Several methods have already been proposed to alleviate MS in these cases but not all of them have been thoroughly researched. Patrao et al. [2] suggest a VR environment should be treated as a theater and not a film; this means camera animations and tricks like involuntary zooming and a lack of a reference can have a massive impact on comfort in VR. A virtual body, on the other hand, has been shown to have a negative impact on motion sickness [17]. Serrano et al. [18] proposed motion parallax to help with MS, but this may not counteract the MS caused by frame delay due to its added computational cost.

The article by Chang et al. [4] provides a significant and influential contribution by presenting a thorough classification system that categorizes the origins of motion sickness induced by VR. The taxonomy identifies three primary domains that contribute to VR-induced motion sickness: hardware-related factors, content-related attributes, and individual-specific human variables. Hardware factors encompass various technical specifications, including display technology, Field of View (FoV) range, latency, and flicker rate. From a content standpoint, various complex factors, such as the dynamics of optical flow, the fidelity of graphical rendering, and the contextually optimized dimensions of the FoV, are identified as crucial precursors. The field of human factors encompasses demographic factors, such as age and gender, as well as experiential factors, with a particular focus on an individual's inherent vulnerability to motion sickness.

Kemeny emphasizes the importance of reducing the discrepancy between visual and vestibular cues as a crucial approach to alleviate motion sickness [19]. The importance of controlling translational and rotational velocities, adopting HMDs with wide fields of view, and avoiding obstructions in the central visual field is supported by the empirical evidence presented in this study.

In their research work, Benz et al. [20] provide significant contributions to the field. Their argument suggests that the use of windscreen projection, which offers a wider field of view both horizontally (40 degrees) and vertically (35 degrees), leads to a reduced likelihood of causing simulator sickness when compared to traditional HMDs.

Modifications to the design of virtual reality content in order to address the issue of motion sickness were proposed by Patrao et al. [2]. The advised methodology promotes the conceptualization of VR experiences in a manner similar to theatrical productions. This involves incorporating static frame references and avoiding involuntary camera movements. These measures collectively contribute to the enhancement of user immersion and the reduction of discomfort.

The study by Cho et al. [21] presents a unique approach that utilizes a distortion technique to align user expectations with the motion experienced in a vehicle, resulting in a decrease in motion sickness.

Lim et al. [22] present an innovative approach that involves modulating the field of view in order to effectively alleviate symptoms of motion sickness. Nevertheless, the research articulates a word of warning regarding the possible consequences of imposing excessive limitations on the field of view, which could negatively impact the perception of presence and lead to a greater dispersion of attentional resources.

Jasper et al. [23] examine the disparities related to gender in terms of susceptibility to and recovery from motion sickness. The results highlight a higher vulnerability but faster recuperation in females compared to males.

Researchers such as Tovsic et al. [5] and Yoon et al. [24] investigate different aspects related to VR. The former examines the influence of latency on simulator sickness in smartphone VR, while the latter explores the effects of extended usage of VR through a smartphone-based HMD on visual parameters. Tovsic et al. find that there is no MS difference between latencies of 18 and 28 ms and Yoon et al. conclude that after 2 h of HMD usage, stereopsis and exophoric deviation significantly worsened.

Lim and Lee present findings that shed light on the intricate relationship between degrees of freedom (DOF) and FoV within the context of virtual reality-induced cybersickness [25]. This study compares the effects of using 3-DOF and 6-DOF HMDs and finds that the latter is associated with a lower likelihood of experiencing cybersickness. The decrease in discrepancy is ascribed to the increased alignment between the visual stimuli and the vestibular information, which is facilitated by the 6-DOF setups. The expansion of the field of view in HMDs is associated with a reduction in instances of cybersickness. This can be attributed to an increased feeling of immersion and a decrease in sensory conflicts.

A recent scholarly publication authored by Rolos and Merchant presents a methodical examination of the distinct vulnerability patterns observed in VR and MR HMDs [26]. It is postulated that the decrease in cybersickness induced by MR can be attributed to the translucent characteristics of MR HMDs, which allow for partial visibility of the real world and facilitate visual-vestibular interactions that enhance congruence.

The study conducted by Goedicke et al. [27] in 2022 presents an innovative approach to a mixed-reality driving simulation known as XR-OOM. The XR-OOM framework has the potential to reduce simulator sickness by integrating virtual elements into real-world driving scenarios using HMDs. This is achieved by ensuring that the visual stimuli presented in the virtual environment align with the vestibular cues experienced by the user.

In their research work, McGill et al. [28] explore the potential of AR/VR/MR technologies in improving passenger comfort during transportation. Through the synchronization of virtual augmentations with real-world surroundings, immersive technologies present a promising approach to mitigate motion-induced distress by harmonizing perceived and actual motion.

3. Study Overview

3.1. Method and Experiment Design

A selection of six different experimental setups were chosen based on parameters that could be changed in a physical VR HMD setup. For all experimental setups, the subjects were assigned the order conditions or changes based on a random distribution. For setups where users were to test only a subset of conditions like the FoV test, users were assigned FoV values based on a uniform distribution.

The experiments were divided into short and long tests. The shorter tests made up the majority of the testing while the longer tests served as a way to properly compare the simulation results to real life results. The shorter SSQ evaluations were conducted at 0, 5, 10, and 15 min. The longer tests were conducted at 0, 5, 10, 20, 30, 45, and 60 min, or when a participant was no longer able to continue being in VR. All new test subjects were administered a MSQ test as these were later used to normalize the results.

Some experiments changed the base nature of the application and may have introduced additional frame delays. Because of this, the experiments that added computationally impacting changes in the simulation were evaluated against additional control tests.

All alterations to the user experience were conducted separately. The visual alterations were not combined in any way so confounding variables were not tested in the frame of this study.

3.1.1. Adaptation

In order to confirm if prior exposure and training on a VR HMD is necessary, users' adaptability to VR had to be validated. The prior literature suggested that users become accustomed to VR over multiple exposure sessions, but the impact of this in relation to other effects has remained unmeasured.

We set up an experiment where we tested users three times on the same system, once with a 1 h and 15 min break between the two sessions and the second time after a several-week-long break. These two different break intervals allowed us to compare the immediate effects of VR exposure and long lasting exposure.

A smaller sample set of users were also tested prior to the repeated tests as that was their first exposure to VR. Their initial results have been marked as "Pre test".

3.1.2. Field of View

Keshavarz et al. [29] showed that a reduced FoV of 32 deg and a reduced visual display angle both contributed to a better motion sickness score. This has also been backed by Zielasko et al. [30] and Bala et al. [31], where their results indicated a FoV reduction had a minor reducing effect on MS. All of the aforementioned sources only tested either very limited FoV ranges or only a couple of selected FoV values.

We set up our experimental scenario with the same conditions, but with nine different FoV settings. The FoV settings tested were {26, 30, 34, 44, 49, 54, 60, 63, 84} degrees. All of the FoV tests were conducted by generating a flat plane in front of the user with a transparent hole in the center with blurred edges; different FoVs were achieved by scaling this plane. This ensured there were no additional distortions caused when the FoV was changed for each of the tests.

Out of the 9 possible FoV settings, 3–4 were chosen at random for each participant.

As the nature of this application comprises a camera, a virtual camera and a HMD, it is important to distinguish between three different types of FoV present in the system. The FoV of the cameras that are projected into the system, the FoV of the virtual camera that is fed into the VR HMD, and the inherent FoV of the headset. A visual representation can be seen in Figure 1. The last link in this pipeline, the FoV of the headset, was the one being measured. The changing the FoV of the virtual camera has already been thoroughly researched [25,32]. In case of the simulated testing setup, the first camera in the pipeline is a simulated camera. Apart from this difference, the two systems were designed to be as similar as possible.

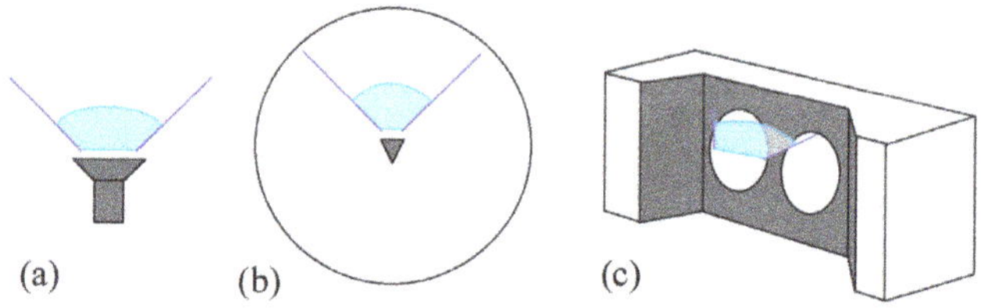

Figure 1. Images showing the 3 types of FoV present in the system. (**a**) is the FoV of the physical camera, the image of which is projected inside a virtual sphere. (**b**) is the FoV of the virtual camera inside the virtual sphere; this camera feed is directly sent to the VR HMD. (**c**) is the FoV inherent to the VR HMD itself.

3.1.3. Blur

The prior literature has suggested that the blurring of the salient area for VR HMD systems increases MS [33]. We wanted to verify the severity of this effect in comparison to other MS-inducing factors as in-field VR equipment may be subjected to less-than-ideal conditions and the lenses may become smudged or blurred as a result.

In order to verify the apparent effect of general blurring on a VR HMD, we set up an experimental scenario where the VR camera had a blurring post-processing (PP) filter applied to it. This PP effect did not visually impact the performance but as a precaution, a secondary test was conducted as a control where the PP effect was still applied but with the blur size being 0. Participants were randomly given one of the two effects to test out and were then asked to test the second setting at a later date.

3.1.4. Vignette

Non-salient area blurring has been shown to reduce MS and test track performance [34]. Non-salient area blurring could not be applied directly to a real camera feed in this application as the added complexity would greatly influence glass-to-glass latency. As such, the effects were applied to the periphery of the VR view, causing what is known as a vignette effect.

An additional test was proposed and conducted after seeing the findings from blurred and FoV tests. The idea was to test blur as a means of reducing FoV artificially while maintaining some semblance of peripheral vision. For this, a new setup was created where the periphery of the VR view was blurred to match the best-performing reduced FoV values. The chosen FoV were 26 deg and 54 deg as they stood out in the raw FoV test data. As this new rendering pipeline introduced additional input delay, new control tests were conducted with the same size black vignettes as the blurred periphery.

For both black and blurred cases for each FoV setting, the same users were tested and the first vignette method was chosen randomly for each user. Additionally, the users were asked to name their preferred vignette for each FoV setting separately.

3.1.5. Distortion

The real use case system uses multiple camera inputs with varying optics to generate the VR display. These cameras' systems come with inherent but subtle image distortions. The effect of these distortions on a VR camera feed are largely unknown and their influence needs to be measured.

The simulation was set up so that each of the cameras capturing feed into the VR view had subtle artificial distortion applied to each of them. The distortions that were tested were common barrel and pincushion distortions as they are the most common and well-known types in image processing applications. The distortion effects were slightly over-exaggerated as compared to the in-field setup. Both distortions were applied at 6% distortion strength in either direction. This was subtle enough to be barely perceivable without directly comparing the two cases and stronger then the distortion observed in the Vegvisir cameras.

The tests were conducted on the same people but they were randomly chosen to have one of the distortions applied to them first and the second one on a later date. As these distortions did not change the render pipeline and thus did not impact performance, completely separate control tests were not needed.

3.1.6. Illumination

Comparing and stitching multiple camera feeds from outdoor conditions highlights the issues with frame exposure and different white balance settings. This was very apparent in field tests and the significance of it on the user had to be verified. Kala et al. [35] has mentioned that decreased brightness on a camera feed has a reducing effect on MS but a high variance on brightness between frames has not been deeply researched.

The simulation was set up to allow for automatic white balance along with different camera illumination settings for each camera separately. The illumination variations between the cameras were greatly over exaggerated as compared to in field setups as the real setup is highly dependent on outside lighting and weather conditions. A secondary control setup was also tested where only automatic white balance was enabled on the simulated cameras. Users were randomly given either the manually exaggerated exposure setting or the default automatic white balance setup. They were then later tested on the second version on a later date.

3.2. *In Field Setup*

The test setup consisted of two separate systems. Namely, the Vegvisir system that used real pass-through live feed cameras and a simulation setup that was deployed on an HTC Vive with simulated camera feeds. A flowchart of how the system operates can be seen in Figure 2. The former was only tested in the field and is referenced in this text as "in field" setup. The latter was a stationary setup and is referenced as the "in simulation" setup.

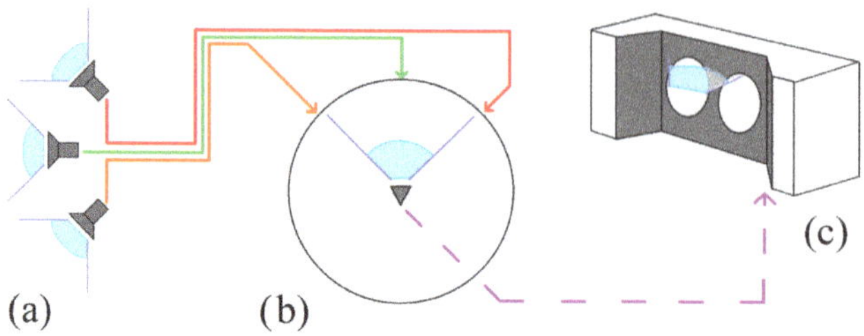

Figure 2. A flowchart of where the camera feed flows in the system based on the 3 components mentioned in Figure 1. The physical (or virtual) cameras render their captured feed onto the virtual sphere (**b**). A VR viewer in the virtual sphere (**b**) captures this feed and sends it to the physical VR HMD in (**c**). In the in field setup, (**a**,**c**) are physical while (**b**) is virtual. In the simulated setup, (**a**,**b**) are simulated while only (**c**) is physical.

In-field tests were conducted on different race tracks, but the test setup and evaluation procedure was the same throughout the tests. The majority of in field tests were conducted as 15 min runs as the longer tests were primarily used to confirm that the simulation and in field systems behave similarly in terms of MS. A more thorough description of the test lengths can be found in the results chapter. A rendering what was displayed for the user can be seen in Figure 3.

Figure 3. A rendering of the VR dome for the Vegvisir in field setup.

For the majority of recorded tests, the test subject was seated at the passenger seat with their view facing the direction of the vehicle. For four cases, the test subject was instead in the drivers seat. Their recorded results were indistinguishable from the other test results.

3.3. VR Simulation Setup

As a proof of concept for the system, a VR simulation in Unity had been developed beforehand. This simulation proved to be ideal for testing different MS-related effects in a controlled environment.

The simulated setup was based around a VR scene of a small town and surrounding fields. The materials of the assets, scene lighting and skybox were adjusted to appear more natural. Most of the similarity was achieved using post-processing effects like depth of field, color balance, bloom, film grain, and others on each of the virtual cameras to replicate the camera feed from the real-life counterpart. Post -rocessing was also used separately on the VR headset camera to replicate the conditions of the Vegvisir system, but these were minor and primarily focused on adjusting bloom and the FoV of the headset.

The VR HMD for the simulation was the HTC Vive Pro Eye instead of the Vegvisir HMD used in the real-life setup. This choice was made as the Vive had a larger viewing angle, allowing tests to exceed the technical specifications of the original Vegvisir system. The simulation was improved based on field test data to achieve as close to real life visual clarity as the final system. The view from the Vive headset was also adjusted in simulation with a blocking vignette to fully mimic the viewing angle of the Vegvisir HMD. The in-simulation camera was adjusted with post-processing effects to adjust the color balance, depth of field, bloom, and grain to mimic the real-life counterpart. A sample of focal depth for the simulated system as well as the real-life counterpart can be seen in Figure 4.

During the tests, the subjects remained seated and they were allowed to move the vehicle using keyboard inputs.

Hardware

For the simulation setup, a HTC Vive Pro was used with a computer comprising of an Nvidia RTX 3060 GPU and an AMD Ryzen 5 5600X CPU. This allowed the simulated setup to closely mimic the real life setup in terms of visual fidelity and motion-to-photo latency. The reported viewing angle for the Vive headset is 107 deg horizontally and the manually recorded FoV was 81 deg.

The Vive HMD and Vegvisir HMD were both manually measured to have a very similar motion-to-photo latency in default conditions. With the blurred vignette setup, the Vive was tested to have a 60 ms higher latency, so all comparison tests with the black vignette were redone on the slower setting.

Figure 4. Images showing a comparison between the simulation (**A**,**C**), and a photo taken with the physical camera (**B**,**D**), the pairs (**A**,**B**) as well as the pair (**C**,**D**) were both taken at roughly the same distances. The center 4 images are the calibration markers from the corresponding images at different distances to calibrate the virtual focal length.

4. Participants

The simulation-side experiment involved 25 individuals. Due to the substantial length of each test and repeated testing on the same individuals to obtain comparable results, it was decided that the small sample size was acceptable. The sample size was also limited due to the specialized nature of tests, which are both time-intensive and physically demanding for participants. Considering the potential discomfort and the need for close monitoring of each participant's well-being, limiting the number of individuals involved was important. The subjects were thirteen females and twelve males—all students and employees of the University of Tartu. The ages of the participants ranged from seventeen to twenty-eight, with no significant prior VR experience. This particular study group was selected because individuals within this age category are the most frequent users of VR. Participants tried different VR parameters and had different numbers of test sessions.

The in-field tests were conducted on 26 test subjects, of which 10 were university students. The participants consisted of three females and twelve males, and the overall distribution was nine females and seventeen males including the university students. The ages of the in-field subjects ranged from 19 to over 60 years. Most of the longer in-field tests were conducted on the overlapping simulation participants, as this made it easier to compare the results of the two systems.

Among the participants, 16 participated in only a single test, 11 of these came from the field tests. The rest of the participants participated, on average, in 5.5 tests. All participant were informed ahead of time about the nature of the experiment and the potential discomfort caused by VR-induced motion sickness. Participants were also allowed to stop the test at any time if they experienced discomfort.

5. Results

5.1. SSQ Normalization Using MSQ

Motion sickness is a variable and often perplexing phenomenon in VR experiences. The individual susceptibility to VR-induced motion sickness can greatly differ among users. The literature has pointed toward a potential solution—a weak correlation between pre-test MSQ scores and subsequent SSQ scores [17]. This revelation opens the possibility of using MSQ scores to normalize SSQ values and establish a more uniform measure of motion sickness severity.

To delve into this correlation, our study initially compiled a wide spectrum of SSQ data, reflecting the diverse reactions people exhibit. Employing linear regression on each participant's SSQ scores, we quantified the spread of the generated lines to derive a baseline score. To bridge the MSQ-SSQ relationship, a series of scaling functions were introduced as $F(x, m, s)$. Here, x signified the raw SSQ input to be scaled, m denoted the corresponding MSQ score for that specific participant, and s was a parameter introduced to optimize the function. Manipulating the s variable and the fundamental function $F(x, m, s)$ led to the identification of the best fit.

The most effective function, derived from this analysis of the raw data, can be expressed as: $F(x, m, s) = x/(1 + m \cdot s)$.

The best s value on all of the data was found to be 1.50, but this varied slightly as newer tests were recorded. Sample best fitting lines can be seen in Figure 5. The first plot depicts the error of the application of the function $F(x, m, s)$ on all recorded data over varying values of s. The x-axis depicts varying values of s and the y-axis depicts the standard deviation for the data at each time step after scaling. The four lines on the plot show where the smallest standard deviation was present for 5, 10, 15, and 60 min after scaling.

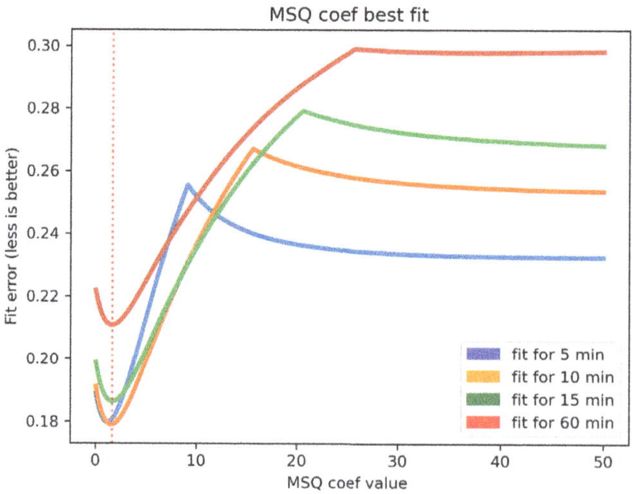

Figure 5. The function parameter s for SSQ normalization using MSQ.

This was a form of focused fitting where the standard deviation of values near the X amount of minutes was used as the fitting error with a varying MSQ coefficient. Each of the lines represents all user data scaled by different MSQ coefficient values (x-axis) and then linearly interpolated to obtain the value at X minutes; all of these values were aggregated and the standard deviation was used as the fit error (y-axis). The values 5, 10, and 15 were chosen as they were the most frequent SSQ sample point values in the dataset and thus the most reliable. The 60 min represents our main target duration but is mostly interpolated.

The function $F(x, m, s)$ was generated in a similar manner. Several functions including linear, polynomial, exponential, and combinations of these were tested. The proposed function of $F(x, m, s) = x/(1 + m \cdot s)$ outperformed the the rest as more complex functions tended to over-fit on a specific time range very quickly and changed drastically when new test results were added for evaluation.

In the context of long-term tests, a distinct pattern emerged. Figure 6 only contains recorded data that exceeded 15 min in test time as this subset proved to be more stable for SSQ curve prediction.

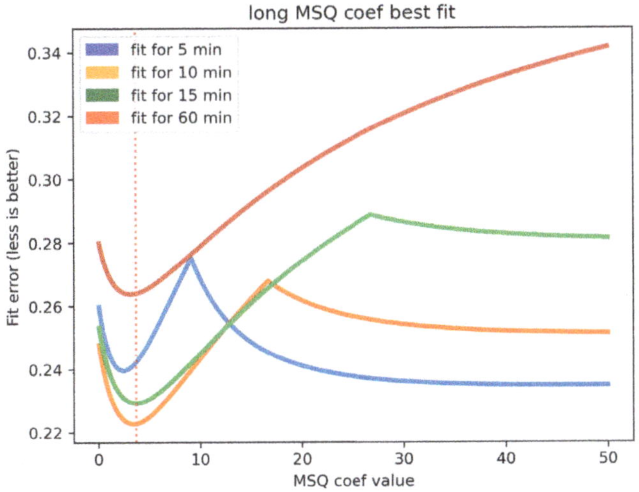

Figure 6. The function parameter s for SSQ normalization using MSQ but only calibrated on experiments that lasted over 15 min.

Fluctuations in the best fit suggested an insufficiency of data, hinting at the challenges of conducting longer-duration tests. Here, the coefficient value ranged from 3.45 to 5.40, with coefficients exceeding 1.00 yielding more stable MSQ scaling results compared to those based on data from all experiment durations. This phenomenon can be attributed to the dominance of 15 min samples, which seemingly lacked sufficient time to allow motion sickness effects to stabilize.

The best s value on the long data was 4.95. Both the initial 1.50 and 4.95 were tested for data normalization and the value 4.95 was found to have a smaller standard deviation on the unaltered data set compared to 1.50. As such, the coefficient of 4.95 was used for MSQ scaling, but this remains to be fully explored.

Exponential and logarithmic functions were also explored but a linear function yielded the most consistent results when the data set was updated. This inconsistency of the logarithmic function was due to the sporadic nature of the SSQ results, causing the function to often fail completely and default to a 0 slope line when there were not enough user samples.

This scaling mechanism was subsequently applied to all SSQ evaluations, thereby creating a theoretically more comparable set of raw values. However, the practical application of MSQ-scaled results proved to be nuanced. While this approach successfully compensated for some outliers, it encountered limitations in cases where subjects either overestimated their susceptibility to motion sickness or lacked prior experience. Nevertheless, the application of MSQ scaling has been included in later analysis as it better addresses outliers who have either very low or very high inherent susceptibility to MS.

To account for inter-individual MSQ variations, a secondary scaling based on the median MSQ was introduced. Using the observed scaling coefficient of 4.95 and targeting a median MSQ value of 5.0, this approach aimed to highlight users who, despite their anticipated low susceptibility, displayed unexpectedly high SSQ scores. Additionally, it aimed to align users with inherently high MSQ scores, expecting them to exhibit correspondingly elevated SSQ scores.

5.2. SSQ Change over Time

Understanding the dynamics of motion sickness and its thresholds is critical for ensuring a comfortable and immersive virtual reality experience. In this section, we delve into the results of long-term tests, focusing on the change of the SSQ scores over time and the establishment of motion sickness thresholds.

The outcomes of extensive long-term tests provide us with valuable insights into the trajectory of SSQ changes over extended durations and the convergence of these changes. By analyzing the baseline curve of SSQ change over time, we gain an overall perspective of how SSQ scores tend to increase over prolonged periods until they level off. This baseline curve, in conjunction with the level-off period and its corresponding value, serves as a means to assess the disparity between simulation and field tests. The adaptation of simulation test results onto field tests is facilitated through this comparative approach.

Across the spectrum of tests, encompassing both field tests and simulated experiments, a notable motion sickness threshold emerges. Approximately at an SSQ score of 75–80, users exhibit signs of motion sickness-related effects. Beyond this threshold, impairing symptoms become prominent. The upper limit of SSQ scores that users can tolerate before discontinuation lies within the range of 100–120 SSQ score. The lower bounds of these lines have been visualized in Figure 7.

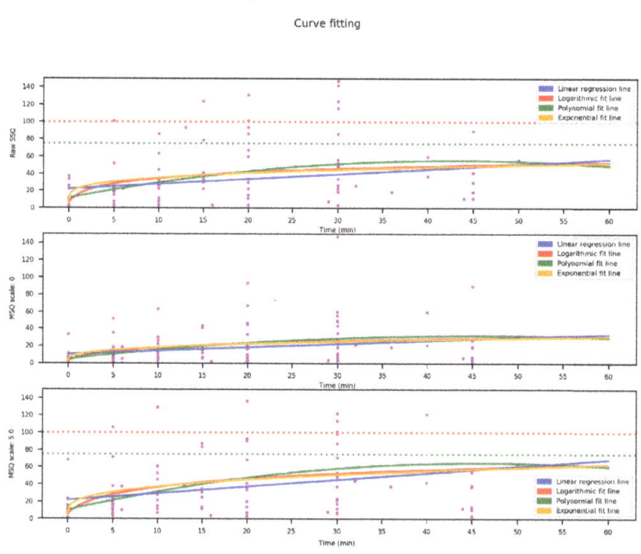

Figure 7. Logarithmic, polynomial, exponential curvature and linear regression lines on the gathered SSQ data. The top plot contains the raw measurements, the bottom two are MSQ scaled with MSQ being set to 0 and 5.0, the median in the data set. The green dotted line denotes the first impairment line and the red dotted line denotes the approximate upper limit. The magenta dots are raw measurement samples. This figure and all following figures have time in minutes as the x-axis.

Visualizing these thresholds, a plotted representation includes user data recorded for durations exceeding 15 min. Instances where users crossed the green dotted threshold line of 75 SSQ score reported symptoms causing impairment. Crossing the more critical red dotted line of 100 SSQ score led users to a point where they were unable to continue. The data encompasses general trends, though it should be noted that two anomalous edge cases were excluded due to their rapid and severe SSQ score increase. These cases were attributed to erroneous tests, showcasing SSQ scores surpassing 100, and resultant impairment. In both cases, the test setup experienced frame-rate drop unrelated to the experiment, causing a rapid increase in SSQ scores.

When searching for the best fitting function for the given dataset, linear regression was first applied. This fit the dataset quite well while the experiments lasted up to 15 min. The root-mean-square deviation (RMSE) for linear regression on all of the data was 27.4, while for the MSQ-scaled results, RMSE was 24.5.

Once longer tests were conducted, it became apparent the SSQ results tended to level off at around 30 to 40 min. The RMSE for a logarithmic curve on all of the data was 23.9 and for the MSQ-scaled results RMSE was 20.1. Both a polynomial as well as an exponential fit were also tested during this time. While the exponential fit had a slightly worse RMSE (24.1 and 20.2), the polynomial fit had a better RMSE (23.5 and 19.8) than the logarithmic function. Manually reviewing the fit, it was apparent the low overall scores of the longest tests greatly impacted the polynomial fit, causing it to slope downwards after the 45 min mark. As this sort of a reduction in MS was not observed on individual test subjects, only on the data as a whole, the fit was discarded as being caused by a lower number of >45 min tests.

For fitting, 61 separate test results were used, which translated into 244 data points with a mean of 24.8 min and Standard Deviation of 18.3 min.

Similar RMSE MSQ scaling patterns were found when only fitting for experiments where the tests lasted longer than 15 min. The >15 min subset contained 27 tests with 125 points with a mean of 38.4 min and Standard Deviation of 19.5 min.

5.3. Comparison of Field Test and Simulation Test Results

Field tests, conducted on race tracks and off-road tracks, mirrored the simulation's predicted curve for the most part. These prediction curves can be observed in Figure 8 as well as the averaged user scores in Figure 9.

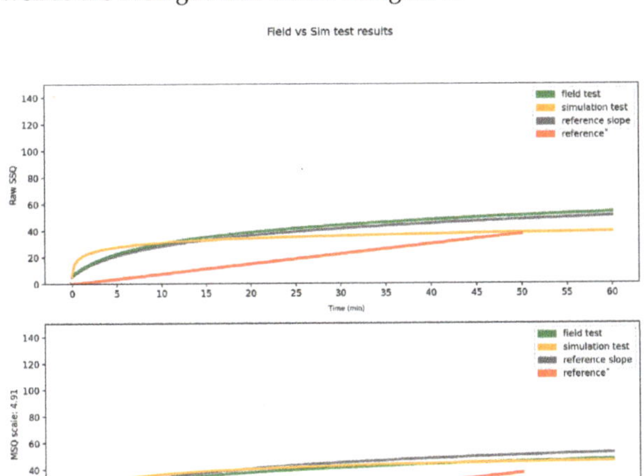

Figure 8. Logarithmic curvature prediction of the SSQ results comparing the in field test to the simulation. The red SSQ line, marked with a ∗, is from a similar experiment by Benz et al. [20].

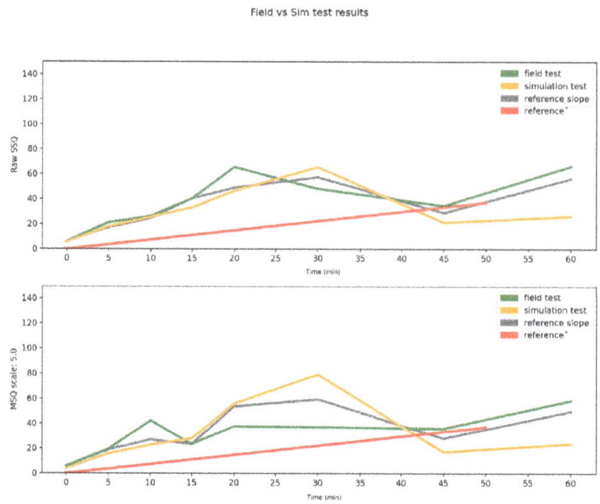

Figure 9. Mean SSQ results comparing the in field test to the simulation. The red SSQ line, marked with a ∗, is from a similar experiment by Benz et al. [20].

In total, 26 control tests were conducted on the unmodified simulated setup. The average duration of the tests was 25.68 min with a standard deviation of 13.03 min. Additionally, 26 in-field tests were conducted on the Vegvisir system. The average duration of the tests was 31.52 min with a standard deviation of 24.5 min. The unmodified simulation setup test results were used as the control value for most of the experimental setups. The blurred vignette did not use these data as the control baseline.

Visually inspecting the results, it is apparent that the raw field test outcomes induce marginally more nausea than the simulated tests. When accounting for MSQ scores, the simulated tests and the field tests were much more similar. The numeric results indicate that the simulator appears to perform poorly in the 20 to 30 min range, but manages to level out at around 40 min. This was backed by post-experiment interviews as users got used to the system and some of their SSQ symptoms were only temporarily heightened.

In the case of in field tests, the SSQ values did not follow a similar spiked pattern at 20 min but tended to be more gradual, especially when normalized by MSQ. These findings can be contextualized against the reference line, which represents aggregated data from all unaltered tests. As an additional reference, the results from a similar experiment by Benz et al. [20], where the SSQ at 50 min was measured, have also been added in red.

Participant feedback from the field tests provides further context. Participants universally reported an improved experience and greater usability of the system after 10–20 min.

All test subjects themselves preferred the field test version to the simulated version in terms of MS. Verbal responses from test subjects have indicated that the heightened realism and detailed view from the camera were prominent factors contributing to this preference. This was also visible when comparing the MS values for the subjects with the highest MSQ values, as they did not feel they crossed the previously noted MS thresholds, even if the SSQ scores indicated otherwise. This could be attributed to the subjectivity of the SSQ tests.

However, it is crucial to acknowledge certain limitations. The feedback indicates concerns related to eye strain and the weight of the VR system on the face. These factors were less prominent in the simulated setup.

The field test setup HMD has a wide focus adjustment range and improper adjustment might have been the primary cause of the eye strain, as most subjects opted not to adjust the focus. The second factor that may have had an effect on eye strain is improper fixation of the headset. If the headset is not properly fixed to the head, the movement and vibrations of the vehicle may move the HMD and cause the user to lose focus. These were known factors, but hard to account for as it requires more experience on the users part to know how to best adjust the headset to fit their needs.

This section offers a comprehensive exploration of SSQ changes over time and the establishment of motion sickness thresholds in VR environments. The findings underscore the significance of adapting tests to real-world conditions for a more accurate assessment of motion sickness responses. Further investigation into anomalies, user feedback, and system ergonomics remains imperative for enhancing the VR experience and ensuring user comfort.

To facilitate better statistical comparisons between the real-world and simulated data, a series of analytical techniques were applied. ANCOVA was employed to ascertain whether a significant difference exists between the field and simulation data sets. Finally, T-tests were conducted to compare individual data slopes, thereby determining the similarity between the two sets of slopes. Both of these tests were used to quantitatively compare the two datasets and if they could be labeled similar regardless of the visual similarity.

The ANCOVA analysis yields a p-value of 0.000112, indicating that the datasets are significantly different. A p-value below 0.05 suggests divergence between the datasets. This divergence may be attributed to a higher percentage of people with high MSQ scores present in the simulation dataset.

Examining the T-test coefficient, a value of -0.39 indicates a negative correlation in slopes, with a p-value of 0.698. This difference in slopes, while notable, is again likely influenced by chance and is thus insignificant.

5.4. Comparison of Long Field and Simulation Test Results

In the pursuit of a comprehensive evaluation of VR motion sickness, it becomes essential to analyze the results of both field tests and unaltered simulation tests. To counterbalance potential biases introduced by an abundance of short-duration tests, a secondary comparison was conducted between field and simulation tests exclusively comprising samples surpassing the 15 min threshold. Both curves can be observed in Figure 10 and in Figure 11, where the scores are averaged.

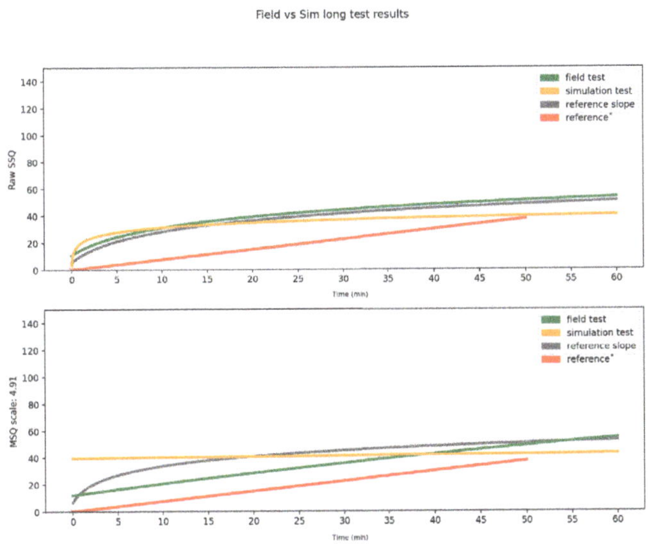

Figure 10. Logarithmic curvature prediction of the SSQ results comparing the in field test to the simulation. Only tests lasting longer than 15 min were included. The red SSQ line, marked with a $*$, is from a similar experiment by Benz et al. [20].

When only comparing the longer experiments, the results become more stable. Unfortunately, with the MSQ-scaled results, the logarithmic prediction for simulated results failed to find a good enough approximation. To compensate for this, a simple linear regression line was plotted instead.

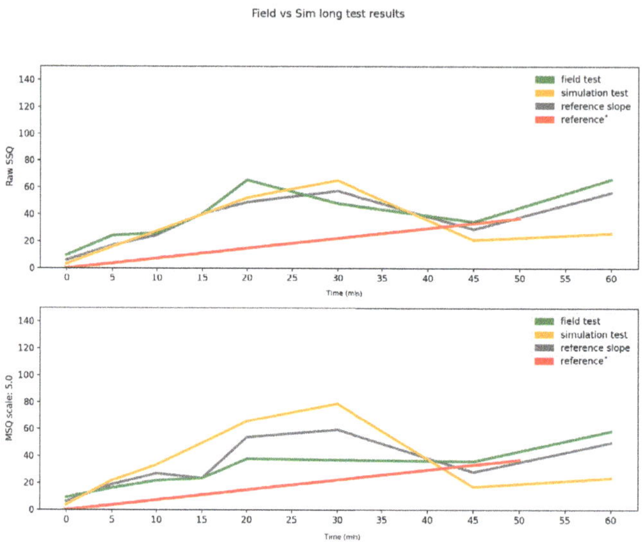

Figure 11. Mean SSQ results comparing the in-field test to the simulation. Only tests lasting longer than 15 min were included. The red SSQ line, marked with a $*$, is from a similar experiment by Benz et al. [20].

Verbal responses by participants mentioned how the in-field setup caused less apparent nausea and was more enjoyable than the simulated setup. The numeric SSQ results for the long field tests were influenced by hot weather during the tests, causing higher reported sweating levels for the SSQ tests than on any previous experiment.

The simulated results and in-field results were also compared using ANCOVA to indicate the presence of a difference between the two data sets and T-test to compare the similarity of the slopes of the two data sets.

The ANCOVA analysis indicates a non-significant difference between the two data sets, with a p-value of 0.1127.

The T-test coefficient, registering at -2.13, denotes a pronounced reversed difference in slopes, with a p-value of 0.06. Due to the high p values in all of the tests, it unfortunately impossible to state that the field and simulated data sets are statistically similar to one another. As such, all of the results obtained on the altered setup experiments cannot be directly compared to the field test results.

5.5. Adaptation

The results of our repeated tests confirm the common understanding that states that people get used to VR over multiple exposure sessions. A total of 26 tests were conducted using the repeat experiment. Each of the three test cases had eight users evaluated, with two extra accounting for the pre-tests. All of these tests were exactly 15 min long.

In our case it was very visible that the users tested were more comfortable during the second round, as seen in Figure 12. Their motion sickness score was reduced. The test also includes the results of two people who were very strongly susceptible and had a single test conducted two weeks prior to the first test. In their case the difference between the pre-test and the later tests was much more significant. The impact of continuous exposure is especially visible when comparing the overall SSQ mean of the test, as seen in Table 1.

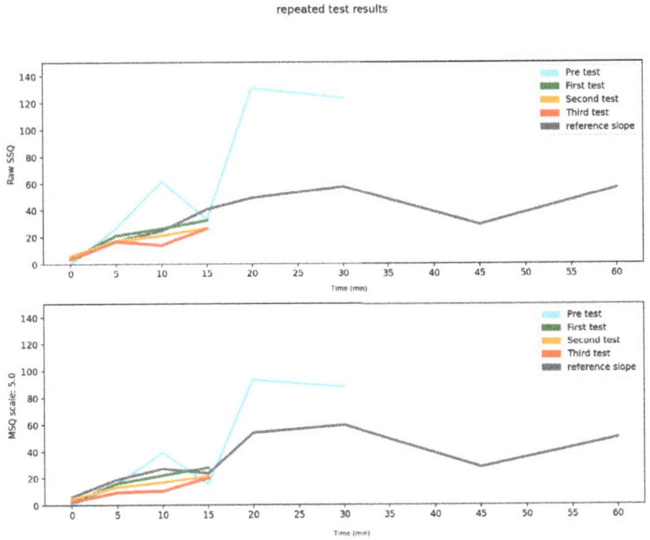

Figure 12. Mean SSQ results for the repeated tests. The pre-test was conducted on users who had no prior experience in VR. The 1st and 2nd tests were 1 h 15 min apart. The 2nd and 3rd tests were more than a week apart.

Table 1. Mean and standard deviation of all recorded SSQ values for unaltered simulator and real life experiments. The field vs. sim tests consisted of 52 samples and the adaptation test consisted of 26 samples.

Test Type	Subtype	RAW Mean	RAW Std	MSQ Mean	MSQ Std
Field vs. Sim	field	23.19	28.26	21.04	21.02
Field vs. Sim	sim	23.3	29.53	22.3	34.39
Field vs. Sim > 15 min	field	30.06	30.97	25.27	21.54
Field vs. Sim > 15 min	sim	31.58	37.6	39.4	55.83
Adaptation	pre	51.52	46.9	34.21	34.68
Adaptation	1st	20.8	23.39	17.06	18.97
Adaptation	2nd	17.77	27.75	14.06	21.59
Adaptation	3rd	15.19	17.37	10.72	12.17

This suggests that first training with a VR device will improve how long a user can withstand using it before becoming uncomfortable or impaired due to motion sickness-related effects. The findings also suggest that the effects of such an exposure will linger for an extended period of time. Based on these findings, first-time VR users will experience significantly worse motion sickness effects than repeat users, indicating that first-time VR users are not a good indicator for how MS-inducing a VR device is.

During our repeated tests, we also noticed short lasting after-effects on users as mentioned by Bles et al. [36]. This was only applicable for people with high MSQ scores during the second round of tests as they had not managed to fully recover before the second experiment. During the second round of adaptation tests and past experience with testing, we have noticed that it takes users between 30 min to 1 h to fully recover from any effects caused by VR. This appeared to depend mostly on the individual users' MS susceptibility.

The third experiment was conducted with a 1-week- to 10-day-long wait period and thus is not influenced by after-effects.

5.6. Field of View

In total, 49 tests were conducted using varying FoV settings. As the subjects were randomly assigned their test FoVs, the number of samples in each category was not equal. All of these tests were exactly 15 min long.

The FoV experiment scores were scaled by MSQ and can be seen in Figure 13. The SSQ scores were also scaled by individual users to remove the effect of random chance causing more susceptible users to be given the same FoV setting.

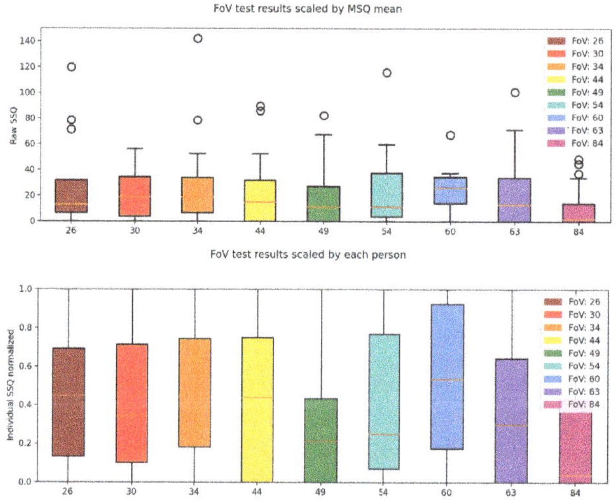

Figure 13. The top figure shows the raw results for each tested FoV setting and the bottom one is individually scaled for each users results.

With the data scaled by users, the results suggest there is a minor local minimum of motion sickness at 49/54 deg and two very significant local minima at 26 deg and 84 deg. The very low FoV performing well is explained by very susceptible users who tended to prefer lower FoV settings in general. This coincides with the findings of Teixeira et al. [37] where FoV reduction on MS susceptible people had a very significant impact on MS reduction. The 84 deg FoV generating a significant minimum is currently unexplained and can only be explained by a lack of data. The primary local minima being between 49 and 54 deg is also similar to the findings of Kala et al. [35], where lower FoV values down to 60 deg were suggested and values below 45 deg proved to be hampering.

The raw SSQ values, shown in Table 2, indicate very similar results but with a larger preference for 49 and 84 deg FoV values than the MSQ-scaled or user-scaled versions.

5.7. Blur

For the blur test, 19 tests were conducted: 10 of these accounted for the actual blurred setup and 9 were separate control conditions. All of the test lengths were 15 min.

The raw SSQ results in Figure 14 indicate that there is no distinguishable impact caused by blur on MS. Scaling the results by MSQ, shows the blurring effect impacted MS negatively. This was backed by user feedback and the results are even more extreme for people with already high MSQ scores.

This indicates that there is a minor impact on the users' motion sickness if their vision is even slightly blurred in the middle of their view.

Table 2. Mean and standard deviation of all recorded SSQ values for HMD-based changes. The FoV tests consisted of 49 samples, blur consisted of 10 + 9 control samples and the vignette test consisted of 16 samples.

Test Type	Subtype	RAW Mean	RAW Std	MSQ Mean	MSQ Std
FoV test	26 deg	28.98	37.09	**15.15**	12.79
FoV test	30 deg	**20.69**	17.96	19.8	27.74
FoV test	34 deg	26.92	32.55	38.59	67.62
FoV test	44 deg	22.17	24.63	18.58	19.78
FoV test	49 deg	**19.87**	24.48	18.81	21.94
FoV test	54 deg	22.6	26.35	**15.17**	17.98
FoV test	60 deg	32.49	36.74	43.25	78.98
FoV test	63 deg	**20.41**	24.69	**14.65**	17.57
FoV test	84 deg	**11.53**	16.65	**8.98**	13.02
Blur	control	**20.67**	26.8	**17.31**	22.11
Blur	blur	20.94	24.42	23.76	42.34
Vignette	black 26 deg	45.35	39.78	-	-
Vignette	black 54 deg	**31.42**	**31.59**	-	-
Vignette	blur 26 deg	46.75	50.47	-	-
Vignette	blur 54 deg	58.09	41.6	-	-

Please note, this test only evaluated blur in the center of vision and not the periphery which has shown to have different results. This means that the HMD screen has to be properly calibrated for each user, any and all blur will have an impact on the user's ability to use the system for extended periods of time. Secondly this also means the lenses of the headset must be kept clean, as smudges and fingerprints are likely in an in-field VR application.

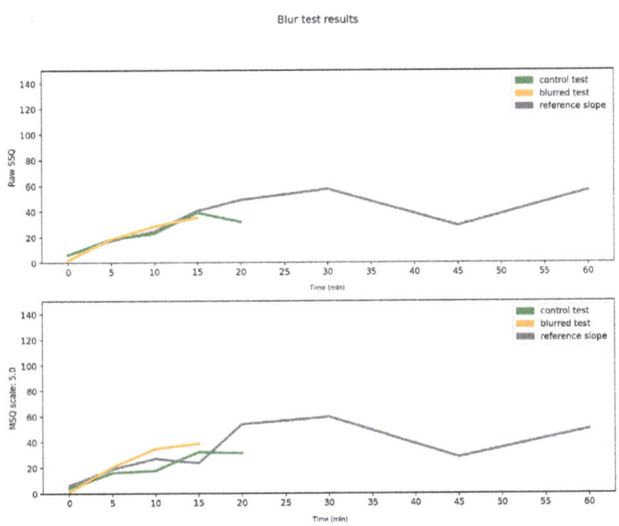

Figure 14. Mean SSQ results for the blur test. Both the new control and general reference line have been included.

5.8. Vignette

For the vignette test, 16 tests were conducted. All of these tests lasted 15 min.

The SSQ scores in Figure 15 indicate that the 54 deg black vignette appeared to perform the best, with the other black vignette and 26 deg blurred periphery following close behind.

Verbally, all participants who tested both setups also indicated their preference to the black vignette over the blurred version.

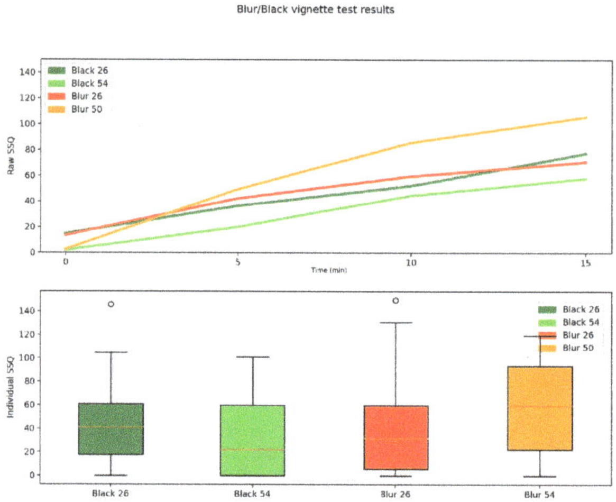

Figure 15. Mean SSQ test results on the top figure and the individually scaled results for each user on the bottom.

5.9. Distortion

With the distortion setup, 14 tests were conducted. All of these tests were exactly 15 min long.

Based on the conducted tests shown in Figure 16, the pincushion effect had more of an effect than the barrel distortion. Neither of the tested effects exhibited very noticeable effects on the test subjects when questioned after the tests. This suggests that if there are distortions present, they do not have to be corrected very precisely in order to preserve computational complexity.

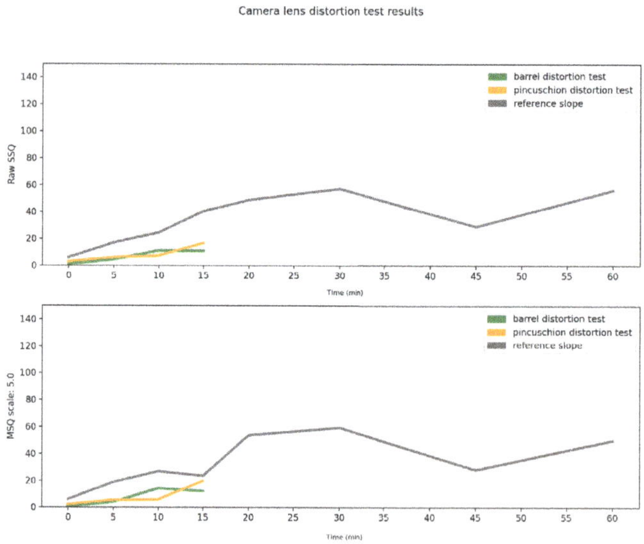

Figure 16. Mean SSQ results for barrel and pincushion distortion effects on the camera feeds.

5.10. Illumination

The illumination test was conducted eight times with four additional control tests with overlapping subjects. The new control tests were conducted to see if a small change in the lighting system had a noticeable impact on our setup but no differences were observed. All of these tests were exactly 15 min long and the control tests had a mean length of 18.75 min with a standard deviation of 2.5 min.

From the tests we conducted, there seems to be a very minor difference between users being exposed to cameras with varying illumination settings and uniformly illuminated cameras. The recorded SSQ scores indicate that illumination variance may have a subtle effect on MS when compared to the control setting. Verbal responses indicated a lack of a visible difference or affirmed the findings of Kala et al. [35] that increased illumination and overexposure increase the effects of MS. The findings of the illumination experiment can be seen in Figure 17 and in the Table 3.

In practice the users tend to focus on a single camera view and rarely view the overlapping area between two cameras. This, along with the users preferring zoomed in view of the sphere, tends to reduce the impact of variance between the different camera feeds. The same user preference was observed with the camera distortion tests where the main impact was near the area where two camera feeds were stitched together.

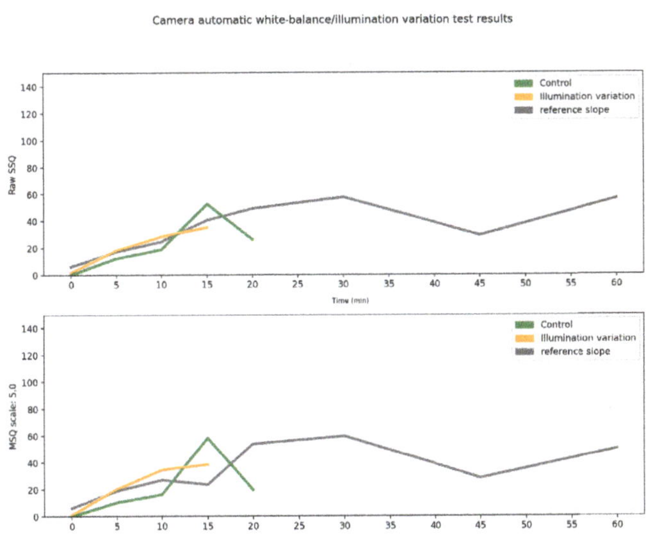

Figure 17. Mean SSQ results for exaggerated over- and under-exposure conditions compared against a control where only automatic white balance was used.

Table 3. Mean and standard deviation of all recorded SSQ values for camera-based distortions. The camera distortion tests consisted of 14 samples and the illumination test consisted of 8 samples with 4 additional control samples.

Test Type	Subtype	RAW Mean	RAW Std	MSQ Mean	MSQ Std
Camera distortion	barrel	**6.9**	8.42	**7.87**	13.37
Camera distortion	pincushion	8.42	10.54	8.4	13.16
Camera illumination	control	**15.9**	20.03	**13.94**	16.78
Camera illumination	illum	20.94	24.42	23.76	42.34

6. Additional Findings

Sakai et al. [38] has mentioned in their findings that a user's sense of speed is greatly influenced by the FoV of the headset. There are also contradictory findings by Perrin et al. [39], where users were able to very accurately estimate their speed in VR.

In our case, users participating in the field tests did encounter the influence of FoV when estimating the speed of the vehicle. A larger FoV caused users to underestimate the actual speed while a lower FoV caused them to overestimate it. The difference was minor compared to Sakai et al. but this was likely affected by other cues like the engine sounds, familiarity to the vehicle and car vibrations that helped the user better estimate their speed.

7. Conclusions

This research was aimed at improving the user experience of a mixed-reality situational awareness system. The results of this study can be applicable for VR applications that display a live pass through camera feed for the user, but, as this research was primarily conducted on a particular system, they cannot be generalized for general-purpose VR systems. Additionally, it should be noted that the simulation results were primarily conducted on 17–28 years old participants. Field test results and their comparisons were conducted on a larger age range.

In order to quantify the effect different changes made to the system, the Simulator Sickness Questionnaire was used as a primary metric with user feedback confirming the results. As SSQ is both a subjective metric and is very dependent on a user's inherent motion sickness susceptibility, the Motion Sickness Susceptibility Questionnaire was used to normalize the results.

Our findings indicate that MSQ values can be used on SSQ recordings in order to normalize the results across users with varying inherent motion sickness susceptibility.

We have also found that a user's motion sickness level increases rapidly in the first 10–20 min of contact with VR and levels out at around 40–50 min. This pattern was best described and approximated with a logarithmic curve.

We observed two motion sickness thresholds. Crossing the first was observed to be detrimental to usability and the second one rendered the user unable to continue. In terms of SSQ scores, these values were found to be at around 75–80 and 100–120, respectively. The specific threshold value varied by user due to the subjective nature of SSQ.

Repeated exposure proved to have a positive impact on motion sickness, even when the time between different VR sessions was less than the general recovery time.

Our tests concluded that a preferable VR setup for an MRSAS application has a FoV between 49–54 deg with a black periphery. Any blurring on the screen or even on the periphery did not have a positive effect on MS.

Camera-based effects on performance caused by lens distortion or automatic white balance were shown to be negligible, as the primary impact of the effects was concentrated on the overlapping or stitched edges of different camera feeds.

Supplementary Materials: The following supporting information can be downloaded at: https://www.mdpi.com/article/10.3390/app14062231/s1.

Author Contributions: Conceptualization, G.A. and R.S.; methodology, R.E.H.; software, R.E.H., V.P., T.L. and A.P.; validation, R.E.H., A.P. and R.S.; formal analysis, N.M. and V.P.; investigation, R.E.H., N.M. and V.P.; resources, R.S., T.L. and A.P.; data curation, R.E.H.; writing—original draft preparation, R.E.H. and G.A.; writing—review and editing, R.E.H.; visualization, R.E.H.; supervision, G.A., R.E.H. and R.S.; project administration, G.A.; funding acquisition, G.A. All authors have read and agreed to the published version of the manuscript.

Funding: This research was partially funded by and in cooperation with Vegvisir (Defensphere OÜ).

Institutional Review Board Statement: Ethical review and approval were waived for this study, due to this article not containing medical research, no personal data nor identifiable medical data being gathered and all subjects consenting to the experiments prior to putting on the headsets.

Informed Consent Statement: Informed consent was obtained from all subjects involved in the study.

Data Availability Statement: The raw data supporting the conclusions of this article will be made available by the authors on request.

Conflicts of Interest: The authors declare no conflict of interest. The research was conducted in order to improve a specific VR system and the results may be beneficial for the development of other similar systems.

Abbreviations

SSQ	Simulator Sickness Questionnaire
MSQ	Motion Sickness Susceptibility Questionnaire
HMD	Head-Mounted Display
MS	Motion Sickness
VR	Virtual Reality
FoV	Field of View

References

1. Hwang, A.D.; Peli, E. Instability of the perceived world while watching 3D stereoscopic imagery: A likely source of motion sickness symptoms. *i-Perception* **2014**, *5*, 515–535. [CrossRef]
2. Patrão, B.; Pedro, S.; Menezes, P. How to Deal with Motion Sickness in Virtual Reality. In *Proceedings of the 22o Encontro Português de Computação Gráfica e Interação 2015*; Dias, P., Menezes, P., Eds.; The Eurographics Association: Eindhoven, The Netherlands, 2020; ISBN 978-3-03868-128-1. [CrossRef]
3. Kennedy, R.S.; Lane, N.E.; Berbaum, K.S.; Lilienthal, M.G. Simulator sickness questionnaire: An enhanced method for quantifying simulator sickness. *Int. J. Aviat. Psychol.* **1993**, *3*, 203–220. [CrossRef]
4. Chang, E.; Kim, H.T.; Yoo, B. Virtual reality sickness: A review of causes and measurements. *Int. J. Hum. Comput. Interact.* **2020**, *36*, 1658–1682. [CrossRef]
5. Tošić, I.; Hoffman, D.; Balram, N. Effect of latency on simulator sickness in smartphone virtual reality. *J. Soc. Inf. Disp.* **2021**, *29*, 561–572. [CrossRef]
6. Keshavarz, B.; Hecht, H. Validating an efficient method to quantify motion sickness. *Hum. Factors* **2011**, *53*, 415–426. [CrossRef] [PubMed]
7. Frank, L.; Kennedy, R.S.; Kellogg, R.S.; McCauley, M.E.; FL, E.C.O. Simulator sickness: A reaction to a transformed perceptual world. 1. scope of the problem. In *Proceedings of the Second Symposium of Aviation Psychology*; Ohio State University: Columbus OH, USA, 1983; pp. 25–28.
8. Kim, S.K.; Kang, S.J.; Choi, Y.J.; Choi, M.H.; Hong, M. Augmented-Reality Survey: From Concept to Application. *KSII Trans. Internet Inf. Syst.* **2017**, *11*, 982–1004.
9. Onsel, I.; Donati, D.; Stead, D.; Chang, O. Applications of virtual and mixed reality in rock engineering. In Proceedings of the ARMA US Rock Mechanics/Geomechanics Symposium, ARMA, Seattle, WA, USA, 17–20 June 2018; p. ARMA-2018-798
10. Hu, H.Z.; Feng, X.B.; Shao, Z.W.; Xie, M.; Xu, S.; Wu, X.H.; Ye, Z.W. Application and prospect of mixed reality technology in medical field. *Curr. Med Sci.* **2019**, *39*, 1–6. [CrossRef] [PubMed]
11. De Armas, C.; Tori, R.; Netto, A.V. Use of virtual reality simulators for training programs in the areas of security and defense: A systematic review. *Multimed. Tools Appl.* **2020**, *79*, 3495–3515. [CrossRef]
12. Coelho, L.P.; Freitas, I.; Kaminska, D.U.; Queirós, R.; Laska-Lesniewicz, A.; Zwolinski, G.; Raposo, R.; Vairinhos, M.; Pereira, E.T.; Haamer, E.; et al. Virtual and augmented reality awareness tools for universal design: Towards active preventive healthcare. In *Emerging Advancements for Virtual and Augmented Reality in Healthcare*; IGI Global: Hershey, PA, USA, 2022; pp. 11–24.
13. Arena, F.; Collotta, M.; Pau, G.; Termine, F. An overview of augmented reality. *Computers* **2022**, *11*, 28. [CrossRef]
14. Kamińska, D.; Zwoliński, G.; Laska-Leśniewicz, A.; Raposo, R.; Vairinhos, M.; Pereira, E.; Urem, F.; Hinic, M.L.; Haamer, R.E.; Anbarjafari, G. Augmented Reality: Current and New Trends in Education. *Electronics* **2023**, *12*, 3531. [CrossRef]
15. Zwoliński, G.; Kamińska, D.; Haamer, R.E.; Coelho, L.F.; Anbarjafari, G. Enhancing empathy through virtual reality: Developing a universal design training application for students. *Med. Pr.* **2023**, *74*, 199–210. [CrossRef]
16. Li, J.; Fan, M.; Wang, G.; Li, X.; Sun, R. Panorama video stitching system based on VR Works 360 video. In Proceedings of the 2018 Chinese Automation Congress (CAC), Xi'an, China, 30 November–2 December 2018; pp. 715–720.
17. Allen, R.C. *The Effect of Restricted Field of View on Locomotion Tasks, Head Movements, and Motion Sickness*; University of Central Florida: Orlando, FL, USA, 2000.
18. Serrano, A.; Kim, I.; Chen, Z.; DiVerdi, S.; Gutierrez, D.; Hertzmann, A.; Masia, B. Motion parallax for 360 RGBD video. *IEEE Trans. Vis. Comput. Graph.* **2019**, *25*, 1817–1827. [CrossRef]
19. Kemeny, A. From driving simulation to virtual reality. In Proceedings of the 2014 Virtual Reality International Conference, Laval, France, 9–11 April 2014; pp. 1–5.

20. Benz, T.M.; Riedl, B.; Chuang, L.L. Projection displays induce less simulator sickness than head-mounted displays in a real vehicle driving simulator. In Proceedings of the 11th International Conference on Automotive User Interfaces and Interactive Vehicular Applications, Utrecht, The Netherlands, 21–25 September 2019; pp. 379–387.
21. Cho, H.j.; Kim, G.J. Roadvr: Mitigating the effect of vection and sickness by distortion of pathways for in-car virtual reality. In Proceedings of the 26th ACM Symposium on Virtual Reality Software and Technology, Ottawa, ON, Canada, 1–4 November 2020; pp. 1–3.
22. Lim, K.; Lee, J.; Won, K.; Kala, N.; Lee, T. A novel method for VR sickness reduction based on dynamic field of view processing. *Virtual Real.* **2021**, *25*, 331–340. [CrossRef]
23. Jasper, A.; Cone, N.; Meusel, C.; Curtis, M.; Dorneich, M.C.; Gilbert, S.B. Visually induced motion sickness susceptibility and recovery based on four mitigation techniques. *Front. Virtual Real.* **2020**, *1*, 582108. [CrossRef]
24. Yoon, H.J.; Moon, H.S.; Sung, M.S.; Park, S.W.; Heo, H. Effects of prolonged use of virtual reality smartphone-based head-mounted display on visual parameters: A randomised controlled trial. *Sci. Rep.* **2021**, *11*, 15382. [CrossRef] [PubMed]
25. Lim, C.H.; Lee, S.C. The Effects of Degrees of Freedom and Field of View on Motion Sickness in a Virtual Reality Context. *Int. J. Hum.—Comput. Interact.* **2023**, 1–13. [CrossRef]
26. Kirollos, R.; Merchant, W. Comparing cybersickness in virtual reality and mixed reality head-mounted displays. *Front. Virtual Real.* **2023**, *4*, 1130864. [CrossRef]
27. Goedicke, D.; Bremers, A.W.; Lee, S.; Bu, F.; Yasuda, H.; Ju, W. XR-OOM: MiXed Reality driving simulation with real cars for research and design. In Proceedings of the 2022 CHI Conference on Human Factors in Computing Systems, New Orleans, LA, USA, 30 April 30–5 May 2022; pp. 1–13.
28. McGill, M.; Li, G.; Ng, A.; Bajorunaite, L.; Williamson, J.; Pollick, F.; Brewster, S. Augmented, Virtual and Mixed Reality Passenger Experiences. In *User Experience Design in the Era of Automated Driving*; Springer International Publishing A&G: Cham, Switzerland, 2022; pp. 445–475.
29. Keshavarz, B.; Hecht, H.; Zschutschke, L. Intra-visual conflict in visually induced motion sickness. *Displays* **2011**, *32*, 181–188. [CrossRef]
30. Zielasko, D.; Meißner, A.; Freitag, S.; Weyers, B.; Kuhlen, T.W. Dynamic field of view reduction related to subjective sickness measures in an HMD-based data analysis task. In Proceedings of the IEEE VR Workshop on Everyday Virtual Reality, Reutlingen, Germany, 18 March 2018; Volume 2.
31. Bala, P.; Dionísio, D.; Nisi, V.; Nunes, N. Visually induced motion sickness in 360 videos: Comparing and combining visual optimization techniques. In Proceedings of the 2018 IEEE International Symposium on Mixed and Augmented Reality Adjunct (ISMAR-Adjunct), Munich, Germany, 16–20 October 2018; pp. 244–249.
32. Bos, J.E.; de Vries, S.C.; van Emmerik, M.L.; Groen, E.L. The effect of internal and external fields of view on visually induced motion sickness. *Appl. Ergon.* **2010**, *41*, 516–521. [CrossRef]
33. Bonato, F.; Bubka, A.; Thornton, W. Visual blur and motion sickness in an optokinetic drum. *Aerosp. Med. Hum. Perform.* **2015**, *86*, 440–444. [CrossRef]
34. Nie, G.Y.; Duh, H.B.L.; Liu, Y.; Wang, Y. Analysis on mitigation of visually induced motion sickness by applying dynamical blurring on a user's retina. *IEEE Trans. Vis. Comput. Graph.* **2019**, *26*, 2535–2545. [CrossRef] [PubMed]
35. Kala, N.; Lim, K.; Won, K.; Lee, J.; Lee, T.; Kim, S.; Choe, W. P-218: An approach to reduce VR sickness by content based field of view processing. In *Proceedings of the SID Symposium Digest of Technical Papers*; Wiley Online Library: San Francisco, CA, USA, 2017; Volume 48, pp. 1645–1648.
36. Bles, W.; Wertheim, A. *Appropriate Use of Virtual Environments to Minimise Motion Sickness*; Human Factors Research Inst TNO Soesterberg: Soesterberg, The Netherlands, 2001; 10p. Available online: https://apps.dtic.mil/sti/pdfs/ADP010785.pdf (accessed on 16 October 2023).
37. Teixeira, J.; Palmisano, S. Effects of dynamic field-of-view restriction on cybersickness and presence in HMD-based virtual reality. *Virtual Real.* **2021**, *25*, 433–445. [CrossRef]
38. Sakai, Y.; Watanabe, T.; Ishiguro, Y.; Nishino, T.; Takeda, K. Effects on user perception of a 'modified' speed experience through in-vehicle virtual reality. In Proceedings of the 11th International Conference on Automotive User Interfaces and Interactive Vehicular Applications: Adjunct Proceedings, Utrecht, The Netherlands, 22–25 September 2019; pp. 166–170.
39. Perrin, T.; Kerhervé, H.A.; Faure, C.; Sorel, A.; Bideau, B.; Kulpa, R. Enactive approach to assess perceived speed error during walking and running in virtual reality. In Proceedings of the 2019 IEEE Conference on Virtual Reality and 3D User Interfaces (VR), Osaka, Japan, 23–27 March 2019; pp. 622–629.

applied sciences

MDPI

Article

Impact of Navigation Aid and Spatial Ability Skills on Wayfinding Performance and Workload in Indoor-Outdoor Campus Navigation: Challenges and Design

Rabail Tahir * and John Krogstie

Faculty of Information Technology and Electrical Engineering, Norwegian University of Science and Technology, 7491 Trondheim, Norway; john.krogstie@ntnu.no
* Correspondence: rabail.tahir@ntnu.no

Abstract: Wayfinding is important for everyone on a university campus to understand where they are and get to where they want to go to attend a meeting or a class. This study explores the dynamics of mobile navigation apps and the spatial ability skills of individuals on a wayfinding performance and perceived workload on a university campus wayfinding, including indoor-outdoor navigation, by focusing on three research objectives. (1) Compare the effectiveness of Google Maps (outdoor navigation app) and MazeMap (indoor-outdoor navigation app) on wayfinding performance and perceived workload in university campus wayfinding. (2) Investigate the impact of participants' spatial ability skills on their wayfinding performance and perceived workload regardless of the used navigation app. (3) Highlight the challenges in indoor-outdoor university campus wayfinding using mobile navigation apps. To achieve this, a controlled experiment was conducted with 22 participants divided into a control (using Google Maps) and an experiment group (using MazeMap). Participants were required to complete a time-bound wayfinding task of navigating to meeting rooms in different buildings within the Gløshaugen campus of the Norwegian University of Science and Technology in Trondheim, Norway. Participants were assessed on spatial ability tests, mental workload, and wayfinding performance using a questionnaire, observation notes and a short follow-up interview about the challenges they faced in the task. The findings reveal a negative correlation between overall spatial ability score (spatial reasoning, spatial orientation, and sense of direction) and perceived workload (NASA TLX score and Subjective Workload Rating) and a negative correlation between sense of direction score and total hesitation during wayfinding task. However, no significant difference was found between the Google Maps and the MazeMap group for wayfinding performance and perceived workload. The qualitative analysis resulted in five key challenge categories in university campus wayfinding, providing implications for designing navigation systems that better facilitate indoor-outdoor campus navigation.

Keywords: indoor-outdoor navigation; wayfinding; campus navigation; mobile navigation app; spatial ability skills

1. Introduction

Wayfinding and navigation are essentials for everyday experiences, whether we need to find our way to university, the grocery store, hiking up a mountain, or going to a new location [1,2]. Wayfinding is the ability to navigate through the environment by identifying one's location and arriving at destinations both cognitively and behaviorally [1]. The wayfinding task can be aided by providing environmental information cues that are helpful in building up important knowledge for the required wayfinding task. Such knowledge is classified into two categories: knowledge in the head and knowledge in the world, and wayfinding aids are useful to acquire information for transforming the knowledge in the world into knowledge in the head [3]. Several theories (such as route-based strategies and landmark-based strategies) and mechanisms (such as images or dual coding) have been studied to understand how knowledge is learned, structured and stored [4].

In general, all research focusing on navigation in the physical environment can be mainly divided into (1) outdoor navigation, (2) indoor navigation, and (3) a combination of both [4,5]. Outdoor navigation systems are already present everywhere and are widely used [6,7]. However, large campus buildings with connected facilities require indoor navigation aids. Therefore, in recent years, indoor navigation has become equally important and is a central challenge as it is a less developed area [6–9]. One of the most popular projects for navigation is Google Maps, but this is limited to outdoor services. This is because digital navigation solutions such as Google Maps are based on GPS signals, aerial photography, and satellite imagery that do not cover the inside of buildings, meaning a different service (indoor positioning system (IPS)) is needed to facilitate indoor navigation. Indoor navigation intends to help users find their way, avoid obstacles, recognize people, and find objects and resources, among others [5]. Several technologies such as Bluetooth, radio-frequency identification, Wi-Fi, magnetic and even LED light-based are used to solve the indoor navigation problem [10]. Some specialized solutions are available for indoor navigation, but the services are only for limited buildings [7]. With the advent of technology, working environments such as organizations or universities are becoming complex and consist of multiple buildings [3]. Therefore, finding one's way around becomes an additional challenge on top of work and study and requires integration of indoor and outdoor navigation [10–13] to provide seamless navigation from the start point to the point of interest within a building, saving time and energy. Although navigation tools have become more prevalent over the last decade as a resource for reliable wayfinding and route planning, finding systems that integrate indoor to outdoor navigation is a challenge in navigational research [14]. MazeMap is one such digital wayfinding solution that seamlessly integrates indoor and outdoor navigation, allowing users to move between buildings, e.g., on a university campus, with ease. Previous research has also focused on the comparison between different navigation aids to examine their effectiveness. For example, a comparison between mobile navigation systems and paper maps [15], a comparison between Electronic Map Interfaces (Google Maps and Yahoo! Map interfaces) [16], Novel map displays such as Mirror-In-The-Sky Map (MitS) and track-up maps [17]. However, among several research studies conducted on navigational aids, only a few are devoted to outdoor vs. indoor-outdoor navigation systems. Although there are several studies on outdoor navigation systems, only a few studies have focused on the human factors of indoor navigation systems, and there is a strong need to evaluate them [6].

Many factors contribute to the wayfinding ability and performance, including both internal (such as age, gender, types of strategies, sense of direction, comprehension, spatial abilities, etc.) and external (density of buildings, availability of meaningful landmarks, etc.) [1,18–20]. However, there is little consensus in the literature regarding the most important factors for predicting wayfinding ability and how to measure it [1]. Many researchers have investigated factors focusing on components of spatial ability, such as mental rotation skill tests, spatial visualization and spatial relations tasks [21]. Spatial ability is the ability to precisely visualize and recognize the visual-spatial world and the cognitive processes involved to achieve transformations (such as locating a target in space, directional relationships, perceiving distance, and mentally transforming objects based on orientation in space) on those perceptions. Spatial abilities are applied in everyday activities while finding a way [22]. According to research, individuals differ in their ability to process and visualize spatial information [23]. Some studies reported that people who have a good sense of direction (SOD) are faster wayfinders, show better performance, make fewer errors and make flexible use of strategies compared to people who have bad SOD [20,24]. Moreover, research has found that self-reported SOD measures quite highly predict the objective measures of these abilities [25]. However, only a few research studies have explored the relationship between spatial ability and wayfinding in the real world, and these studies have shown mixed results [21,26]. Overall, these studies can be grouped into self-assessment (in which researchers ask participants about their ability) studies and behavioral studies (in which researchers ask participants to solve tasks), with very few studies that have examined wayfinding tasks in the real environment [2]. More detailed research is required to investigate different variables in spatial ability skills and wayfinding and understand their influence on wayfinding performance to design more legible wayfinding systems [22].

Due to technological advancements, there is an increased reliance on navigation systems for wayfinding, but research focusing on the use, effects and evaluation of these approaches is still limited [22]. Therefore, in this study, we aim to examine Google Maps (an outdoor navigation app) in comparison with MazeMap (an indoor-door navigation app) for wayfinding on a university campus. The study focuses on understanding the role of navigation aids and spatial ability skills on perceived workload and wayfinding performance in indoor-outdoor campus navigation tasks. Moreover, the study also aims to identify the challenges in the use of current navigation systems for wayfinding in complex environments such as large university campuses requiring indoor-outdoor navigation. These results could provide empirical evidence and insights supporting the future design of integrated indoor-outdoor navigation systems. To achieve this, the study aims at answering the following research questions.

RQ1. Does indoor-outdoor navigation aid improve university campus wayfinding performance and reduce workload in comparison with outdoor-only navigation systems?
RQ2. Do spatial ability skills of wayfinders affect workload and wayfinding performance in university campus navigation?
RQ3. What are the main challenges in indoor-outdoor university campus navigation using mobile navigation aids?

The rest of this paper is structured as follows: Section 2 presents the related work. Section 3 describes the method used for this study. Section 4 presents the results from the wayfinding experiment concerning the three research questions. Section 5 discusses the results, presents design implications, and highlights the limitations of the study. Lastly, Section 6 concludes the paper and gives directions for future research.

2. Related Work

In this section, we provide an overview of relevant studies on wayfinding systems, indoor-outdoor navigation and the role of spatial ability and workload in wayfinding tasks that ground and guide our research. As far as we know, there are no other studies published that look at the relation between spatial ability, workload, and performance in wayfinding tasks for indoor-outdoor navigation. Moreover, there are only a few studies that compare the performance of outdoor-only navigation systems with indoor-outdoor navigation systems for wayfinding in urban environments, such as university campuses combining indoor and outdoor spaces.

2.1. Wayfinding and Navigation Systems

Wayfinding and navigation systems are important for improving people's quality of life. A number of applications are used by people on a daily basis to help them in moving across the urban environment and within buildings by providing them with useful information and detecting their position [27]. Outdoor navigation systems use global positioning systems (GPS) to track the location of the user, which are not effective in an indoor environment; therefore, other technologies (such as WiFi, Bluetooth, Beacons or RFID, etc.) are needed to track the location of the user in indoor areas [11,28].

A large body of research has been conducted to study navigation in outdoor environments. While outdoor navigation systems have been used extensively and are already represented everywhere, navigation is much less developed in the enclosed space and therefore, indoor navigation systems are still in the early stages of development [6,7,29]. Huang and Gartner [29] proposed an evaluation framework combining the key aspects of indoor navigation to investigate mobile indoor navigation systems. The study also presents a comparison of current mobile indoor navigation systems based on the proposed framework and concludes that "indoor navigation systems are still in an early development stage". Rehman and Cao [6] presented an augmented-reality application to assist people with indoor navigation. This research study particularly examined a wearable head-mounted device (Google glass) compared to handheld navigation aid such as a paper map and smartphone, focusing on technical assessment (including reliability and feasibility of the system) and human factors (including navigation time, route memory retention, perceived accuracy, subjective workload, and subjective comfort). The results indicated that the digital navigation aids (both wearable devices and smartphones) were better than the paper map with regard to lower workload and shorter navigation time. However, they were worse in terms of route retention. Moreover, the wearable device was not significantly different from the smartphone based on performance and workload but was perceived to be more accurate.

2.2. Wayfinding Focusing Indoor-Outdoor Navigation

Complex working environments, such as organizations with large campuses and multiple buildings, require the integration of indoor and outdoor navigation [10–12]. Although several studies have been conducted on navigation and wayfinding, more research focusing on combined indoor-outdoor navigation is required to meet the needs of wayfinding in complex urban environments [29].

Only recently, researchers have started to focus on combined indoor-outdoor navigation. For instance, Giudice et al. [9] reported on the development of a research plan for integrating outdoor and indoor spaces and discussed potential application domains and a variety of models (from formal models to functional models) related to the usability of indoor and indoor-outdoor information systems. Prandi et al. [27] presented a systematic mapping study focusing on software applications and devices intended to foster accessible wayfinding and navigation in outdoor and indoor environments [27]. Ren et al. [30] proposed an integrated navigation method (using differential GNSS and Lidar SLAM) to meet the needs of indoor and outdoor navigation. Similarly, Jiang et al. [31] described an integrated navigation system to support seamless indoor-outdoor navigation based on Global Navigation Satellite Systems (GNSS), Inertial Navigation Systems (INS), and Terrestrial Ranging Techniques. The proposed system was evaluated by conducting an indoor-outdoor test, and results showed that it could provide reliable and continuous position and attitude solutions. Moreover, Vanclooster and Maeyer [14] studied the impact of combining indoor and outdoor navigation on route planners by reviewing several case studies in multiple route planners. The study revealed different aspects and requirements, such as data constraints (preventing optimal use of navigation routes) and addressed matching methodologies (the exit influencing the exit choice of the building) for integration of indoor with outdoor connection in wayfinding. Furthermore, Pfaff et al. [32] proposed an approach based on multi-level surface maps for navigation in combined outdoor and indoor environments. The approach does not rely on GPS information, and the presented technique allows the robot to actively explore the environment [32]. Nikander et al. [11] presented a prototype for an indoor and outdoor mobile navigation system using floor plans together with street maps for route finding and displaying a hybrid map for a university campus. The prototype is implemented on iPad, but the work is still under process; GPS is used for outdoor positioning, and QR codes are used for indoor positioning.

2.3. University Campus Navigation

The route planning functions in current navigation systems in the market do not efficiently process the outdoor and indoor localization information simultaneously. Moreover, these systems also have limitations by the service platforms which compromises the user interface (UI) experiences [12]. Torres-Sospedra et al. [10] focused on mobility in smart campuses and developed a novel combined indoor and outdoor positioning system to support wayfinding with seamless indoor-outdoor navigation for the Universitat Jaume I

(Spain). They further developed and evaluated two mobile applications (SmartUJI APP and SmartUJI AR) that allow users to get map-based information about different campus facilities and provide an augmented reality interface to interact with the campus. The applications were tested by university staff, students and visitors, and they found it useful for locating university facilities and improving spatial orientation. Similarly, Zhou et al. [12] presented an augmented reality-based campus navigation system and conducted indoor and outdoor navigation experiments at the campus of Shanghai University of Electric Power to compare the proposed system with traditional systems, such as Internet media or Gaode map. The results showed an increased precise outdoor localization and improved the user's interactive experience facilitating the usability and effectiveness of learning on campus. The Norwegian University of Science and Technology (NTNU) partnered with Wireless Trondheim to develop MazeMap (earlier known as CampusGuide [33]) to provide seamless indoor-outdoor navigation initially for its main Gløshaugen campus [8], which is now used on a large number of campuses and universities [34]. Biczok et al. [8] conducted a campus case study presenting insights from MazeMap reflecting on the spatial and temporal distribution of MazeMap user requests (geo-location and wayfinding), mapping user mobility patterns of the campus, and spatial-logical connection between different locations in different buildings. The results from their study highlighted the potential of such systems for usage on campus and for other venues such as hospitals, shopping malls, and emergency preparedness.

Several researchers have focused on the use of different navigation aids for wayfinding across campus environments. Schnitzler and Hölscher [18] compared user experience and wayfinding strategy choices for navigation through a complex building using three different types of wayfinding aid (without any map, with a printed map, and with a digital map). According to the results, there was a significant difference between the two map groups and the no map group with respect to search time. However, there were no significant differences between the two map groups in how efficiently they found destinations. Moreover, wayfinding strategy choices did not differ between the groups, but the choices were shaped by the participants' individual preferences and the building with its signage. Cheung [35] investigated the use of different representational options for providing visitors with navigational cues to the site (Auckland University Campus). Wayfinding performance data was obtained from participants for prototyped navigation systems of different representations. The advantages and disadvantages of different systems were identified to find appropriate approaches for designing an interactive system. Kim et al. [36] explored the use of digital signage for improved wayfinding performance. The study evaluated the campus information digital signage, which can be useful for campus visitors as a wayfinding tool. The results showed that digital signage is better than traditional or paper signage and indicated that the integration of these methods could highlight effective content and design features to help design user-centered wayfinding digital signage. Asif and Krogstie [37] investigated and evaluated the use of a mobile student information system (MSIS) to provide relevant information to students to enhance their campus life. The results showed a high intention to use such services, especially providing information based on students' schedules, interests, and localization. Gao et al. [38] explored the adoption of indoor location-based services in Norway using Technology Acceptance Model (TAM). The results of the survey showed a strong influence of perceived usefulness on the intention to use, whereas no positive effect of trust was found on intention to use.

2.4. Spatial Ability and Workload in Wayfinding Task

Different researchers have investigated the relationship between spatial ability and wayfinding, focusing on either outdoor or indoor navigation. Malinowski [21] investigated the relationship between mental rotation skills (using the Vandenburg Kuse Mental Rotations test) and college students' wayfinding performance in an outdoor navigation task. The results showed a significant correlation between mental rotation skills and wayfinding performance on a 6-km orienteering task. The study also highlighted sex differences in spatial ability. Similarly, Nori et al. [26] examined the impact of visuo-spatial ability (V-SA) on wayfinding performance. The wayfinding task focused on a route (approximately 360 m long) in the botanical garden of the university. According to the results of their study, participants with high V-SA had better wayfinding performance and different wayfinding behavior than participants with low V-SA. Participants with low V-SA hesitated more during traveling even if traveling time was the same. Moreover, participants with low V-SA adopted a landmark strategy to navigate, whereas people with high V-SA adopted a route or survey strategy. Verghote et al. [23] conducted an experiment to analyze the influence of spatial cognition and different information formats (two-dimensional (2D) drawings or a three-dimensional (3D) model) on individual wayfinding in an unfamiliar environment. The wayfinding task focused on indoor navigation from the starting point to checkpoint A and then to checkpoint B within the building. The results showed that 3D models have a positive impact on individual wayfinding. Moreover, participants with a low spatial cognition had a significantly improved success rate when using a 3D model but the success rate of participants with a high spatial cognition was not affected by the information format. However, more research is needed to confirm whether information format such as 3D effectively reduces cognitive demand [23]. Although there are many studies that examine the relationship between spatial ability and wayfinding performance, only a few focus on complex environments with combined indoor-outdoor navigation [39].

Mental workload is another important measurement and subject of interest for wayfinding and navigation tasks. Researchers have investigated the effects on users' mental workload and task performance when navigating through indoor or outdoor environments using different technologies or evaluating the usability of technologies in many application areas [17,40]. Vasquez et al. [17] investigated the effect of practice with Mirror-in-the-Sky (MitS) and track-up maps on navigation performance and mental workload (MWL) in a simple navigation task in a virtual environment. According to the results of a three-session experiment, secondary task performance improved, and task completion times, collisions and subjective MWL decreased with practice. The decrease in MWL was larger with MitS as compared with the track-up map. However, the differences between the track-up map and MitS decreased and were negligible as sessions progressed. Lin and Wang [16] studied the impact of four designing factors (map type, map size, zoom function and direction key) as well as gender differences and sense of direction (SOD) ability on the visual wayfinding performance for a simulated electronic map interface for three wayfinding tasks. The research showed that participants with higher SOD scores had faster response times and lower overall workload for the target task. As the response time of participants increases, they experience a higher workload. Moreover, for target and direction tasks, the interaction effect of map type and SOD affect the mean response time. The male participants with good SOD, using a mixed map, and good SOD using E-map had faster mean response times. Whereas females using E-map had faster mean response time. Bradley and Dunlop [4] investigated the difference in mental and physical demands between a group of sighted and visually impaired participants when provided with two different verbal instruction sets guiding them to four different landmarks. The two verbal instructions were a set of route descriptions: the first derived from sighted people, and the second set derived from visually impaired people. The study focused on navigation time, number of deviations and workload. The results showed that instructions derived from visually impaired people resulted in quicker time, fewer deviations and lower workload scores for visually impaired participants but caused higher workload scores for the sighted participants. It is important to understand if some navigation aids are more intuitive than others, but it is also crucial to investigate if the effects on task performance are resultant of the effect on mental workload, and this area is less explored [17,40].

3. Materials and Methods

This section presents the participants of this study, the experimental design describing research context and procedures, the digital navigation platforms used for the experiment, the data sources, and the methods used for data analysis in this wayfinding study.

Wayfinding is important for students, faculty, staff, visitors, and anyone else on a university campus to understand where they are and where they want to go to attend a meeting or a class. Navigation systems are ubiquitous tools to assist wayfinders with various navigational tasks, thus reducing human effort for navigating. In this study, we conducted a field experiment using control and experiment groups to evaluate users' wayfinding performance and workload in an indoor-outdoor campus navigation when using Google Maps (an outdoor-only navigation app) in comparison with MazeMap (an indoor-outdoor navigation app) for wayfinding on a university campus. In addition to understanding the role of navigation aids, this study also explores the effects of spatial ability skills on perceived workload and wayfinding performance in campus wayfinding tasks and identifies challenges for designing efficient indoor-outdoor navigation systems for wayfinding in complex environments such as large university campuses. This study uses both quantitative and qualitative methods to address the three research questions presented in the introduction. The following hypotheses were formulated for the first two research questions:

Hypothesis (H1). *Indoor-outdoor navigation app group will show better wayfinding performance as compared to the outdoor-only navigation app group in the university campus wayfinding task.*

Hypothesis (H2). *Indoor-outdoor navigation app group will experience less workload as compared to the outdoor-only navigation app group in the university campus wayfinding task.*

Hypothesis (H3). *Participants who have better spatial ability skills will experience less workload in the university campus wayfinding task.*

Hypothesis (H4). *Participants who have better spatial ability skills will have better wayfinding performance in the university campus wayfinding task.*

3.1. Participants

For this experiment, twenty-two participants were recruited during March and April 2022. Out of the 22 subjects, 5 were males (23%) and 17 females (77%) who ranged in age from 24 to 43 (mean = 30.8 years, SD = 5.9). This research focused on wayfinding tasks, including indoor-outdoor navigation at the Gløshaugen campus of the Norwegian University of Science and Technology (NTNU) in Trondheim, Norway. The Gløshaugen campus is the largest campus of NTNU and covers 350,000 sqm with more than 60 buildings and 13,000 rooms [8]. The participants for this experiment were students, permanent and temporary staff, visitors, and others on the university campus, as wayfinding is important for them to understand where they are and how to get to where they want to go, such as attending a meeting or a class. The participants included 14 temporary staff (63.6%), 5 students (22.7%), and one of each: permanent staff, visitor and other (4.5% each). The participants had different academic backgrounds; 13 had a master's degree (59.1%), 5 had a bachelor's degree (22.7%), and 4 had a PhD (18.2%). All participants had experience using outdoor navigation apps, e.g., Google Maps, and most of them (95.5%) reported frequent use. On the contrary, all except one participant had experience using indoor-outdoor navigation apps, e.g., MazeMap, but most of them (77.3%) reported that they rarely use them, and only 4 participants (18.2%) reported frequent use. None of the participants had already seen the rooms mentioned in the wayfinding task in the experiment.

The participants were contacted through the university mailing list and intranet channel. This study was organized in multiple sessions conducted over two weeks, from 24 March 2022 to 7 April 2022. The researcher explained the study and obtained consent from participants for data collection. They were told that their participation was voluntary, that they could withdraw their consent at any time without giving a reason, and that all information about them would then be made anonymous. The participants were compensated with a gift card as a token of appreciation for their time and effort. The experiment was conducted on the Gløshaugen campus at the Norwegian University of Science and Technology (NTNU) Trondheim, Norway. No prior knowledge or preparation was required; participants just came with their smartphones or were provided with one.

3.2. Experiment Design

The wayfinding experiment design is illustrated in Figure 1. The experiment took approximately one hour and had three phases (as shown in Figure 1):

1. Pre-Wayfinding Task Phase (15–20 min): Before the task, all participants read and signed a consent form and filled out a demographic questionnaire. The participants were also required to complete the Santa Barbara Sense of Direction Scale (SBSOD) [25] and two spatial ability tests to evaluate the sense of direction, spatial orientation and spatial reasoning (see Appendix A) of each individual.
2. Wayfinding Task (30–35 min): This phase focuses on performing the actual wayfinding task. The study used an independent group design, and participants were divided into two groups (control and experiment) of eleven participants each. Participants were randomly assigned to either the control or the experiment group and were required to complete a wayfinding task using either an outdoor or indoor-outdoor navigation app, depending on the assigned group. Both groups were provided with a clear set of instructions concerning the experiment and performed the same wayfinding task. The only difference was that the control group used an outdoor navigation app (Google Maps), and the experiment group used an indoor/outdoor navigation app (MazeMap) to complete the same wayfinding task. Each participant was tested individually.

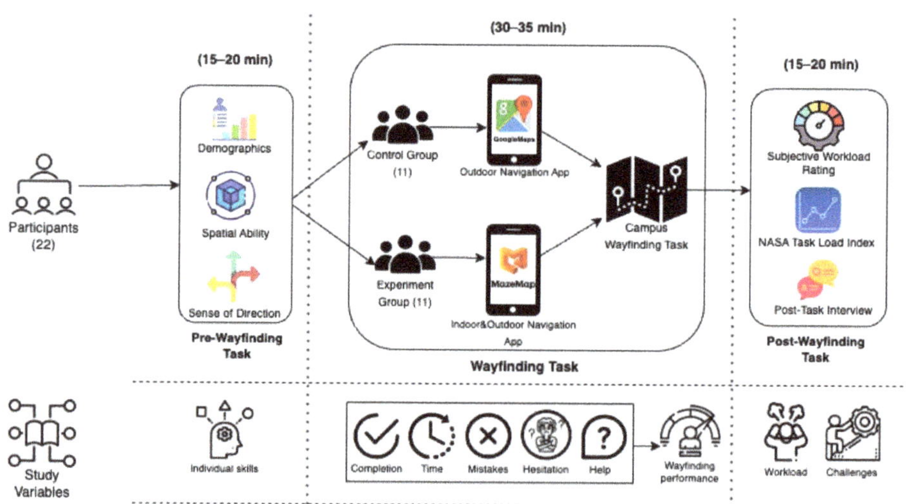

Figure 1. Wayfinding experiment design.

The wayfinding task selected for the experiment included navigating to two different meeting rooms in a specified order within the campus (see Appendix B for details of the task). The wayfinding task routes were all located on the Gløshaugen campus of the Norwegian University of Science and Technology (NTNU) in Trondheim, Norway (shown in Appendix B). The starting points and destinations were pre-defined. The participants were instructed to find their way to the destinations from the starting point by referring to the mobile navigation app (Google Maps or MazeMap), depending on their designated group.

The task finished when the participant reached the second venue. Each participant individually performed this task where an observer followed them silently (without intervening) to take field notes (recording wayfinding performance parameters, i.e., task completion, time, mistakes, hesitation, and help) and observations. The participants had 30 min to complete this task. The task scenario and route maps are presented in Appendix B.

3. Post-Wayfinding Task Phase (15–20 min): After completing the wayfinding task, participants were asked to complete mental workload questionnaires (NASA-TLX rating and subjective workload questionnaire) and a short follow-up interview.

3.3. Digital Mobile Navigation Apps Used

Two digital navigational aids were used for the experiment: Google Maps was adopted as an outdoor-only navigation aid, and MazeMap was used as an indoor-outdoor navigation aid. Participants used a smartphone to run the application for navigation during the wayfinding task.

Google Maps is one of the most popular navigation aids and enables people to navigate to their desired destination by providing the shortest and most convenient route. However, Google Maps is only limited to outdoor navigation. It has many helpful features, such as street view and the location of nearby cafes, hospitals and police-stations [41]. MazeMap is an indoor-outdoor positioning and navigation system. It can locate the position of the user on campus and provides room-level wayfinding and object search capabilities. Moreover, when the user is outdoors, MazeMap is also able to use standard GPS positioning, and the system is also able to combine WiFi and GPS signals if available. The system uses construction drawings on the user interface side to present the users with readable maps and interprets them to recognize different objects. The user can choose what to hide or show so the maps are not overloaded with unnecessary technical data [8].

3.4. Data Sources

The data sources used in this study included the following:

- Demographic questionnaire: This included questions related to bio-demographics (such as age, gender, education, and role) and experience using outdoor and indoor-outdoor navigation tools.
- Spatial ability test: The measurement of spatial ability skills depends on multiple factors focusing on multiple spatially relevant abilities. While different researchers have focused on different factors, the two that are most often cited include spatial orientation (also referred to as spatial relations) and spatial visualization (Spatial reasoning) [42]. Therefore, we included two tests, a "spatial reasoning test" and a "spatial orientation test", to assess participants' visuo-spatial ability. A spatial reasoning test (shown in Appendix A) was used to determine the participant's ability to manipulate 2D and 3D objects, visualize movements and changes between shapes, and spot patterns between those shapes. This test consisted of 9 questions, each consisting of abstract shapes, which participants had to mentally manipulate and choose the correct option out of the four choices. There is a limit of 40 s for each question. The spatial orientation test was used to test participants' ability to imagine different perspectives or orientations in space. The test used for this study was adopted from the version used by [43]. There were 12 items in this test, one on each page. Each question contained a picture of an array of objects and an "arrow circle" with a question about the direction between some of the objects. The task was to draw an arrow from the center object showing the direction to a third object from the facing orientation. Participants had 5 min to complete this test.
- Santa Barbara Sense of Direction scale (SBSOD): According to [44], three substantially correlated factors emerged after examining the structure of spatial ability, and this included navigation in addition to object manipulation and visualization. Therefore, we used the Santa Barbara Sense of Direction Scale (SBSOD) scale [25] to access navigation. This is a self-report scale of environmental spatial ability and is a useful instrument for

measuring the construct of Sense of Direction (SOD). It contained 15 statements asking about participants' spatial and navigational abilities, experiences, and preferences. The participants had to rate each statement that describes the participant's sense of direction on a 7-point Likert scale from "1" (strongly agree) to "7" (strongly disagree).

- NASA Task Load Index (TLX) rating: NASA-TLX [45,46] is a multi-dimensional scale that provides an overall workload score using six workload-related factors consisting of mental demand, physical demand, temporal demand, performance, effort and frustration). This study aimed to use NASA-TLX as an objective measure of the workload of the wayfinding task [47].
- Subjective workload rating (SWR): In addition to the NASA-TLX, a second instrument was used as a subjective measure of workload. A Subjective Workload Rating (SWR) questionnaire developed by [42] was administered to all participants to evaluate how they felt about the wayfinding task on a scale of 1 to 10. This instrument consists of the following sub-scales: Easy/difficult, relaxing/stressful, required minimal effort/labor intensive, pleasant/annoying, calming/frustrating and simple/complex.
- Field Notes: This was used to record the parameters of wayfinding performance [48] and observation notes. The following parameters were recorded for the wayfinding task: start time, end time, number of times asked for help/assistance, took a wrong direction (e.g., turned back after taking a direction), frequency of hesitations (i.e., every time participant stopped to think about the way to take), task completion and observation notes regarding campus navigation during the task.
- Post-Task Interview: The interview consisted of three questions related to their experience, challenges they faced, and suggestions to mitigate such challenges and improve wayfinding applications. The questions were:
    1. How was your experience with this wayfinding task? (e.g., were you stressed/relaxed, was it easy/difficult, asked for help for navigating to the meeting rooms, etc.)
    2. Describe the challenges you faced during the wayfinding task?
    3. What would have helped you in completing this wayfinding task more efficiently? Give suggestions for improving the navigation tool.

3.5. Data Analysis

The IBM SPSS Statistics v28 software was used for the quantitative data analysis (RQ1–2) and NVivo software was used for the qualitative data analysis (RQ3). We conducted a *t*-test to examine any potential difference in the Wayfinding performance (Task time, task completion, mistakes, hesitation, help) and workload of participants in the control and experiment groups. Spearman correlation was used to identify any potential correlation between spatial ability skills, workload, and wayfinding performance. The qualitative data collected through interviews and observation notes was analyzed and interpreted through grounded theory approach proposed by Gioia et al. [49]. This resulted in a data structure for challenges and identified key issues that hindered the successful and satisfactory usage of navigational apps for indoor-outdoor navigation in large university campuses. Finally, based on the findings, some design recommendations were proposed for designing efficient navigation tools for campus wayfinding.

4. Results

This section presents the results from the wayfinding experiment to answer the three research questions presented in the introduction section. The first two questions are addressed with quantitative data analysis using statistical tests. We reported the differences in the wayfinding performance and workload between the control and experiment groups. Secondly, we looked at the effect of spatial ability skills on the workload and wayfinding performance of participants regardless of their group in indoor-outdoor navigation tasks. We will explore four hypotheses (H1–H4 presented in Section 3) in this study. The first two are linked to research question one, and the next two are related to research question two.

The third question is addressed with qualitative data analysis and identifies the challenges and issues faced by participants when performing the indoor-outdoor university campus wayfinding task using mobile navigation apps.

4.1. RQ1: Indoor-Outdoor vs. Outdoor Only Mobile Navigation Aid in Campus Wayfinding (Performance and Perceived Workload)

This research question investigated how the participants' wayfinding performance and perceived workload in campus navigation tasks using the provided navigation aid (outdoor navigation app vs. indoor-outdoor navigation app) differs between the control (using Google Maps) and the experiment group (using MazeMap). This research question focused on hypotheses H1 and H2.

The null hypothesis for H1 was that there was no difference between the wayfinding performance of participants in the control (outdoor navigation app: Google Maps) and experiment (indoor-outdoor navigation app: MazeMap) groups.

The wayfinding performance is measured in terms of total task duration, total help, total wrong directions, total hesitation, and total task completion. An independent *t*-test was conducted to compare the wayfinding performance between the control and experiment groups. The independent variable was the groups with different navigation aid: outdoor navigation app vs. indoor-outdoor navigation app, and the dependent variables were wayfinding performance parameters: total task duration, total help, total hesitation, total wrong directions, and total task completion. The *t*-test is appropriate for our experiment as it is used to study if there is a statistical difference between two independent groups. The results from the *t*-test are shown in Table 1.

Table 1. Independent *t*-test to compare wayfinding performance between the control and experiment groups.

				Levene's Test for Equality of Variances		*t*-Test for Equality of Means						
											95% Confidence Interval of Difference	
		Mean	SD	F	*p*-Value	t	df	*p*-Value	Mean Diff.	Std. Error Diff.	Lower	Upper
Total task duration	CG EG	24.82 22.00	7.561 8.390	0.015	0.902	0.828	20	0.418	2.818	3.405	−4.285	9.922
Total help	CG EG	1.18 0.55	1.168 0.934	1.107	0.305	1.411	20	0.174	0.636	0.451	−0.304	1.577
Total wrong directions	CG EG	2.00 2.19	1.414 1.514	0.107	0.747	−0.146	20	0.886	−0.091	0.625	−1.394	1.212
Total hesitation	CG EG	2.27 1.64	1.555 1.120	3.805	0.065	1.101	20	0.284	0.636	0.578	−0.569	1.842
Total task completion	CG EG	1.64 1.73	0.674 0.647	0.217	0.646	−0.323	20	0.750	−0.091	0.282	−0.678	0.497

Significance level, $p < 0.05$. Mean for total task duration is in minutes and number of times for others. For task completion, a score of 0 was given if no meeting room was found, 1 if one meeting room was found and 2 if both meeting rooms were found.

According to the results, no significant difference was found in wayfinding performance between the control and experiment groups. There were no significant differences in total task duration (t (20) = 0.828, $p = 0.418$); total help (t (20) = 1.411, $p = 0.174$); total wrong directions (t (20) = −0.146, $p = 0.886$); total hesitation (t (20) = 1.101, $p = 0.284$); and total task completion (t (20) = −0.323, $p = 0.750$) between the two groups ($p \geq 0.05$), failing to reject the null hypothesis. Despite this, the mean task duration (M = 24.82, SD = 7.561) for the control group (CG) was higher than the experiment group (EG) (M = 22.00, SD = 8.390). Similarly, the mean help (M = 1.18, SD = 1.168) and hesitation (M = 2.27, SD = 1.555) for the control group (CG) was higher than the mean help (M = 0.55, SD = 0.934) and hesitation (M = 1.64, SD = 1.120) for the experiment group (EG). However, the magnitude of the difference in the means for total task duration (mean difference = 2.818, 95% CI: [−4.285, 9.922]), total help (mean difference = 0.636, 95% CI: [−0.304, 1.577]), and total hesitation (mean difference = 0.636, 95% CI: [−0.569, 1.842]) was not significant. Hence, H1 was not supported.

The null hypothesis for H2 was that there was no difference between the perceived workload of participants in the control (Google Maps) and experiment (MazeMap) groups.

The perceived workload is measured using the NASA TLX score and subjective workload rating (SWR) score. Again, an independent *t*-test was conducted to compare the perceived workload between the control and experiment groups as it is used to study if there is a statistical difference between the two independent groups. The independent variable was the groups with different navigation aid (outdoor navigation app vs. indoor-outdoor navigation app). The dependent variables were the NASA TLX score and SWR score. The results from the *t*-test for H2 are shown in Table 2. According to the results, there was no significant difference in NASA TLX score (t (20) = −0.429, $p = 0.673$) and SWR score (t (20) = 0.706, $p = 0.488$) between the two groups ($p \geq 0.05$), failing to reject the null hypothesis. Hence, H2 was also not supported.

Table 2. Independent *t*-test to compare perceived workload between control and experiment group.

				Levene's Test for Equality of Variances		*t*-Test for Equality of Means						
		Mean	**SD**	**F**	***p*-Value**	**t**	**df**	***p*-Value**	**Mean Diff.**	**Std. Error Diff.**	**95% Confidence Interval of Difference**	
											Lower	**Upper**
NASA TLX score	CG EG	40.23 43.71	22.064 15.498	0.616	0.442	−0.429	20	0.673	−3.485	8.130	−20.443	13.473
SWR score	CG EG	5.02 4.45	1.849 1.874	0.007	0.932	0.706	20	0.488	0.561	0.794	−1.095	2.216

Significance level, $p < 0.05$.

4.2. RQ2: Effect of Spatial Ability Skills on Campus Wayfinding (Performance and Perceived Workload)

This research question investigated the effect of spatial ability skills on the perceived workload and wayfinding performance of all participants in indoor-outdoor campus navigation tasks regardless of their group (i.e., used navigation aid). The spatial ability skills of all participants were assessed before conducting the navigation task using mobile navigation aid. After the first phase (pre-task phase, see Figure 1), participants were randomly assigned to the control and experiment groups. The descriptive statistics presented in Table 3 show that the spatial ability skill parameters of participants in both groups are fairly equal, which shows that results in the previous section were not affected by the spatial ability skills of participants. The spatial ability skills were measured on three parameters spatial reasoning test, spatial orientation test and navigation (using the Santa Barbara Sense of Direction (SBSOD) scale). The scores of these three factors were added to get an overall score for spatial ability skills.

Table 3. Descriptive statistics for control and experiment groups.

Descriptive Statistics	Control Group	Experiment Group
Age	29.54	31.64
Male	2	3
Female	9	8
Spatial reasoning	6.45	6.73
Spatial orientation	7.27	8.09
SBSOD	3.96	4.16

A series of Pearson's correlations were conducted to identify any potential correlation between the participant's spatial ability skills parameters (spatial reasoning score, spatial orientation score and SBSOD score), their perceived workload (NASA TLX score and SWR score), and wayfinding performance parameters (task duration, task completion, total help, total hesitation, and total wrong directions). Pearson's correlation measures the strength and direction of the linear relationship between two variables. Pearson's correlation was used to determine whether the variables of interest are correlated or related to each other, and the data met all the assumptions required for using Pearson's correlation. All results for Pearson's correlation are presented in Tables 4 and 5. This research question focused on hypotheses H3 and H4.

Table 4. Pearson's correlation between spatial ability skills and perceived workload.

		NASA TLX Score	SWR Score
Overall Spatial Ability Skill	Pearson Correlation *p*-Value N	−0.559 ** 0.007 22	−0.466 * 0.029 22
Spatial reasoning	Pearson Correlation *p*-value N	−0.404 0.062 22	−0.287 0.195 22
Spatial orientation	Pearson Correlation *p*-value N	−0.500 * 0.018 22	−0.348 0.113 22
SBSOD	Pearson Correlation *p*-value N	−0.104 0.644 22	−0.351 0.109 22

* Correlation is significant at the 0.05 level. ** Correlation is significant at the 0.01 level.

Table 5. Pearson's correlation between spatial ability skills and wayfinding performance.

		Total Task Duration	Total Help	Total Wrong Directions	Total Hesitation	Total Task Completion
Overall Spatial ability skill	Pearson Correlation *p*-value N	−0.342 0.119 22	−0.079 0.726 22	−0.195 0.385 22	−0.266 0.231 22	0.365 0.095 22
Spatial reasoning	Pearson Correlation *p*-value N	−0.040 0.860 22	−0.133 0.556 22	−0.062 0.786 22	−0.035 0.877 22	0.112 0.619 22
Spatial orientation	Pearson Correlation *p*-value N	−0.330 0.134 22	−0.161 0.474 22	−0.157 0.485 22	−0.151 0.501 22	0.308 0.164 22
SBSOD	Pearson Correlation *p*-value N	−0.239 0.285 22	0.346 0.115 22	−0.183 0.415 22	−0.498 * 0.018 22	0.306 0.166 22

* Correlation is significant at the 0.05 level.

The third hypothesis focuses on the relationship between spatial ability skills and workload in campus wayfinding. The null hypothesis for H3 was that the spatial ability skills of participants will not impact the workload they experience in the university campus wayfinding task.

Pearson's test (see Table 4) revealed a significant negative relationship between the overall score for spatial ability skill and perceived workload parameters: NASA TLX and SWR. It means that the spatial ability skills of participants affect the workload they experience in indoor-outdoor campus wayfinding tasks using mobile navigation apps. It shows that participants who had higher spatial ability skills experienced less workload during campus navigation using mobile apps. However, if we look further in detail at the three parameters of spatial ability skill, only spatial orientation has a significant relationship with the NASA TLX score. There was no significant relationship between any of the other two spatial ability skill parameters (Spatial reasoning and *SBSOD*) and perceived workload (NASA TLX score and SWR score).

Pearson's correlation of overall spatial ability skill and NASA TLX score was found to be a strong correlation and statistically significant ($r = -0.559$, $p < 0.01$), rejecting the null hypothesis. Similarly, the correlation between the spatial orientation and NASA TLX score was also found to be a strong correlation and statistically significant ($r = -0.500$, $p < 0.05$). At the same time, a medium correlation ($r = -0.466$, $p < 0.05$) was found between the overall spatial ability skill and SWR score. Hence, H3 was supported. This show that participants with higher spatial ability skills would experience less workload in the indoor-outdoor campus wayfinding task.

The fourth hypothesis focuses on the relationship between spatial ability skills and wayfinding performance in campus wayfinding. The null hypothesis for H4 was that the spatial ability skills of participants will not impact their wayfinding performance in the university campus wayfinding task.

Pearson's test (see Table 5) revealed a significant negative relationship only between one of the three parameters of spatial ability skills, i.e., navigation (SBSOD score) and total hesitation. It means that participants' sense of direction affects one of the five parameters of wayfinding performance, i.e., total hesitation. It shows that participants with a better sense of direction experienced less hesitation during wayfinding in indoor-outdoor campus navigation using the mobile app. However, none of the other two parameters of spatial ability nor the overall spatial ability skill had a significant relationship with any of the five wayfinding performance parameters (total task duration, total help, total hesitation, total wrong directions, and total task completion). Pearson's correlation of SBSOD and total hesitation was found to be a medium correlation and statistically significant ($r = -0.498$, $p < 0.05$), rejecting the null hypothesis. Hence, H4 was partially supported.

4.3. RQ3: Challenges in University Campus Wayfinding Using Mobile Navigation Apps

The last research question investigated the challenges faced by participants during indoor-outdoor university campus wayfinding task using mobile navigation apps. As the participants were divided into control and experiment groups, two navigation apps were used. The control group used Google Maps (an outdoor-only navigation app), and the experiment group used MazeMap (an indoor-outdoor navigation app). The qualitative data was collected through observation notes from the wayfinding task and short post-task interview responses from participants of both groups regarding their experience and challenges in university campus wayfinding. According to [50], qualitative data analysis is important to deepen the understanding of research participants. Therefore, the procedure described by Gioia et al. [49] was followed for analyzing, interpreting, and presenting data. The results highlight the issues in indoor-outdoor campus wayfinding using the two navigation apps: Google Maps and MazeMap. It also provides insight into challenges related to the specific technology that they used and how these navigation applications can be further improved to facilitate university campus wayfinding that combine indoor and outdoor navigation.

The analysis followed the grounded theory approach described by [51,52] and resulted in a data structure with three levels, as shown in Figure 2:

(1) First column—the first level highlighted the challenges from participant's feedback, i.e., raw data.
(2) Second column—the second level identified the themes for these challenges.
(3) Third column—the third level aggregated the created themes into categories.

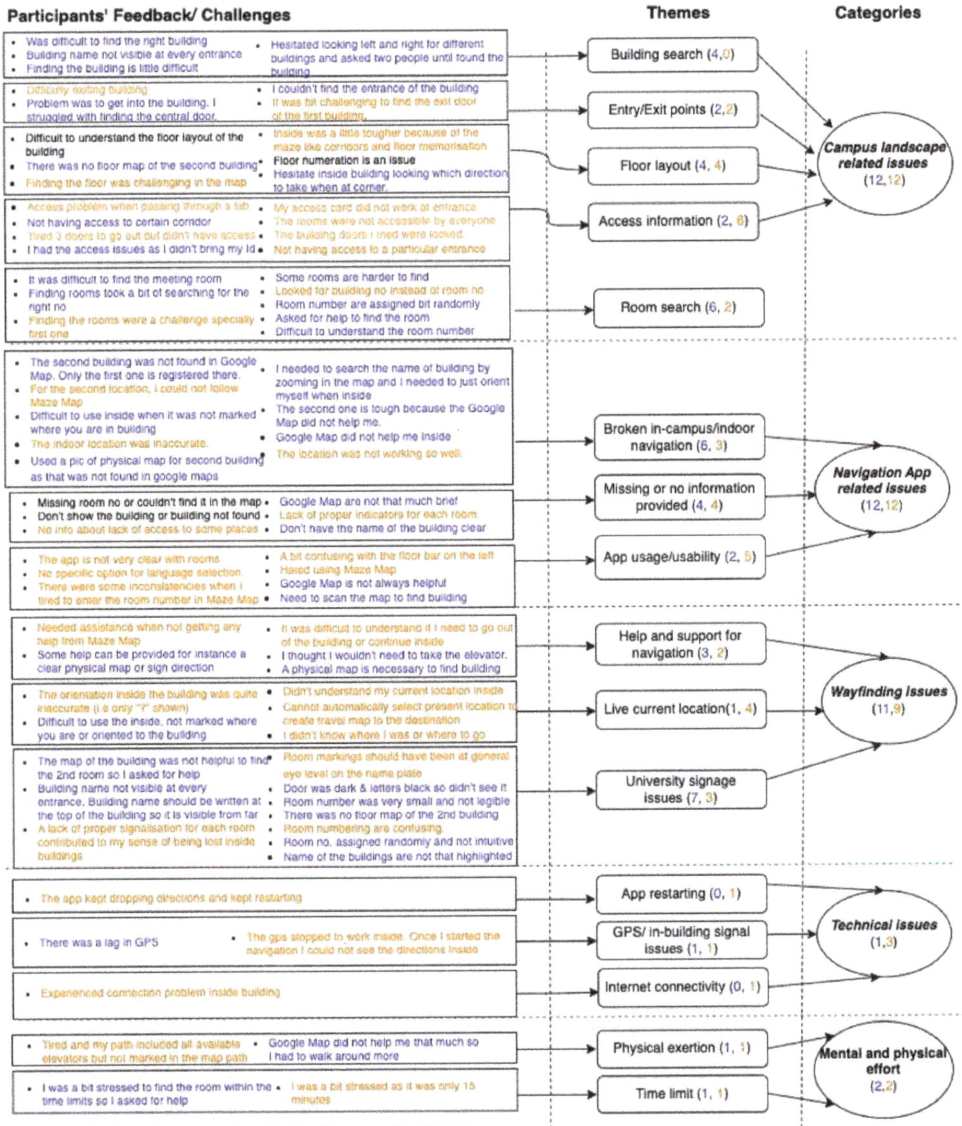

Figure 2. Challenges in indoor-outdoor campus wayfinding using mobile navigation apps (Google Maps and Maze Map).

For simplicity of representation, in the first column in Figure 2, the participants' feedback is encoded using the colors orange, blue, and black. The "blue" color represents the feedback from participants using Google Maps, the "orange" color for feedback from participants using MazeMap, and the "black" color infers that the same feedback is from both groups. Moreover, the frequency of occurrence of challenges for each theme is reflected in the second column with numbers in round brackets using blue color for Google Maps and orange for Maze Map, and cumulatively the parameters of all themes are presented in the categories respectively in column three of Figure 2.

Our analysis resulted in five categories that characterize challenges in indoor-outdoor university campus wayfinding using mobile navigation apps. The categories include campus landscape-related issues, navigation app-related issues, wayfinding issues, technical issues, and mental and physical effort. According to the qualitative analysis (presented in Figure 2), the category "campus landscape related issues" was based on five themes related to challenges linked to the university campus landscape. The themes include building search, entry/exit points, floor layout, assess information and room search. It was difficult for some participants to find the right building. Moreover, participants found it challenging to find the entry and exit points to get into or out of the building. Similarly, finding room numbers and understanding the floor map by navigating corridors was troublesome for some participants. Lastly, lack of access to some places made it difficult to navigate to the desired location as certain corridors and rooms were not accessible by everyone and required card access to open doors. As shown in Figure 2, participants in both groups (Google Maps and MazeMap) reported equal challenges in this category. However, participants using Google Maps faced more difficulty in building and room searches as compared to participants using MazeMap. This is evident from the fact that in addition to outdoor navigation MazeMap also provides indoor navigation, which is lacking in Google Maps. Moreover, the experiment group (participants using MazeMap) reported more access problems. This is because they were following the route provided by the app but had access issues to certain places within that route, which was more frustrating as they were following the directed path. Whereas for the control group (participants using Google Maps), the reported access issues were comparatively less as participants were not using Google Maps inside and orienting themselves, so they considered it as the wrong route instead of an access issue.

The category "navigation app related issues" was based on three themes that included challenges linked to the used technology (Google Maps and MazeMap). The themes include broken in-campus/indoor navigation, missing or no information provided and app usage/usability. For "broken in-campus/indoor navigation", participants using Google Maps experienced more challenges compared to the MazeMap group as it did not provide indoor navigation. Most participants reported that they did not use Google Maps for indoor navigation as it was not helpful and pointless to use inside when it is not marked where you are in the building. In addition to this, some of the campus buildings were not registered in Google Maps, and participants needed to 'zoom in' the map to search for the building. On the contrary, participants using MazeMap did not have complaints about finding the building on the map. However, the issues they experienced were either the indoor location given in the app was not very accurate, or it was difficult to follow the map when the indoor positioning was not working so well. For "missing or no information provided", most participants using Google Maps complained that the app is not complete and detailed for campus wayfinding as no information was provided for indoor navigation, such as rooms and buildings are either not found in the map or the name is not clear. On the contrary, the issues faced by the MazeMap group were more related to missing information, such as missing room numbers, lack of proper indicators for each room or missing information related to lack of access to some places. Lastly, for "App usage/usability", the group using MazeMap reported more issues as compared to the group using Google Maps. The usability issues related to MazeMap included inconsistencies with room search, confusing floor bar on the left, did not find any option for language selection, or participants did not like the overall experience of using MazeMap. Whereas participants using Google Maps only had an issue that they needed to scan the map to find the building, and they were not satisfied with the app for campus navigation as it was not helpful for indoor navigation.

The "wayfinding issues" category was related to the following themes: help and support for navigation, live current location, and university signage issues. Overall, the Google Maps group experienced more wayfinding issues as compared to the MazeMap group. For "help and support for navigation", most participants using Google Maps relied more on receiving as much help as possible from physical maps and directions within the university campus. This is because indoor navigation was not provided, and even for outdoor navigation, some building names were not clear or helpful. On the contrary, based on the qualitative analysis, the participants using MazeMap only asked for assistance when they got stuck or when certain information was not provided in the app. For example, when it was difficult to understand whether to go out of the building or continue inside while following the route in the app because either the signal dropped or the door was not accessible, or the current location was not clear. For "live current location", most issues were reported by the MazeMap group. Overall, the majority of participants were satisfied with the live location feature provided by Google Maps, except that it was not available for indoor navigation. The MazeMap group also referred to it when giving feedback "MazeMap should also provide live location like Google Maps". Although MazeMap provides the current location, most participants faced issues with accuracy. Most of the time, only "?" was shown instead of the exact current location when inside the building, which made orientation difficult as the participants could not understand where they were and where to go. Moreover, participants also complained that MazeMap does not automatically select the current location to create the route map to the destination; instead, you must enter the current location yourself, which is sometimes difficult when you do not know your exact location. Lastly, for "university signage issues", most challenges were reported by participants using Google Maps. This is because participants in this group relied more on university signage and physical maps on campus, which is also in line with the theme "help and support for navigation" and "broken in-campus/indoor navigation". The university signage issues included problems such as building names not being visible at the entrance or not highlighted, room numbers being randomly assigned and not intuitive, font size and color contract issues with the displayed room labels, no floor maps in some buildings, the physical maps inside the building not helpful to find the meeting rooms and lack of proper directions for each room. These problems are in the real world and not inside the map of navigation apps.

The category "technical issues" were related to app restarting, GPS/in-building signal issues and internet connectivity. The participants using MazeMap reported more technical issues. There were problems with the app restarting and dropping directions, the GPS stopped working inside, and the indoor positioning system had some issues as participants experienced connection problems inside buildings. On the other hand, participants using Google Maps only had a few technical issues related to lag in GPS.

Lastly, the "mental and physical effort" category included challenges related to physical exertion and time limit. Both groups (Google Maps and MazeMap) experienced similar challenges related to these themes. Participants were tired as they had to walk more when the app was not helping them or giving them information for quicker or easier paths (such as the availability of elevators instead of taking stairs). Similarly, most participants were stressed due to the time limit for finding a location which is similar to when you are running late for a meeting.

Overall, from the qualitative analysis presented in Figure 2, analyzing the parameters in the category column (third column) shows that both the Google Maps and MazeMap groups had equal challenges for "campus landscape related issues", "navigation app related issues" and "mental and physical effort". However, the analysis highlights that participants using Google Maps experienced more "Wayfinding issues" and participants using MazeMap experienced more "technical issues".

5. Discussion

This section will first discuss some more general conclusions based on results (presented in the previous section) from the university campus wayfinding experiment. The study examined the impact of navigation aid and spatial ability skills on workload and performance in university campus wayfinding contexts using quantitative and qualitative data. Subsequently, some of the implications and design recommendations will be presented for indoor-outdoor campus navigation technology, and some limitations of this study will be discussed.

5.1. Impact of Workload on University Campus Wayfinding Performance

We found significant negative correlations between participants' spatial ability skills and their workload and wayfinding performance. The participants who had higher overall spatial ability skills had lower perceived workload and less frequency of hesitation in the wayfinding task. In this study, the participants in the experiment group had a slightly lower mean SWR score (4.45) than participants in the control group (5.02). Similarly, the mean total task duration (22.00), mean total help (0.55), and mean total hesitation (1.64) for the experiment group were also slightly lower than the mean total task duration (24.82), mean total help (1.18), and mean total hesitation (2.27) for the control group. Therefore, we again used Pearson's correlations to identify any potential relation between the participant's perceived workload and wayfinding performance parameters (See Table 6). It is interesting to investigate if the effects on task performance are resultant of the effect on mental workload [17,40]. Pearson's test verified a statistically significant strong positive relationship between the SWR score and the two parameters of wayfinding performance (total task duration and total wrong directions) and a statistically significant strong negative relationship between the SWR score and total task completion. Similarly, participants' NASA TLX score was positively related to total task duration and negatively related to total task completion. On the other hand, no significant relation was found between any of the two workload parameters (SWR score, NASA TLX score) and total help and total hesitation.

Table 6. Pearson's correlation between workload and wayfinding performance.

		Total Task Duration	Total Help	Total Wrong Directions	Total Hesitation	Total Task Completion
SWR score	Pearson Correlation	0.593 **	0.109	0.518 *	0.299	−0.742 **
	p-value	0.004	0.631	0.014	0.176	<0.001
	N	22	22	22	22	22
NASA TLX score	Pearson Correlation	0.491 *	0.334	0.411	0.211	−0.764 **
	p-value	0.020	0.129	0.058	0.346	<0.001
	N	22	22	22	22	22

* Correlation is significant at the 0.05 level. ** Correlation is significant at the 0.01 level.

It means that participants who experienced a higher workload took more time to complete the task and took more wrong directions. Overall, the results showed that the task completion rate was higher for participants who experienced less workload. The qualitative data from observation showed that participants were mostly stressed and exhausted because of time and more physical exertion. Some participants complained that app did not provide additional information about the availability of escalators and no access to certain doors, due to which they had to walk more. Others also suggested that the navigation app should provide options for different routes (shortest, accessible, etc.) so they can decide which path to take based on their time and accessibility; this could reduce stress. Moreover, participants who experienced more workload also recommended that the navigation aid should provide notifications for route directions. For instance, if we take the wrong direction, the app should immediately notify us and accurately guide us. To conclude, it appears that the workload experienced by the wayfinders affects their wayfinding performance in university campus navigation. This is in line with previous research [53] that conflicting cognitive demand in the navigational task affects navigation performance. Therefore, it is important to consider cognitive load when designing navigation technologies and systems, as navigation applications with low cognitive load will ensure better performance and safety [53]. Although no significant difference was found in workload and wayfinding performance between the two groups using different navigation apps in this study. However, it is important to design options in the navigation aids that reduce the cognitive load of wayfinders.

5.2. Impact of Navigation Aid: Cognitive Load and Wayfinding Strategy

Outdoor navigation apps and indoor-outdoor navigation apps are two common mobile-based navigation aids, and one of the objectives of this research study was to explore the benefits of using these two aids in university campus wayfinding that includes both indoor and outdoor navigation.

5.2.1. No Effect on Performance and Workload

It was assumed that MazeMap should have provided better results as it provides seamless outdoor to indoor navigation, which is required in university campus wayfinding. Surprisingly this was not the case. According to the results from the conducted experiment, no significant differences were found between the two aids in workload and wayfinding performance. There were no differences in NASA TLX score, SWR score, frequency of wrong directions and total task completion. Although the findings showed differences in mean task duration, frequency of help, and hesitation, which was higher for outdoor navigation app. However, these differences were also not significant. Many researchers have focused on the use of different navigation aids for wayfinding and have found similar results. Dong et al. [48] found no radical difference in the effectiveness and efficiency of wayfinding time, errors, and interaction with tools between AR users and 2D electronic map users. Similarly, Xu et al. [54] compared voice-assisted digital maps with paper maps and found that regardless of the navigation aid used, the cognitive workload and search efficiency of environmental information are quite similar, and no significant difference was found in landmark knowledge acquisition and distance and time estimation. Rehman and Cao [6] also found that digital navigation aids (wearable devices and smartphones) were not significantly different in performance and workload. However, the wearable device was perceived to be more accurate. According to the results of Schnitzler and Hölscher [18], there were no significant differences between the two wayfinding aids (printed map, and digital map) in how efficiently they found destinations. However, these non-significant results may have several causes. First the small sample size (N = 22) could be one of the reasons as statistically significant findings are harder to detect with small sample sizes. The other possible reasons are presented in the next section.

5.2.2. Previous Experience, Usability, and Technical Issues: Cognitive Load and Multimodal Navigation

The quantitative data from the demographic questionnaire showed that all the participants had frequently used Google Maps, whereas most of the participants had rarely used MazeMap before. This difference in *previous experience* led to a difference in familiarity with apps between the groups, which could have affected the cognitive load and, thus, wayfinding performance. According to Ishikawa et al. [15], novelty could be one reason for the ineffectiveness of GPS-based navigation systems. Therefore, it is interesting to see if the wayfinding performance improves after multiple uses. Another research by Ishikawa [55] reported that the accumulated experience of navigation tool use affects wayfinding and spatial orientation. The participants using MazeMap also faced more *usage/usability issues* (floor understanding, room search, and overall experience as highlighted by the qualitative data) which added to the cognitive load and, in turn, increased the workload for this group. As highlighted by Albers [56], usability issues increase the user's cognitive load, thus increasing the cognitive resources required for the primary task. According to Georgsson [57], NASA TLX determines the cognitive load of the user in performing a task. Therefore, the results could be different if participants had more experience and fewer usability issues. However, Google Maps does not provide indoor navigation, so participants in this group were on their own when inside the building; this could offset the ease of using Google Maps compared with the ease of using MazeMap. Third, the qualitative analysis from the interview revealed that participants using MazeMap faced more *technical issues* related to internet and GPS/IPS connectivity which also affected live or current location inside the building and could have affected the workload. Regardless the experiment group performed better although not significant.

According to [58,59], multimodality can enhance the usability of applications and provide better results in complicated tasks compared to unimodal interaction. Liljedahl et al. [60] reported that careful design of multimodal interfaces can provide good user experiences. Multimodal technology can also improve accessibility in user environments such as navigation systems and mobile devices [61]. Therefore, multimodal navigation has the potential to produce good user experiences for various user groups by integrating several modalities and facilitating learning of the application [60].

5.2.3. Navigation Aid and Wayfinding Strategy: Importance of Landmarks

The qualitative data from the interview and observation notes showed that most participants in the control group (using Google Maps) focused more on landmarks. Google Maps did not provide indoor navigation at all, and the outdoor navigation within the campus was not great as most building names were not provided on the map, so they relied on direct experience. Whereas participants using MazeMap did not focus on landmarks around them and relied on the app. According to qualitative data, participants who did not focus on landmarks thought they were lost and made more wrong turns. This difference in focus on landmarks is also highlighted in research by Dong et al. [48]. They found that participants using smartphone-based AR aids paid less attention to environmental objects such as buildings compared to participants using 2D maps. Total wrong directions are the only parameter where the control group performed slightly better than the experiment group (see Table 1). Although it is a very minor difference, a large sample size might have produced different results. This finding is consistent with [15,62,63], who found that participants using GPS-based mobile navigation systems made more mistakes/direction errors and provided fewer details of landmarks than participants who used local signs or direct experience for wayfinding. Moreover, Ishikawa et al. [15] found that participants using mobile navigation apps rated wayfinding tasks as more difficult compared to direct-experience participants. The finding from our experiment also showed that the workload parameter (NASA TLX scores) for the experiment group (mean = 43.71) was slightly higher as compared to the control group (mean = 40.23). However, based on results from our experiment there was not much difference in total wrong directions taken by the control (mean = 2.00) and experiment groups (mean = 2.19). Moreover, results from our experiment also showed a strong and statistically significant correlation between total wrong direction and SWR score ($r = 0.518$, $p < 0.05$). Based on previous research and our findings, wayfinding strategy affects the workload and wayfinding performance. Although, more quantitative research is required to prove it. However, it is important to explore and improve wayfinding strategies as participants who focus on landmarks perform better.

To conclude, in this study, the mobile-based outdoor and indoor-outdoor navigation aids did not affect the perceived workload and wayfinding performance of participants in the university campus wayfinding task, but it impacted their cognitive load and wayfinding strategy (RQ1). To summarize the above discussion, the navigation aid for university campus wayfinding can be improved by improving usability and technical issues in the app and following design guidelines to reduce the cognitive load of users. We know from Section 5.1 that workload is related to wayfinding performance; thus, higher performance can be achieved by designing navigation aids that reduce the cognitive load of participants. The integration of appropriate landmarks in university campus navigation apps can decrease the cognitive load on the wayfinders [64,65] and help users increase their environmental awareness and spatial recognition ability [65,66]. Landmark-based campus wayfinding solutions can increase the efficiency and reliability of such applications [65].

5.3. Impact of Spatial Ability Skills: Technology and Design for Spatial Knowledge Acquisition

The findings from this research study showed a negative correlation between participants' overall spatial ability skills and their workload, consistent with the previous research in this area. According to Nori et al. [26], basic spatial abilities are related to navigation in a real-world environment. Lin and Wang [16] found in their study that participants with a better sense of direction (SOD) had a lower overall workload for the target task. However, in our study, only spatial orientation out of the three individual parameters of overall spatial ability had a significant relationship with the NASA TLX score (see Table 4). With respect to wayfinding performance, this study outlined that the spatial ability skill parameter (sense of direction) is negatively related to the frequency of hesitation. However, total task duration, total help, total wrong directions, and task completion are not associated with spatial ability skills. Other researchers have also found an association between sense of direction and wayfinding performance. A study by Brunyé et al. [67] found that a higher spatial sense of direction was related to higher path efficiency. Lin and Wang [16] indicated that participants with a better SOD had faster response time.

To conclude, the spatial ability skills of wayfinders can be associated with their perceived workload and wayfinding performance (RQ2). The wayfinders with higher spatial ability skills experience less workload and are more confident navigating indoor-outdoor campus wayfinding. According to the research investigating navigation assistance and wayfinding, navigation aid can play a role in shaping human spatial ability as it influences human wayfinding and navigation [54]. Therefore, it is important to design navigation aids that improve spatial ability as it affects the workload.

Navigation assistance systems should be improved in regard to the acquisition of spatial knowledge. Navigation applications should include features to improve users' skills and knowledge instead of taking on all cognitive functioning. In this way, users will make informed choices when using the navigation aid, which provides optimal support for the acquisition of spatial knowledge [68]. Modern digital technology can provide novel options for enhancing spatial learning with promising transformative impacts [69]. Researchers have explored different aspects of navigation aids that can influence spatial knowledge acquisition. Xu et al. [54] found that the use of voice-assisted digital maps may result in time efficiency, but it can reduce the spatial knowledge acquisition of pedestrians. According to Münzer et al. [68], appropriate visualization can support the acquisition of self-to-object spatial knowledge. The findings by Verghote et al. [23] showed significant improvement in the success rate of the participants with low spatial cognition when using 3D models instead of 2D drawings. However, the format of information did not affect the participants with high spatial cognition. Similarly, Qiu et al. [70] found that AR maps are better than 2D maps in assisting individuals in acquiring spatial knowledge. AR displays provide comprehensive information about the environment highlighting the landmark locations and providing exocentric spatial information, thus helping in acquiring landmark and route knowledge. Moreover, landmarks are important for wayfinding and navigation [66]; they can assist in spatial knowledge acquisition during navigation [71] and can support correct decision-making [72]. According to Zheng and Hsu [66], highlighting buildings at intersections can aid wayfinders in increasing their intuitive spatial ability by helping them to use intersections as references.

5.4. Implications for Indoor-Outdoor Campus Navigation App Design

In this section, design implications are presented based on both the qualitative analysis of interview data and observation notes that identified the main challenges in indoor-outdoor campus wayfinding using a mobile navigation app (RQ3) and the lessons learned from the main findings of the quantitative data (RQ1–2). The design recommendations are summarized as follows.

5.4.1. Landmarks and Additional Route and Location Information

Landmarks play an essential role in wayfinding and navigation tasks [66] and can aid in the acquisition of spatial knowledge without much workload [71]. Landmarks can be used by wayfinders as a point of reference for the identification of directions and for making route decisions which can reduce confusion and hesitation [66]. Therefore, navigation systems should use physical landmarks as a directional aid in wayfinding directions because they help with orientation [73,74]. Previous research has provided different landmark suggestions. According to Ishikawa et al. [15], a global landmark in the area does not help navigation if they are not shown on the map as a navigation clue. According to Qiu et al. [70], AR displays are useful for supporting individuals in acquiring landmark and route knowledge. In the present study, participants who focused on landmarks and the surrounding environment were more confident navigating in the wayfinding task.

It was also found from qualitative data that most participants faced challenges related to route and location. Participants in both groups (Google Maps and MazeMap) reported issues related to missing or no information provided (see Figure 2). Some of the participants' suggestions to improve the navigation app included: "Provide information about access to rooms", "Maybe show which doors I have access or not", "Clearly marked elevators", "Building/door opening hours", "Info telling me where public access doors are", "Clearly marked entrances, including which are for common use and which require special access", "Providing options of accessible entrance (updates)". Previous research also shows that route and location information is important and can improve navigation aid for users [59]. Mallik et al. [75] found that floor identification is important where a user can travel between floors, and it is crucial to provide floor numbers in multi-floor structures.

According to Sato et al. [76], the surrounding information, such as elevator locations, access information, and floor changes, is useful to aid in the orientation and mobility of users and increase confidence during navigation by confirming their location. Vainio [59] concluded that navigation aids could be improved for users by designing predictive clues and rhythm in mobile-based navigation applications. Mallik et al. [75] suggested that sensor-based 3D localization strategies should be used for localizing a person traveling through elevators, stairs, or escalators. Huang et al. [77] suggested adding landmark objects into real-world environments in order to enhance the user experience of indoor navigation systems.

5.4.2. Real-Time Location

There is a high demand to provide real-time indoor navigation solutions as most location-based service users spend the majority of their time (70–90%) indoors, increasing public attention to these potential applications [78]. In the present study, most participants faced wayfinding issues related to live current location, especially during indoor navigation with MazeMap (see Figure 2). Based on interview results, some of the suggestions for improvement included: "Indicate my current position", "Location dot moving in real time", "It would be helpful if the navigation feature in the map was real time showing the path you are following", "Updating real time location of the person", "It could be helpful to highlight both the start and target floor in the bar so that one knows where to navigate and from where". Zheng and Chen [79] suggested that floor indication design in mobile-based indoor navigation interfaces should mark the current floor and be arranged according to the sequence of floors from the bottom up. MazeMap provides such floor indication; however, according to qualitative analysis in the present study, some participants felt that the floor bar in the app was confusing and should be improved in design with respect to marking the current floor.

A lot of research is focusing on indoor location determination technologies for making indoor positioning systems reliable and highly accurate [80]. An ideal navigation application should present real-time location and direction [81]. According to Harper et al. [73], it is important that navigation apps are designed with knowledge of the user's current location to help them identify their location in the environment, as they have expectations that maps should operate from their viewpoint.

5.4.3. Improved Location Search

Navigation systems should provide different options for improved location search. Some of the options for improved location search suggested by participants in this study include: "Option to automatically put current location as starting point instead of a manually entering the location", "Floor maps within the building and option to find the building by name", "If there was room search option in Google Maps it would have saved some time".

Wayfinding focuses on searching the route from the user's current position to the target destination. Therefore, it is not only important to detect the user's position, but providing different options related to route guidance and search is also important [82]. Zheng and Chen [79] suggested providing a self-locating icon and using a dialogue box with text showing "your current location" in the beginning to prompt users to use the self-locating icon. According to Ng and Lim [82], AR technology can be used in the mobile-based indoor navigation system to detect the user's position and display AR navigation directions on top of real-world camera view to provide an immersive navigation experience and guide the user to the destination. Guo et al. [83] focused on providing a personalized, customized navigation menu bar, one click to open and search on demand functions for 3D visualization indoor navigation system. Harper et al. [73] proposed including a universal search function in wayfinding systems to enable users to search for things specifically instead of by category, which may be difficult to use and frustrating.

Zheng and Chen [79] indicated that smartphone-based indoor navigation interfaces must also provide options such as multiple choices of route planning and reminders. Moreover, in the present study, participants suggested that the navigation app should provide alternate route options. Some of the comments include: "Show alternate routes", "Quickest path illustrating on MazeMap", "It should have alternative options as well. I could not find the second building, so I think it should be more specified towards the building name".

5.4.4. Improved Location Visualization

Appropriate visualization can support the acquisition of spatial knowledge [68]. Previous research has highlighted the benefits of using 2D, 3D and augmented reality (AR) technology for visualization in indoor and outdoor navigation apps. The results from the research by Torres-Sospedra et al. [10] indicated the usefulness of visualizations (AR viewer and intuitive map-based view) in locating university facilities and improving spatial orientation. They developed and evaluated two mobile-based indoor-outdoor navigation applications: SmartUJI APP, which uses an intuitive map-based view to visualize information, and SmartUJI AR, which uses an AR interface with interactive icons placed over campus facilities allowing users to interact with the campus and visualize the location of the nearby facilities and points of interest with an AR viewer through the mobile device camera. Sandoval et al. [84] used 3D-Enhanced Facility Management System to assist visually impaired users in indoor navigation and achieved satisfactory preliminary results regarding user experience.

In the present study, participants also suggested improving the visualization for indoor-outdoor campus navigation systems. Some of the comments from participants interviews include: "Visualising buildings by their photos", "Street view/corridor view inside the building", "Show picture of target room (sometimes no room number is given at the door, or it is hard to find)", "Different angles to see where I am waiting", "It should be more visualised with the 3D view and updated always".

5.4.5. Notifications and Alerts

Navigation systems should provide notifications and alerts to guide users in finding their destination [85–87]. Previous research has shown that a lack of visual or audio notification upon arrival at a destination can lead to failure in recognizing the target location by participants [88]. Therefore, one useful feature of wayfinding technology is providing notification when the user is approaching the destination [87,88]. Moreover, notification of obstacles or road closures in the path can lead to further improvement in wayfinding aid [89].

However, sudden notifications can disturb users in uncertain situations [90]. According to Huang et al. [77], the responsiveness of the notification system is important to create a better user experience. It is useless to notify the user when they are not in the area or notify when they have already passed the area. Alnabhan and Tomaszewski [86] highlighted the importance of the accuracy of displayed notifications. They developed an AR-based indoor navigation system, "INSAR", that issues a 2 s notification vibration when it finds the destination and displays a message telling the user that they have arrived at their destination [86]. Hall [91] used a map feature to support users in campus navigation which is enabled by sending notifications when they are within 60 m of the target destination.

The qualitative data also highlighted the importance of notifications and alerts. Many participants suggested that the navigation app must provide relevant notifications about destinations and warnings for wrong directions for further improvement as a campus wayfinding aid. Some comments include: "More obvious notification when a room is inside another room, is behind a locked/special access door, etc.", "Pop-up notification when you are taking the wrong direction", "Warning if you are turning towards wrong directions", "I think a beep or a sound to inform me when I need to go upstairs or downstairs of the building", "A ring when the location is closer."

5.4.6. Assistance or Help: Voice, Text, Symbolic Notations

It is important to design adequate help and assistance in campus navigation applications to support users in wayfinding. However, not much navigation assistance is currently available for indoor navigation [92]. In the present study, many participants faced challenges related to help and support for navigation and suggested that campus wayfinding apps can be improved by providing path directions with arrows, symbolic notations, text for description of route, turn-by-turn instructions and voice or audio instructions to make the navigation easier. Some of the participants' comments include: "Show path directions maybe with arrows", "Symbolic notations to make the navigation easier for all in any condition", "Some text may help the navigation. For example, buildings like Realfagbygget, some suggestions about which entrance to take which block, etc. can be helpful", "Offer caption to get description of route (similar to Google Maps) so that one knows i.e., to which floor to navigate. Especially helpful if connection is lost", "Turn by turn written instructions (e.g., turn left, go 100 m, turn right)", "Audio of the written instructions", "Voice commentary should be there", "Would be cool to have a tool that guide in a smart way with voice hints", "Voice instruction".

Previous research has also explored different ways to provide assistance and help in navigation applications to aid users in wayfinding and should be used in designing better campus navigation applications. Xu et al. [54] found that voice-assisted digital maps can increase participants' comfort levels and help them reach their destination more quickly compared to the other types of maps. Vainio [59] demonstrated the potential of multidisciplinary approaches and multimodal navigation aids to support the navigation and wayfinding of mobile users [59]. Liu and Meng [93] suggested, based on their study focusing on augmented reality-based indoor navigation applications, that adding more semantic icons with text to an AR real-time scene interface can serve as virtual landmarks and enhance users' spatial learning [48]. Harper et al. [73] suggested using relatable and simple instructions for users; for example, instead of telling how many steps or feet to walk, using time such as 15 min walk is more beneficial as it is not easy for users to estimate distance by steps. Sato et al. [76] suggested augmenting navigation aids and providing more instructions when the user is standing and only offering little and essential guidance when walking so as not to hinder awareness of their surroundings.

Harper et al. [73] recommended that a wayfinding system should provide users an option to request assistance through human-human contact if they are unable to find the location. This is because sometimes users just want to speak to a person instead of interacting with a system. Helmi et al. [94] suggested improving indoor campus navigation applications by adding features such as arrow-based direction for giving indoor direction pathways or adding a Live chat page so users can contact the admin if needed.

5.4.7. Consistency

The consistency between the user interface (UI) and the real world is crucial and should be addressed when designing maps and UI for navigation systems [95]. High interface consistency is expected to result in higher user satisfaction, fewer errors, and faster task completion [96]. To ensure consistency, it is important to match the elements of the application to those in the environment by understanding the naming conventions, landmarks, labeling, zones, icons, color codes, images and signage of buildings and places [73]. According to [95,97], more attention should be paid to designing consistency between the real world and digital presentation when designing indoor maps because the shapes, colors, and materials of the environment vary from building to building.

However, consistency in navigational signage used in the real world is also important to aid wayfinding in complex environments with multiple buildings. Apelt et al. [98] also recommended using consistent naming protocols for places with multiple buildings, such as educational institutions, although it is suggested to avoid alphanumeric coding as it is less memorable as compared to place names. However, it is also important to provide consistency within the alpha-numeric coding system (such as Room B3.7) when using it as a naming protocol. Moreover, ensuring consistency in sign types is also important to identify and recognize signage, for example, consistent colors, logos, and typeface; consistent graphic layouts; consistent materials and construction; and consistent overall appearance [98]. In the present study, participants reported some challenges related to university signage issues highlighting inconsistency in naming, numbering, and labeling for rooms, floors, and buildings. The suggestions for improvement focused on the need to ensure consistency, some of the comments included: "There should be consistent numbering of floors", "Same buildings names in different platforms/navigation tools", "It is possible to mark every building by their name and number both".

5.4.8. Easy to Understand

It is important that the campus navigation system is easy to use for a truly seamless behavior. However, the ease of use and user experience of such applications depends not only on user-friendly interface design but many other factors, such as intermittent internet connectivity or storage limitation of mobile phones for location prediction and delivering a quality experience [75]. According to Harper et al. [73], it is important to reduce unnecessary on-screen information so users can focus on important pieces of information. This can be done by avoiding clutter and adhering to platform standards instead of making users learn new methods of interaction [73]. Similarly, designing less detailed maps in navigation apps can make orientation easier [95,97]. Information and wording should be classified according to user expectations, and it is important to consider accessibility in on-screen interactions and hand positioning for touchscreen systems [73].

In the present study, participants reported some issues related to app usage and usability that negatively impacted their experience of using the navigation app for campus wayfinding. Some participants using MazeMap were frustrated as they thought it was a bit confusing and not easy to use. From the qualitative analysis, one of the suggestions for improvement was that the campus navigation app should provide a "simple layout".

5.4.9. Update Location Changes

It is crucial for the navigation app to provide updated location information and maps. Participants are interested and want to be aware of landmarks and points of interest, such as cafes, escalators or stairs, that can help them in confirming their location or aid in navigating the environment [76]. However, as the layout and appearance of the facilities change greatly, and the navigation system must be able to rescan and post-process the new data in a short time in order to minimize the drop in user experience due to layout change [83].

In the present study, one of the participants pointed out that location changes were not updated in MazeMap, which caused confusion in indoor navigation. However, this could be due to delays in updates by the organization rather than the application and taps into the issue of how often universities (in this case, NTNU) update MazeMap. The participants further suggested that the navigation app needs to be regularly updated to improve campus wayfinding. The comment made was as follows: "MazeMap scenery description needs to be regularly updated as it conveyed wrong surroundings. The cafe area with seating arrangement now in realfag building is not the same as shown in MazeMap. It is on the opposite side and not as many tables as shown in MazeMap".

5.5. Limitations of the Study

This work entails several limitations. One such limitation is that the gender distribution was not equal. Overall, the study included more female participants (77%) than male participants (23%). It is possible that results could have been different if there were more male participants. Some of the previous research has shown gender differences in spatial ability skills [99], where males significantly outperformed females in overall spatial ability. However, the gender distribution within both groups was fairly equal and thus reducing gender bias. The control group had two male and nine female participants, whereas the experiment group had three male and eight female participants. Therefore, the analysis was not affected due to overall unequal gender distribution. Moreover, there were no differences in the navigation task and how the two mobile navigation apps were used. Moreover, the spatial ability skills of the two groups were relatively similar.

Another limitation of this study was its size. The small sample size makes it harder to detect statistically significant findings and even difficult to generalize the resulting findings. No statistically significant difference was found in wayfinding performance and workload of the control and experiment groups which could be potentially due to the small sample size. One of the concerns with a small sample size is more chances of producing type II errors (beta errors) and failing to reject the null hypothesis (H0) when the null is false. Therefore, these initial findings and the outlined effects need to be replicated with additional testing using a larger and more diverse sample of participants and with different navigation apps. However, we also collected qualitative data from participants to triangulate the findings, thus increasing the validity.

The results reported in this paper should apply to other mobile navigation apps and different venues and scenarios requiring combined indoor and outdoor navigation. However, we acknowledge that the results might not be transferable to any navigation app, especially for different platforms and other indoor-outdoor wayfinding scenarios, such as hospitals or large organizations with multiple buildings. The results presented in this study may be only relevant in the context of the use of university campus wayfinding and navigation systems that provide navigation features like Google Maps and MazeMap. The average age of participants in this study is 30.8, and it is possible that similar analyses with other user groups (elderly or children) might produce different results due to age differences.

6. Conclusions and Future Work

This study explored the effects of two different navigation aids (outdoor vs. indoor-outdoor navigation app) and spatial ability skills on university campus wayfinding performance and perceived workload. This paper also identified the challenges in indoor-outdoor campus wayfinding using mobile navigation aids. An experiment was conducted with twenty-two participants divided into control and experiment groups using Google Maps and MazeMap, respectively, to complete a time-bound university campus wayfinding task.

Both quantitative and qualitative data was collected using multiple methods, including spatial ability test, questionnaires, interview, and observation/field notes. Based on the findings, the following conclusions and design implications are presented for facilitating university campus wayfinding by improving navigation systems:

- Seamless indoor-outdoor navigation is crucial for university campus wayfinding, and navigation apps must provide support for indoor navigation.
- The spatial ability skills of participants affect their perceived workload and wayfinding performance. Moreover, the mental workload experienced by the wayfinders also impacts their wayfinding performance in university campus navigation. Therefore, it is important to consider cognitive load and spatial knowledge acquisition when designing campus navigation systems.
- The employed navigation aids do not affect the workload and wayfinding performance, but they can impact the cognitive load and wayfinding strategy of participants in university campus wayfinding. Therefore, navigation aid can play a role in reducing cognitive load and shaping human spatial ability to improve wayfinding and navigation.
- The app usage experience, usability and technical issues in the navigation app can affect the cognitive load of users. Thus, the design of indoor-outdoor campus navigation applications should focus on reducing the cognitive load of wayfinders by improving usability and technical issues in the app and catering to users with different prior experiences (novice and experienced). The effective use of multimodal technology integrating multiple modalities can enhance the usability and user experience of navigation systems and facilitate learning the application.
- The campus navigation apps should help wayfinders in spatial knowledge acquisition. The wayfinders with higher spatial ability skills experience less workload and are more confident navigating indoor-outdoor campus wayfinding. Modern digital technology, such as 3D models, AR maps and appropriate visualization, can assist individuals in acquiring spatial knowledge.
- Landmarks can decrease the cognitive load on the wayfinders and assist them in spatial knowledge acquisition. Therefore, the integration of appropriate landmarks is important in designing university campus navigation systems.
- Some other design recommendations for indoor-outdoor campus navigation applications include additional route and location information, real-time location, improved location search, improved location visualization, notifications and alerts, assistance or help, consistency, easy to understand and updated location changes.

Some of the directions for future work include focusing on conducting experiments on a larger scale to compare the differences between navigation aids with different user groups and on multiple campuses, exploring the impact of different wayfinding strategies on the workload and wayfinding performance with navigational aids, and exploring the impact of gender and age difference on wayfinding performance. Future studies could also investigate topics such as how user's spatial knowledge can be improved through navigation aids, how to adapt navigation aids for different types of wayfinders and their strategies and investigate the effectiveness of multimodal instructions on wayfinding performance in navigation tasks with different cognitive loads.

Author Contributions: For Conceptualization, R.T.; methodology, R.T.; validation, J.K.; analysis, R.T.; investigation, R.T.; data curation, R.T.; writing—original draft preparation, R.T.; writing—review and editing, R.T. and J.K.; visualization, R.T. All authors have read and agreed to the published version of the manuscript.

Funding: This research was funded by the Research Council of Norway (NFR) through the Smart Workspace past COVID-19 project, project number: 317725. Available online: https://shorturl.at/bdmU3 (accessed on 18 July 2023).

Informed Consent Statement: Informed consent was obtained from all subjects involved in the study.

Data Availability Statement: The data presented in this study are available on request from the corresponding author. Additional meta-data is needed for the data to made use of in another context, thus we have not made them openly available.

Conflicts of Interest: The authors declare no conflict of interest. The funders had no role in the design of the study; in the collection, analyses, or interpretation of data; in the writing of the manuscript; or in the decision to publish the results.

Appendix A Spatial Reasoning Test

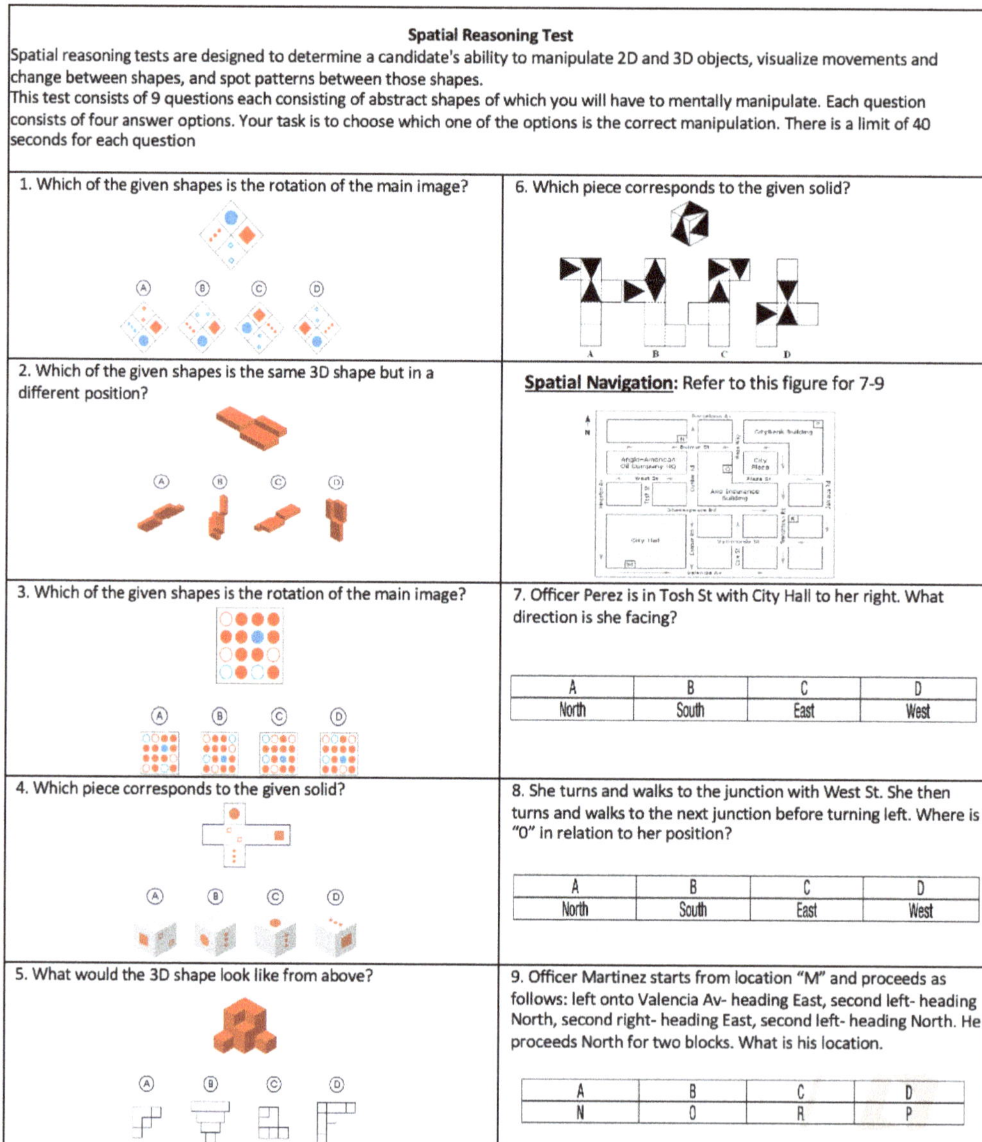

Spatial Reasoning Test

Spatial reasoning tests are designed to determine a candidate's ability to manipulate 2D and 3D objects, visualize movements and change between shapes, and spot patterns between those shapes.

This test consists of 9 questions each consisting of abstract shapes of which you will have to mentally manipulate. Each question consists of four answer options. Your task is to choose which one of the options is the correct manipulation. There is a limit of 40 seconds for each question

1. Which of the given shapes is the rotation of the main image?

(A) (B) (C) (D)

2. Which of the given shapes is the same 3D shape but in a different position?

(A) (B) (C) (D)

3. Which of the given shapes is the rotation of the main image?

(A) (B) (C) (D)

4. Which piece corresponds to the given solid?

(A) (B) (C) (D)

5. What would the 3D shape look like from above?

(A) (B) (C) (D)

6. Which piece corresponds to the given solid?

A B C D

Spatial Navigation: Refer to this figure for 7-9

7. Officer Perez is in Tosh St with City Hall to her right. What direction is she facing?

A	B	C	D
North	South	East	West

8. She turns and walks to the junction with West St. She then turns and walks to the next junction before turning left. Where is "O" in relation to her position?

A	B	C	D
North	South	East	West

9. Officer Martinez starts from location "M" and proceeds as follows: left onto Valencia Av- heading East, second left- heading North, second right- heading East, second left- heading North. He proceeds North for two blocks. What is his location.

A	B	C	D
N	O	R	P

Figure A1. Spatial reasoning test.

Appendix B Navigation Task and Route Maps Generated from MazeMap and Google Map

You are in Gløshaugen campus on a normal workday. This is your office room (Gløshaugen IT-bygget (317): Room number 122, floor 1) and you have two upcoming consecutive meetings. You have already booked two different meeting rooms (see venue below) for these meetings.

Venue 1: The room booked for the first meeting is: **Building name & code**: Gløshaugen Byggteknisk (337): **Room number:** 1-001, **Floor** 1
Venue 2: The room booked for the second meeting is **Building name & code**: *Gløshaugen Hovedbygningen (301):* **Room number:** *228,* **Floor** *2*

Your task is to navigate from your office room (IT-bygget:122, floor 1) to the venue 1 (Byggteknisk: Room number 1-001, Floor 1 to attend the first meeting and then from this place navigate to the venue 2 (Hovedbygningen: Room number 228, Floor 2) to attend the second meeting. You do not need to stop at venue 1, as soon as you reach the room continue to navigate to the venue 2. Your task finishes when you reach venue 2. You have 30 minutes to complete this task.

Control Group: You will only use **Google maps** to complete this navigation task. *Please do not use any other tool.*
Experiment Group: You will only use **Mazemap** to complete this navigation task. *Please do not use any other tool.*

Figure A2. Navigation task.

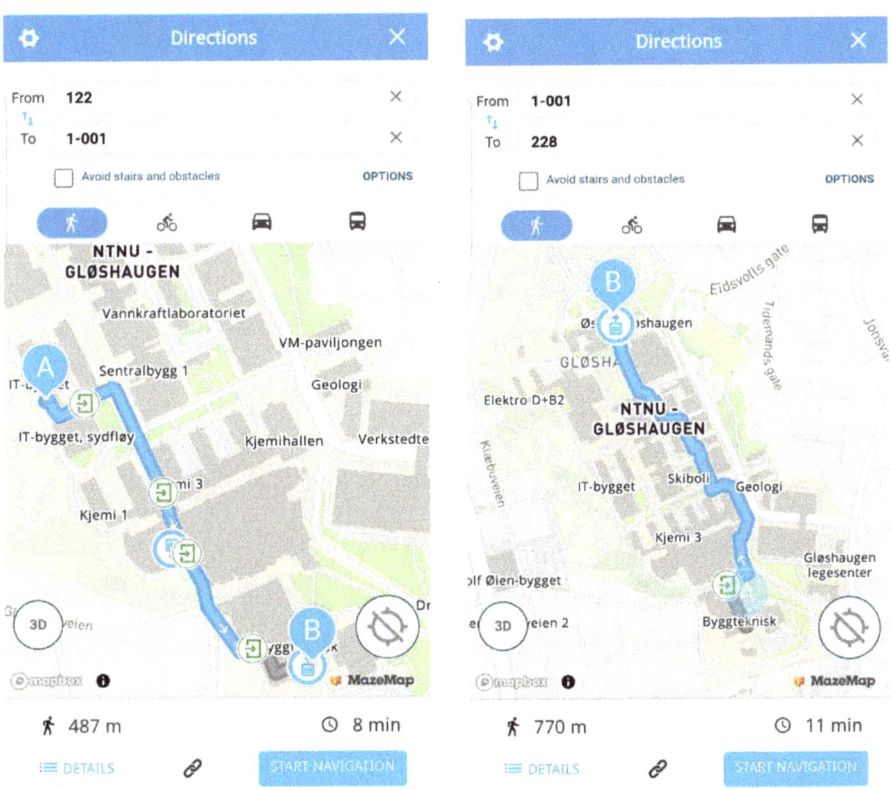

Figure A3. MazeMap: Start point to 1st meeting room (**left image**); 1st meeting room to 2nd meeting room i.e., end point (**right image**).

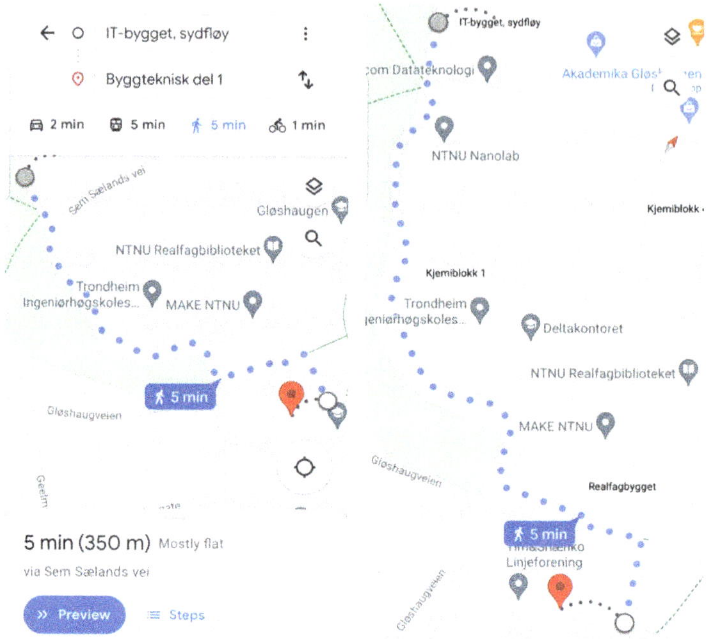

Figure A4. Google Map: Start point to 1st meeting room (right image is enlarged view).

Figure A5. Google Map: 1st meeting room to 2nd meeting room i.e., end point (right image is enlarged view).

References

1. Prestopnik, J.L.; Roskos–Ewoldsen, B. The relations among wayfinding strategy use, sense of direction, sex, familiarity, and wayfinding ability. *J. Environ. Psychol.* **2000**, *20*, 177–191. [CrossRef]
2. Munion, A.K.; Stefanucci, J.K.; Rovira, E.; Squire, P.; Hendricks, M. Gender differences in spatial navigation: Characterizing wayfinding behaviors. *Psychon. Bull. Rev.* **2019**, *26*, 1933–1940. [CrossRef]
3. Iftikhar, H.; Asghar, S.; Luximon, Y. The efficacy of campus wayfinding signage: A comparative study from Hong Kong and Pakistan. *Facilities* **2020**, *38*, 871–892. [CrossRef]
4. Bradley, N.A.; Dunlop, M.D. An experimental investigation into wayfinding directions for visually impaired people. *Pers. Ubiquitous Comput.* **2005**, *9*, 395–403. [CrossRef]
5. Martínez-Cruz, S.; Morales-Hernández, L.A.; Pérez-Soto, G.I.; Benitez-Rangel, J.P.; Camarillo-Gómez, K.A. An outdoor navigation assistance system for visually impaired people in public transportation. *IEEE Access* **2021**, *9*, 130767–130777. [CrossRef]
6. Rehman, U.; Cao, S. Augmented-reality-based indoor navigation: A comparative analysis of handheld devices versus google glass. *IEEE Trans. Hum.-Mach. Syst.* **2016**, *47*, 140–151. [CrossRef]
7. Schuldt, C.; Shoushtari, H.; Hellweg, N.; Sternberg, H. L5IN: Overview of an indoor navigation pilot project. *Remote Sens.* **2021**, *13*, 624. [CrossRef]
8. Biczok, G.; Martínez, S.D.; Jelle, T.; Krogstie, J. Navigating MazeMap: Indoor human mobility, spatio-logical ties and future potential. In Proceedings of the 2014 IEEE International Conference on Pervasive Computing and Communication Workshops (PERCOM WORKSHOPS), Budapest, Hungary, 24–28 March 2014; IEEE: New York, NY, USA, 2014; pp. 266–271.
9. Giudice, N.A.; Walton, L.A.; Worboys, M. The informatics of indoor and outdoor space: A research agenda. In Proceedings of the 2nd ACM SIGSPATIAL International Workshop on Indoor Spatial Awareness, San Jose, CA, USA, 2 November 2010; pp. 47–53.
10. Torres-Sospedra, J.; Avariento, J.; Rambla, D.; Montoliu, R.; Casteleyn, S.; Benedito-Bordonau, M.; Gould, M.; Huerta, J. Enhancing integrated indoor/outdoor mobility in a smart campus. *Int. J. Geogr. Inf. Sci.* **2015**, *29*, 1955–1968. [CrossRef]
11. Nikander, J.; Järvi, J.; Usman, M.; Virrantaus, K. Indoor and outdoor mobile navigation by using a combination of floor plans and street maps. In *Progress in Location-Based Services*; Springer: Berlin/Heidelberg, Germany, 2013; pp. 233–249.
12. Lu, F.; Zhou, H.; Guo, L.; Chen, J.; Pei, L. An ARCore-Based Augmented Reality Campus Navigation System. *Appl. Sci.* **2021**, *11*, 7515. [CrossRef]
13. Pawar, U. Design and implementation of outdoor and indoor campus navigation system. *Int. J. Mod. Trends Eng. Res.* **2016**, *3*, 8–10.
14. Vanclooster, A.; De Maeyer, P. Combining indoor and outdoor navigation: The current approach of route planners. In *Advances in Location-Based Services, Proceedings of the 8th International Symposium on Location-Based Services, Vienna, Austria, 21–23 November 2011*; Springer: Berlin/Heidelberg, Germany, 2012; pp. 283–303.
15. Ishikawa, T.; Fujiwara, H.; Imai, O.; Okabe, A. Wayfinding with a GPS-based mobile navigation system: A comparison with maps and direct experience. *J. Environ. Psychol.* **2008**, *28*, 74–82. [CrossRef]
16. Lin, Y.-L.; Wang, C.-H. Evaluation of wayfinding performance and workload on electronic map interface. In Proceedings of the Human-Computer Interaction. Towards Mobile and Intelligent Interaction Environments: 14th International Conference, HCI International 2011, Orlando, FL, USA, 9–14 July 2011; Proceedings, Part III 14. Springer: Berlin/Heidelberg, Germany, 2011; pp. 311–320.
17. Vasquez, H.M.; Hollands, J.G.; Jamieson, G.A. The Effects of Practice on Navigation Performance and Mental Workload with a Mirror-In-The-Sky Map. In Proceedings of the Human Factors and Ergonomics Society Annual Meeting, Baltimore, MD, USA, 4–7 October 2021; SAGE Publications Sage CA: Los Angeles, CA, USA, 2021; Volume 65, pp. 1546–1550.
18. Schnitzler, V.; Hölscher, C. User Experience and Strategy Choices During Navigation: A Content Analysis of Navigators Using Different Types of Wayfinding Devices. In Proceedings of the EAPCogSci, Torino, Italy, 25–27 September 2015.
19. Chien, Y.C.; Tsai, P.A.; Lin, Y.T.; Wu, S.M.; Chen, K.T.; Han, Y.T.; Hwang, P. Uncertainty and mental workload among wayfinding strategies. In Proceedings of the Universal Access in Human-Computer Interaction. Users and Context Diversity: 10th International Conference, UAHCI 2016, Held as Part of HCI International 2016, Toronto, ON, Canada, 17–22 July 2016; Proceedings, Part III 10. Springer: Berlin/Heidelberg, Germany, 2016; pp. 556–565.
20. Jamshidi, S.; Ensafi, M.; Pati, D. Wayfinding in interior environments: An integrative review. *Front. Psychol.* **2020**, *11*, 549628. [CrossRef] [PubMed]
21. Malinowski, J.C. Mental rotation and real-world wayfinding. *Percept. Mot. Ski.* **2001**, *92*, 19–30. [CrossRef] [PubMed]
22. Shamsuddin, N.A.A.; Din, S.C. Spatial ability skills: A correlation between Augmented Reality (AR) and conventional way on wayfinding system. *Environ.-Behav. Proc. J.* **2016**, *1*, 159–167. [CrossRef]
23. Verghote, A.; Al-Haddad, S.; Goodrum, P.; Van Emelen, S. The effects of information format and spatial cognition on individual wayfinding performance. *Buildings* **2019**, *9*, 29. [CrossRef]
24. Kato, Y.; Takeuchi, Y. Individual differences in wayfinding strategies. *J. Environ. Psychol.* **2003**, *23*, 171–188. [CrossRef]
25. Hegarty, M.; Richardson, A.E.; Montello, D.R.; Lovelace, K.; Subbiah, I. Development of a self-report measure of environmental spatial ability. *Intelligence* **2002**, *30*, 425–447. [CrossRef]
26. Nori, R.; Grandicelli, S.; Giusberti, F. Visuo-spatial ability and wayfinding performance in real-world. *Cogn. Process.* **2006**, *7*, 135–137. [CrossRef]

27. Prandi, C.; Barricelli, B.R.; Mirri, S.; Fogli, D. Accessible wayfinding and navigation: A systematic mapping study. *Univers. Access Inf. Soc.* **2021**, *22*, 185–212. [CrossRef]
28. Kunhoth, J.; Karkar, A.; Al-Maadeed, S.; Al-Attiyah, A. Comparative analysis of computer-vision and BLE technology based indoor navigation systems for people with visual impairments. *Int. J. Health Geogr.* **2019**, *18*, 29. [CrossRef]
29. Huang, H.; Gartner, G. A survey of mobile indoor navigation systems. In *Cartography in Central and Eastern Europe*; Springer: Berlin/Heidelberg, Germany, 2009; pp. 305–319.
30. Ren, G.; Ai, C.; Xu, Q.; Wang, Z.; Wang, Z.; Geng, D. Research on Indoor and Outdoor Navigation Technology Based on the Combination of Differential GNSS and Lidar SLAM. In Proceedings of the 2020 IEEE International Conference on Real-Time Computing and Robotics (RCAR), Asahikawa, Japan, 28–29 September 2020; IEEE: New York, NY, USA, 2020; pp. 134–139.
31. Jiang, W.; Li, Y.; Rizos, C.; Cai, B.; Shangguan, W. Seamless indoor-outdoor navigation based on GNSS, INS and terrestrial ranging techniques. *J. Navig.* **2017**, *70*, 1183–1204. [CrossRef]
32. Pfaff, P.; Kümmerle, R.; Joho, D.; Stachniss, C.; Triebel, R.; Burgard, W. Navigation in combined outdoor and indoor environments using multi-level surface maps. In Proceedings of the WS on Safe Navigation in Open and Dynamic Environments, IROS, San Diego, CA, USA, 29 October–2 November 2007; Volume 7.
33. Krogstie, J. Bridging research and innovation by applying living labs for design science research. In Proceedings of the Nordic Contributions in IS Research: Third Scandinavian Conference on Information Systems, SCIS 2012, Sigtuna, Sweden, 17–20 August 2012; Proceedings 3. Springer: Berlin/Heidelberg, Germany, 2012; pp. 161–176.
34. MazeMap Website. Available online: http://mazemap.com (accessed on 22 March 2023).
35. Cheung, A.K.L. Representational issues in interactive wayfinding systems: Navigating the Auckland University Campus. In Proceedings of the Web and Wireless Geographical Information Systems: 6th International Symposium, W2GIS 2006, Hong Kong, China, 4–5 December 2006; Proceedings 6. Springer: Berlin/Heidelberg, Germany, 2006; pp. 90–101.
36. Kim, S.; Park, E.; Hong, S.; Cho, Y.; del Pobil, A.P. Designing digital signage for better wayfinding performance: New visitors' navigating campus of university. In Proceedings of the 4th International Conference on Interaction Sciences, Busan, Republic of Korea, 16–18 August 2011; IEEE: New York, NY, USA, 2011; pp. 35–40.
37. Asif, M.; Krogstie, J. Mobile student information system. *Campus-Wide Inf. Syst.* **2011**, *28*, 5–15. [CrossRef]
38. Gao, S.; Krogstie, J.; Thingstad, T.; Tran, H. An empirical study of the adoption of an indoor location-based service: Finding reading rooms. *Int. J. Technol. Hum. Interact.* **2017**, *13*, 70–88. [CrossRef]
39. Castelli, L.; Corazzini, L.L.; Geminiani, G.C. Spatial navigation in large-scale virtual environments: Gender differences in survey tasks. *Comput. Hum. Behav.* **2008**, *24*, 1643–1667. [CrossRef]
40. Jeffri, N.F.S.; Rambli, D.R.A. A review of augmented reality systems and their effects on mental workload and task performance. *Heliyon* **2021**, *7*, e06277. [CrossRef]
41. Mehta, H.; Kanani, P.; Lande, P. Google maps. *Int. J. Comput. Appl.* **2019**, *178*, 41–46. [CrossRef]
42. Jong, S. Rethinking Spatial Abilities: A Multimethod Examination of Its Context-Dependent Nature and Whether Tests Require Increased Conceptual, Contextual, and Perceptual Similarity to Usage Context. Ph.D. Thesis, Carleton University, Ottawa, ON, Canada, 2015.
43. Hegarty, M.; Waller, D. A dissociation between mental rotation and perspective-taking spatial abilities. *Intelligence* **2004**, *32*, 175–191. [CrossRef]
44. Malanchini, M.; Rimfeld, K.; Shakeshaft, N.G.; McMillan, A.; Schofield, K.L.; Rodic, M.; Rossi, V.; Kovas, Y.; Dale, P.S.; Tucker-Drob, E.M. Evidence for a unitary structure of spatial cognition beyond general intelligence. *NPJ Sci. Learn.* **2020**, *5*, 9. [CrossRef]
45. Hart, S.G. NASA-task load index (NASA-TLX); 20 years later. In Proceedings of the Human Factors and Ergonomics Society Annual Meeting, San Fransisco, CA, USA, 16–20 October 2006; Sage publications Sage CA: Los Angeles, CA, USA, 2006; Volume 50, pp. 904–908.
46. Hart, S.G.; Staveland, L.E. Development of NASA-TLX (Task Load Index): Results of empirical and theoretical research. In *Advances in Psychology*; Elsevier: Amsterdam, The Netherlands, 1988; Volume 52, pp. 139–183.
47. Bell, S.W.; Kong, J.C.; Clark, D.A.; Carne, P.; Skinner, S.; Pillinger, S.; Burton, P.; Brown, W. The National Aeronautics and Space Administration-task load index: NASA-TLX: Evaluation of its use in surgery. *ANZ J. Surg.* **2022**, *92*, 3022–3028. [CrossRef]
48. Dong, W.; Wu, Y.; Qin, T.; Bian, X.; Zhao, Y.; He, Y.; Xu, Y.; Yu, C. What is the difference between augmented reality and 2D navigation electronic maps in pedestrian wayfinding? *Cartogr. Geogr. Inf. Sci.* **2021**, *48*, 225–240. [CrossRef]
49. Gioia, D.A.; Corley, K.G.; Hamilton, A.L. Seeking qualitative rigor in inductive research: Notes on the Gioia methodology. *Organ. Res. Methods* **2013**, *16*, 15–31. [CrossRef]
50. Creswell, J.W.; Creswell, J.D. *Research Design: Qualitative, Quantitative, and Mixed Methods Approaches*; Sage Publications: New York, NY, USA, 2017.
51. Hannula, O.; Irrmann, O. Played into collaborating: Design games as scaffolding for service co-design project planning. *Simul. Gaming* **2016**, *47*, 599–627. [CrossRef]
52. Lichtman, M. *Qualitative Research in Education: A User's Guide*; Sage Publications: New York, NY, USA, 2012.
53. Zhang, X.; Jin, L.; Zhao, J.; Li, J.; Luh, D.-B.; Xia, T. The Influences of Different Sensory Modalities and Cognitive Loads on Walking Navigation: A Preliminary Study. *Sustainability* **2022**, *14*, 16727. [CrossRef]

54. Xu, Y.; Qin, T.; Wu, Y.; Yu, C.; Dong, W. How do voice-assisted digital maps influence human wayfinding in pedestrian navigation? *Cartogr. Geogr. Inf. Sci.* **2022**, *49*, 271–287. [CrossRef]
55. Ishikawa, T. Satellite navigation and geospatial awareness: Long-term effects of using navigation tools on wayfinding and spatial orientation. *Prof. Geogr.* **2019**, *71*, 197–209. [CrossRef]
56. Albers, M.J. Tapping as a measure of cognitive load and website usability. In Proceedings of the 29th ACM International Conference on Design of Communication, Pisa, Italy, 3–5 October 2011; pp. 25–32.
57. Georgsson, M. NASA RTLX as a novel assessment tool for determining cognitive load and user acceptance of expert and user-based usability evaluation methods. *Eur. J. Biomed. Inform.* **2020**, *261*, 185–190.
58. Jöst, M.; Häußler, J.; Merdes, M.; Malaka, R. Multimodal interaction for pedestrians: An evaluation study. In Proceedings of the 10th International Conference on Intelligent User Interfaces, San Diego, CA, USA, 10–13 January 2005; pp. 59–66.
59. Vainio, T. Exploring multimodal navigation aids for mobile users. In Proceedings of the Human-Computer Interaction—INTERACT 2009: 12th IFIP TC 13 International Conference, Uppsala, Sweden, 24–28 August 2009; Proceedings, Part I 12. Springer: Berlin/Heidelberg, Germany, 2009; pp. 853–865.
60. Liljedahl, M.; Lindberg, S.; Delsing, K.; Polojärvi, M.; Saloranta, T.; Alakärppä, I. Testing two tools for multimodal navigation. *Adv. Hum.-Comp. Int.* **2012**, *2012*, 15. [CrossRef]
61. Kuriakose, B.; Shrestha, R.; Sandnes, F.E. Multimodal navigation systems for users with visual impairments—A review and analysis. *Multimodal Technol. Interact.* **2020**, *4*, 73. [CrossRef]
62. Dillemuth, J. Map design evaluation for mobile display. *Cartogr. Geogr. Inf. Sci.* **2005**, *32*, 285–301. [CrossRef]
63. Chang, H.H. Which one helps tourists most? Perspectives of international tourists using different navigation aids. *Tour. Geogr.* **2015**, *17*, 350–369. [CrossRef]
64. Caduff, D.; Timpf, S. A framework for assessing the salience of landmarks for wayfinding tasks. *Cogn. Process.* **2006**, *7*, 23. [CrossRef]
65. Golestanha, Z.; Satterfield, D. Immersive AR Landmark-Based Campus Wayfinding Solution with Focus on People with Navigation Difficulties. *Hum. Side Serv. Eng.* **2022**, *62*, 67.
66. Zheng, M.-C.; Hsu, Y.-W. How 2.5 D Maps Design Improve the Wayfinding Performance and Spatial Ability of Map Users. *Informatics* **2021**, *8*, 88. [CrossRef]
67. Brunyé, T.T.; Gardony, A.L.; Holmes, A.; Taylor, H.A. Spatial decision dynamics during wayfinding: Intersections prompt the decision-making process. *Cogn. Res. Princ. Implic.* **2018**, *3*, 13. [CrossRef]
68. Münzer, S.; Lörch, L.; Frankenstein, J. Wayfinding and acquisition of spatial knowledge with navigation assistance. *J. Exp. Psychol. Appl.* **2020**, *26*, 73. [CrossRef]
69. Kuhl, P.K.; Lim, S.-S.; Guerriero, S.; van Damme, D. Technologies for Spatial Learning. In *Developing Minds in the Digital Age: Towards a Science of Learning for 21st Century Education*; OECD: Paris, France, 2019.
70. Qiu, X.; Yang, Z.; Yang, J.; Wang, Q.; Wang, D. Impact of AR Navigation Display Methods on Wayfinding Performance and Spatial Knowledge Acquisition. *Int. J. Hum.–Comput. Interact.* **2023**, *27*, 1–21. [CrossRef]
71. Liu, B.; Ding, L.; Meng, L. Spatial knowledge acquisition with virtual semantic landmarks in mixed reality-based indoor navigation. *Cartogr. Geogr. Inf. Sci.* **2021**, *48*, 305–319. [CrossRef]
72. Snopková, D.; Švedová, H.; Kubíček, P.; Stachoň, Z. Navigation in indoor environments: Does the type of visual learning stimulus matter? *ISPRS Int. J. Geo-Inf.* **2019**, *8*, 251. [CrossRef]
73. Harper, C.; Duke, T.; Avera, A.; Crosser, A.; Jefferies, S.; Klisans, D.V. Exploring hospital wayfinding systems: Design guidelines for wayfinding interfaces. In *Advances in Human Factors and Ergonomics in Healthcare and Medical Devices: Proceedings of the AHFE 2020 Virtual Conference on Human Factors and Ergonomics in Healthcare and Medical Devices, Virtual Conference, USA, 16–20 July 2020*; Springer: Cham, Switzerland, 2020; pp. 30–36.
74. Foltz, M.A. Designing Navigable Information Spaces. Master's Thesis, Massachusetts Institute of Technology, Department of Electrical Engineering and Computer Science, Cambridge, MA, USA, 1998.
75. Mallik, M.; Panja, A.K.; Chowdhury, C. Paving the way with machine learning for seamless indoor–outdoor positioning: A survey. *Inf. Fusion* **2023**, *94*, 126–151. [CrossRef]
76. Sato, D.; Oh, U.; Guerreiro, J.; Ahmetovic, D.; Naito, K.; Takagi, H.; Kitani, K.M.; Asakawa, C. NavCog3 in the wild: Large-scale blind indoor navigation assistant with semantic features. *ACM Trans. Access. Comput. (TACCESS)* **2019**, *12*, 1–30. [CrossRef]
77. Huang, B.-C.; Hsu, J.; Chu, E.T.-H.; Wu, H.-M. Arbin: Augmented reality based indoor navigation system. *Sensors* **2020**, *20*, 5890. [CrossRef] [PubMed]
78. Li, Y.; Zhang, P.; Niu, X.; Zhuang, Y.; Lan, H.; El-Sheimy, N. Real-time indoor navigation using smartphone sensors. In Proceedings of the 2015 International Conference on Indoor Positioning and Indoor Navigation (IPIN), Banff, AB, Canada, 13–16 October 2015; IEEE: New York, NY, USA, 2015; pp. 1–10.
79. Zheng, M.-C.; Chen, C.-I. Designing indoor navigation interfaces on smartphones compatible with human information processing in an emergency evacuation scenario. *J. Asian Archit. Build. Eng.* **2019**, *18*, 599–616. [CrossRef]
80. Meliones, A.; Sampson, D. Blind MuseumTourer: A system for self-guided tours in museums and blind indoor navigation. *Technologies* **2018**, *6*, 4. [CrossRef]
81. Harper, C.; Duke, T.; Crosser, A.; Avera, A.; Jefferies, S. Designing hospital wayfinding systems, touchscreen kiosks, environmental cues and mobile apps: An evaluation of a mobile wayfinding application. In *Advances in Human Factors and Ergonomics in Healthcare*

and Medical Devices, Proceedings of the AHFE 2019 International Conference on Human Factors and Ergonomics in Healthcare and Medical Devices, Washington DC, USA, 24–28 July 2019; 10; Springer: Berlin/Heidelberg, Germany, 2020; pp. 89–96.
82. Ng, X.H.; Lim, W.N. Design of a mobile augmented reality-based indoor navigation system. In Proceedings of the 2020 4th International Symposium on Multidisciplinary Studies and Innovative Technologies (ISMSIT), Istanbul, Turkey, 22–24 October 2020; IEEE: New York, NY, USA, 2020; pp. 1–6.
83. Guo, C.; Liu, C.; Wei, E. Design of three-dimensional visualization indoor navigation system in smart city construction. In Proceedings of the E3S Web of Conferences, Odesa, Ukraine, 16 April 2021; EDP Sciencesp: Les Ulis, France, 2021; Volume 283, p. 02025.
84. Sandoval, E.B.; Li, B.; Diakite, A.; Zhao, K.; Oliver, N.; Bednarz, T.; Zlatanova, S. Visually Impaired User Experience using a 3D-Enhanced Facility Management System for Indoors Navigation. In Proceedings of the Companion Publication of the 2020 International Conference on Multimodal Interaction, Virtual, 25–29 October 2020; pp. 92–96.
85. Zheng, X.; Maredia, A.; Zahabi, M. A Scoping Literature Review and Content Analysis of Navigation Apps for Blind and Visually Impaired Users. In Proceedings of the Human Factors and Ergonomics Society Annual Meeting, Atlanta, GA, USA, 10–14 October 2022; SAGE Publications Sage CA: Los Angeles, CA, USA, 2022; Volume 66, pp. 2295–2299.
86. Alnabhan, A.; Tomaszewski, B. INSAR: Indoor navigation system using augmented reality. In Proceedings of the Sixth ACM SIGSPATIAL International Workshop on Indoor Spatial Awareness, Dallas/Fort Worth, TX, USA, 4 November 2014; pp. 36–43.
87. Swobodzinski, M.; Parker, A.T. *A Comprehensive Examination of Electronic Wayfinding Technology for Visually Impaired Travelers in an Urban Environment*; Transportation Research and Education Center (TREC): Portland, OR, USA, 2019.
88. Sanford, J.A.; Melgen, S.; Guhathakurta, S.; Zhang, G. *Utility of a Mobile Route Planning App for People Aging with Disability*; IARIA: Indianapolis, IN, USA, 2018.
89. Swobodzinski, M.; Parker, A.T. *Electronic Wayfinding for Visually Impaired Travelers: Limitations and Opportunities*; Transportation Research and Education Center (TREC): Portland, OR, USA, 2019.
90. Flores, J.Z.; Charbonnier, E.; Cornus, S.; Rasseneur, L. Navi Campus: Quantitative Methodology for Evaluating the User Interface of a Navigation App Using Eye Tracker and Smartphone. In Proceedings of the Sixth International Conference on Informatics and Assistive Technologies for Health-Care, Medical Support and Wellbeing, Barcelona, Spain, 3–7 October 2021; pp. 37–40.
91. Hall, A.J. *A Computer Vision and Maps Aided Tool for Campus Navigation*; Texas A&M University: College Station, TX, USA, 2022.
92. Liu, B.; Ding, L.; Wang, S.; Meng, L. Designing Mixed Reality-Based Indoor Navigation for User Studies. *KN-J. Cartogr. Geogr. Inf.* **2022**, *72*, 129–138. [CrossRef]
93. Liu, B.; Meng, L. Doctoral colloquium—Towards a better user interface of augmented reality based indoor navigation application. In Proceedings of the 2020 6th International Conference of the Immersive Learning Research Network (iLRN), San Luis Obispo, CA, USA, 21–25 June 2020; IEEE: New York, NY, USA, 2020; pp. 392–394.
94. Helmi, R.A.A.; Ravichandran, H.A.; Jamal, A.; Mohammed, M. Design and Development of Indoor Campus Navigation Application. In Proceedings of the 2022 IEEE 10th Conference on Systems, Process & Control (ICSPC), Malacca, Malaysia, 17 December 2022; IEEE: New York, NY, USA, 2022; pp. 77–82.
95. Puikkonen, A.; Sarjanoja, A.-H.; Haveri, M.; Huhtala, J.; Häkkilä, J. Towards designing better maps for indoor navigation: Experiences from a case study. In Proceedings of the 8th International Conference on Mobile and Ubiquitous Multimedia, Cambridge, UK, 22–25 November 2009; pp. 1–4.
96. Mendel, J.; Pak, R. The effect of interface consistency and cognitive load on user performance in an information search task. In Proceedings of the Human Factors and Ergonomics Society Annual Meeting, San Antonio, TX, USA, 19–23 October 2009; SAGE Publications Sage CA: Los Angeles, CA, USA, 2009; Volume 53, pp. 1684–1688.
97. Zulkiflie, S.A.; Kamaruddin, N.; Wahab, A. Dynamic navigation indoor map using Wi-Fi fingerprinting mobile technology. *Bull. Electr. Eng. Inform.* **2020**, *9*, 739–746. [CrossRef]
98. Apelt, R.; Crawford, J.; Hogan, D.J. *Wayfinding Design Guidelines*; CRC for Construction Innovation: Boca Raton, FL, USA, 2007.
99. Yuan, L.; Kong, F.; Luo, Y.; Zeng, S.; Lan, J.; You, X. Gender differences in large-scale and small-scale spatial ability: A systematic review based on behavioral and neuroimaging research. *Front. Behav. Neurosci.* **2019**, *13*, 128. [CrossRef]

applied sciences

MDPI

Article

Presenting Job Instructions Using an Augmented Reality Device, a Printed Manual, and a Video Display for Assembly and Disassembly Tasks: What Are the Differences?

Halimoh Dorloh, Kai-Way Li * and Samsiya Khaday

College of Management, Chung Hua University, Hsin-Chu 30012, Taiwan
* Correspondence: kai@chu.edu.tw; Tel.: +86-25-84395366

Abstract: Components assembly and disassembly are fundamental tasks in manufacturing and the product service industry. Job instructions are required for novice and inexperienced workers to perform such tasks. Conventionally, job instructions may be presented via printed manual and video display. Augmented reality (AR) device has been one of the recent alternatives in conveying such information. This research compared the presentation of job instruction via AR display, video display, and a printed manual in performing computer component assembly and disassembly tasks in terms of efficiency, quality, and usability. A Microsoft® HoloLens 2 device and a laptop computer were adopted to present the job instruction for the AR and video conditions, respectively. A total of 21 healthy adults, including 11 males and 10 females, participated in the study. Our findings were that AR display led to the least efficiency but the best quality of the task being performed. The differences of the overall usability scores among the three job instruction types were insignificant. The participants felt that support from a technical person for the AR device was significantly more than the printed manual. More male participants felt the AR display was easier to use than their female counterparts.

Keywords: augmented reality; job instruction; assembly and disassembly; usability

1. Introduction

Manufacturing is a complex process to produce goods. It is the process of changing raw materials into finished products with specific functions or values to fulfill the needs of customers. Automatic productions are widely used in contemporary manufacturing systems. Even in the highly industrialized era, workers are still required to perform certain tasks which may not be easily accomplished by machines [1]. With the advancement of science and technology, workers have opportunities to use sophisticated digital tools to accomplish their tasks more easily. One of the advancements in technology that facilitates the progress of industrialization is augmented reality (AR) [2].

AR is the technology to superimpose computer-generated images onto a real environment [3]. It is characterized by combining virtual objects in a real environment and allowing human interaction with the virtual objects in a real time manner [4,5]. AR is achievable in different ways. However, the most common approach to realize this technology is via the use of AR devices such as AR glasses [6]. A set of AR glasses is a transparent displayer which may display both 2D and/or 3D images in the visual field of the users. The images do not block the vision of the real world so that the users may see both what actually exists and those virtual visions generated by the AR device. The displayed images of the device may be assistive and supplemental to the users when incorporated into the real world. They may also be independent objects for special purposes.

AR devices have been applied for various purposes. They have been used as guidance for wire assembly tasks [7], product planning and design [8,9], logistics [10,11], maintenance and repairment [12,13], assembly and disassembly tasks [14–17], training and practices in both dentistry [18] and surgery [19,20], and so on. Empirical studies have shown the advantages of AR in training automotive service technicians [21,22] and medical students [19].

Assembly tasks are fundamental in both manufacturing and maintenance operations [14]. They may include alignment and fastening to put two or more parts together [23,24]. Disassembly tasks, on the other hand, refer to separating and removing parts in order to repair or replace them. They are fundamental in maintenance operations. Both assembly and disassembly are performed following specific standard operating procedures (SOPs) [25]. Assembly in mass productions is normally accomplished using automatic or semi-automatic machines unless they are technically unfeasible. Small lot size assembly and many disassembly tasks, on the other hand, are commonly performed manually. The SOPs for assembly and disassembly tasks are established by engineers and designers. They are established so that workers may follow them to perform the tasks successfully. Experienced and skilled workers may perform the tasks without reading the SOPs if they are very familiar with them. Novice or inexperienced workers, however, need to read those procedures either step by step or occasionally in order to complete their tasks.

Traditionally, hard copies of SOPs, such as printed manuals, are provided at workplaces. Inexperienced workers may read them step by step to accomplish their tasks. The procedures may also be presented on a computer monitor. Both printed manual and video presentation of job instructions have been used for a long time. Presenting a SOP in an AR device has been a relatively new approach in providing support to workers. There are studies discussing the use of an AR device at workplaces [26–28]. The literature has shown inconsistent results in using an AR device to convey job instructions in facilitating the assembly and disassembly tasks. For example, Tang et al. [29] has shown that using an AR device in assembly tasks reduced 82% of error rate. Several other authors, such as Fiorentino et al. [30], Wang et al. [16], and Masood and Egger [31], have shown that an AR device may improve worker efficiency and reduce the cost in both assembly and disassembly tasks. On the contrary of these findings, the literature [32,33] has also shown that the performance of wearing a set of AR glasses was not significantly higher than those without using such a device. There are even reports that AR devices could cause safety and health problems for workers including musculoskeletal problems [34] and motion sickness [35]. It is apparent that more studies are required to discover the benefits and applicability of using AR technology in assisting workers to perform assembly and disassembly tasks.

The purpose of this study was to compare the difference between using an AR device to convey job instructions and using traditional job instructions, including paper manual and computer monitor display, in assembly and disassembly tasks as performed in common maintenance and repairment services. In addition to the time to complete the task, the quality of the tasks being performed was assessed in terms of the number of errors made by the participants. Our hypothesis was that using an AR device to present job instructions may either improve the efficiency or the quality of the tasks being performed. The usability of the types of job instructions were also analyzed and discussed.

2. Materials and Methods

In this section, we introduce the participants, hand dexterity test, steps of assembly and disassembly tasks, types of job instruction, and experimental design and data analyses of this study.

2.1. Participants

A total of 21 participants, including 11 males and 10 females, were recruited. All the participants were college students without computer assembly and disassembly experience. Fifteen (nine males and six females) of them were right-handed. The other six (four females and two males) were left-handed. A Landolt C chart and a color book were used to examine the visual acuity and color deficiency of the participants. The mean ($\pm$SD) of the age and decimal visual acuity (both eyes) were 20.70 ($\pm$1.34) yrs and 0.91 ($\pm$0.27), respectively, for females and 19.91 (±2.16) yrs and 0.95 (±0.22), respectively, for males. The stature and body weight of female participants were 158.91 (±5.53) cm and 56.30 (±12.22) kg, respectively, and 172.55 (±5.84) cm and 66.73 (±12.56) kg, respectively, for male participants. None of the participants had color deficiency. Each participant received a cash reward of TWD 600 for their participation in the experiment.

2.2. Hand Dexterity Tests

We measured the hand dexterity of the participants using both the Purdue Pegboard Test (PPT) and the Minnesota Dexterity Test (MDT) (see Figure 1). The test kits were placed on a table 75 cm high. The PPT required the seated participant to place as many pegs into the holes in 30 s, first with each hand separately and then together, followed by the assembly of as many pegs, washers, and collars as possible in 1 min [36–39]. We recorded the average numbers of pegs completed for two trials both using dominant hand in 30 s and assemblies of pegs, washers, and collars using both hands in 1 min for each trial. The MDT test included a turning test and a placing test [40,41]. The participants stood in front of the table to perform the tests. We recorded the average times of two trials to complete the placing test using dominant hand and the time to complete the turning test using both hands.

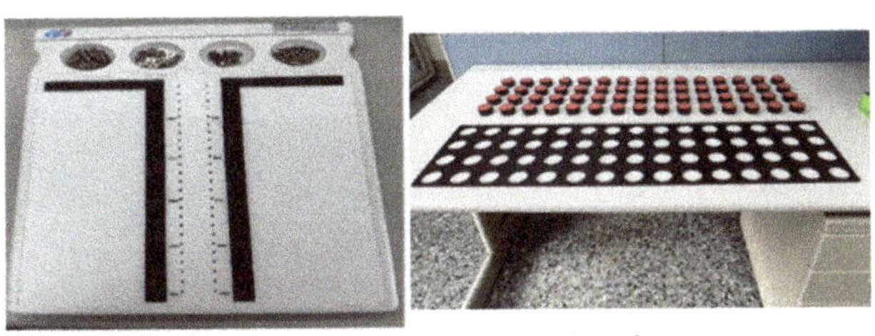

(a) (b)

Figure 1. Manual dexterity test kits. (**a**) Purdue Pegboard Test (PPT); (**b**) Minnesota Dexterity Test (MDT).

2.3. Assembly and Disassembly Tasks

A desktop computer (ASUS®, ASD760, Taipei, Taiwan; see Figure 2) was prepared for the assembly and disassembly tasks on a table 75 cm above the floor. In computer maintenance and repairment services, only partial components are disassembled and assembled. In this study, we requested our participants to disassemble and assemble the following components: a fan unit, a power supply, a DVD drive, and a motherboard (see Figure 3).

Figure 2. A desktop computer.

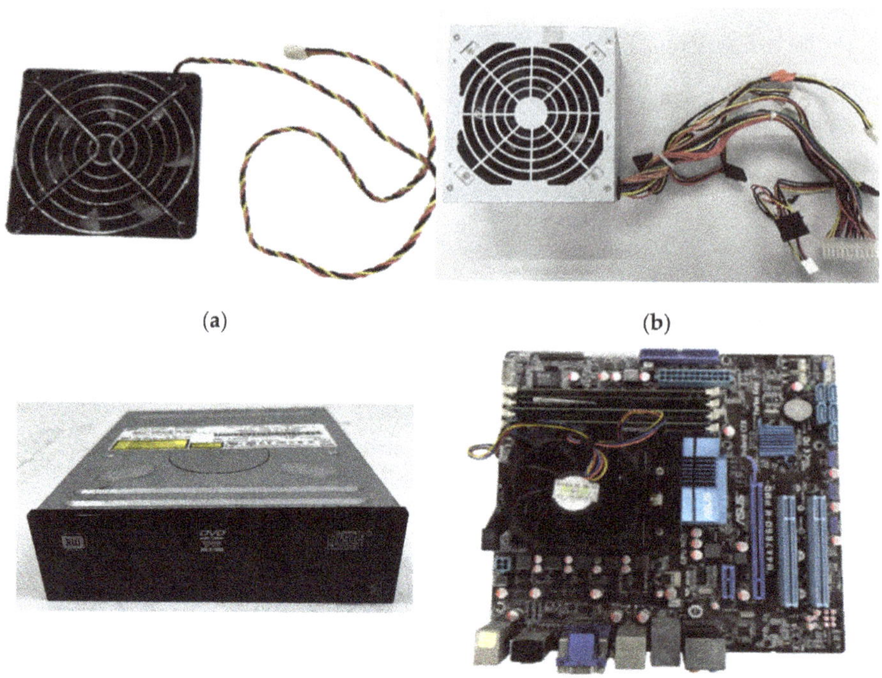

(**a**) (**b**) (**c**) (**d**)

Figure 3. Components in the assembly/disassembly tasks. (**a**) Fan; (**b**) power supply; (**c**) DVD drive; (**d**) motherboard.

The steps of assembly and disassembly tasks are shown in Table 1.

For each assembly and disassembly task, a total of 20 screws needed to be tightened or unscrewed and 4 cable connectors needed to be plugged or unplugged.

2.4. Job Instructions

The job instructions or the steps of assembly and disassembly of the computer components were presented in three different ways, including a paper manual, a video displayed in a laptop computer, and an AR display. For all these job instructions, there were no vocal explanations.

For the printed manual, Chinese versions of both the disassembly and assembly steps are printed as two 6-page manuals. The steps in these manuals are illustrated using text accompanied with a photo highlighting the locations of the screws and cable connections (see Figure 4). For video display on a laptop computer (Acer® Swift 3, SF314-42, New Taipei, Taiwan), the steps were shown in a video without text explanation (see Figure 5). The participant could click to pause, reverse, and continue to see what he or she needed to do. The locations and connections were not highlighted in this condition because they were obvious in the video.

Table 1. Steps of assembly and disassembly tasks.

Assembly Task	Disassembly Task
1. Pick up the DVD drive from the table and put it into the DVD slot of the case.	1. Use a screwdriver to unscrew two screws of the panel. Put the screws in grid 1 in the container on the right.
2. Pick up the screwdriver and two screws in grid 5 of the container. Tighten the screws to fix the DVD drive.	2. Pull and remove the panel and put it in the assigned location on the table.
3. Pick up the motherboard on the table using both hands and put it in the assigned area inside the computer case.	3. Unplug the fan cable connected to the motherboard.
4. Pick up the screwdriver and 8 screws in grid 4 of the container. Tighten the screws to fix the motherboard.	4. Use a screwdriver to unscrew 4 screws of the fan unit and put them in grid 2 in the container. Put the fan unit in the assigned location on the table.
5. Pick up the power supply unit on the table and put it in the assigned location inside the computer case.	5. Use a screwdriver to unscrew 4 screws of the power supply unit and put the screws in grid 3 of the container.
6. Pick up the screwdriver and 4 screws in grid 3 of the container. Tighten the screws to fix the power supply unit.	6. Unplug the two cables of the power supply connected to the motherboard. Unplug the 3rd cable of the power supply connected to the DVD drive.
7. Grab a cable connector attached to the power supply unit (see the job instruction) and plug it into the designated slot on the motherboard. Press it firmly. Grab another cable connector attached to the power supply (see the job instruction) and plug it into another designated slot on the motherboard. Press it firmly. Grab the third cable connector attached to the power supply (see the job instruction) and plug it into the designated slot on the DVD drive. Press it firmly to ensure successful connection.	7. Remove the power supply unit and put it in the assigned location on the table.
8. Pick up the fan unit on the table and put it in the assigned area inside the computer case.	8. Use a screwdriver to unscrew 8 screws fixing the motherboard and put the screws in grid 4 in the container.
9. Pick up the screwdriver and 4 screws in grid 2 of the container. Tighten the screws to fix the fan unit.	9. Take the motherboard out using both hands and put it in the assigned location on the table.
10. Grab the connector of the fan cable and plug it into the designated slot on the motherboard. Press it firmly to ensure successful connection.	10. Use a screwdriver to unscrew the two screws of the DVD drive and put them in grid 5 in the container.
11. Pick up the panel on the table using both hands and put it on the top of the computer case.	11. Push then pull out the DVD drive from its slot and put it in the assigned location on the table.
12. Pick up the screwdriver and 2 screws in grid 1 of the container. Tighten the screws to fix the panel.	

1. Panel

Take out 2 screws at the back, and put the screws into slot 1 of the box at the right of PC.

Figure 4. Step 1 of disassembly task in the printed manual.

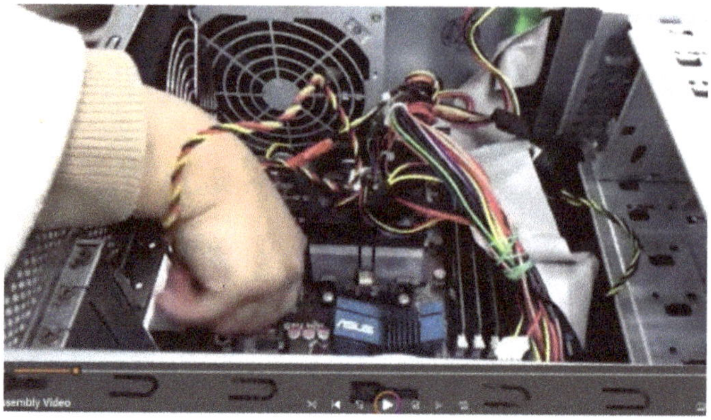

Figure 5. Video display showing how to unplug the cable of the fan connected to the motherboard in the disassembly task (step 3).

For AR-displayed condition, a set of smart glasses (HoloLens 2, Microsoft® Inc., Redmond, WA, USA) was adopted as the AR device in presenting the instructions of the tasks (see Figure 6). This device encompasses see-through holographic lenses. Users may see both virtual instruction and the components in the computer. The Dynamics 365 Guides app (Microsoft® Inc., Redmond, WA, USA) in the HoloLens 2 was adopted to present the job instructions. The text (in Chinese) of each step for both assembly and disassembly tasks were incorporated into the Dynamic 365 app. Virtual circles and squares were added using the AR toolkit to highlight the locations of the screws and cable connections, respectively (see Figure 7), so as to provide supplemental information. When a cable needed to be plugged or unplugged, a virtual hand appeared to highlight the cable (dis)connection operation (see Figure 8). Virtual hand(s) also appeared when pushing or pulling a component was required (see Figures 9 and 10). The virtual hand(s) disappeared after the participant had followed this instruction. The participant might click on a virtual arrow on the top right corner of the instruction to go to the next step when he/she has completed one step using his/her index finger.

Figure 6. Microsoft® HoloLens 2 AR glasses.

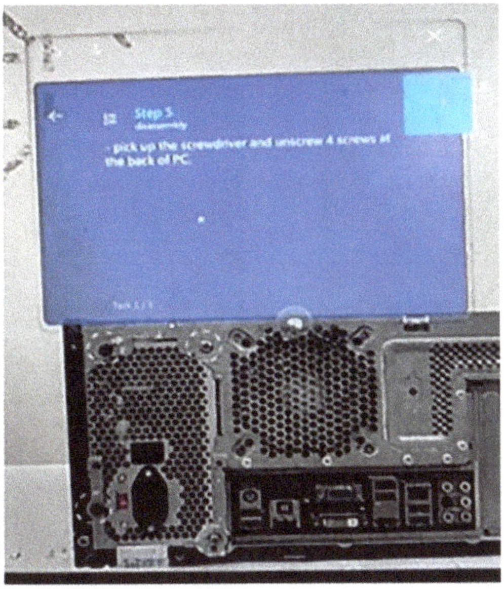

Figure 7. Virtual image in the HoloLens 2 showing both the text explanation to unscrew the power supply unit and the highlighted (dotted circles) screws to be unscrewed and removed (step 5 of disassembly task).

Figure 8. A virtual hand highlights the cable needed to be unplugged.

Figure 9. Virtual hands show how to grab the motherboard to move it out.

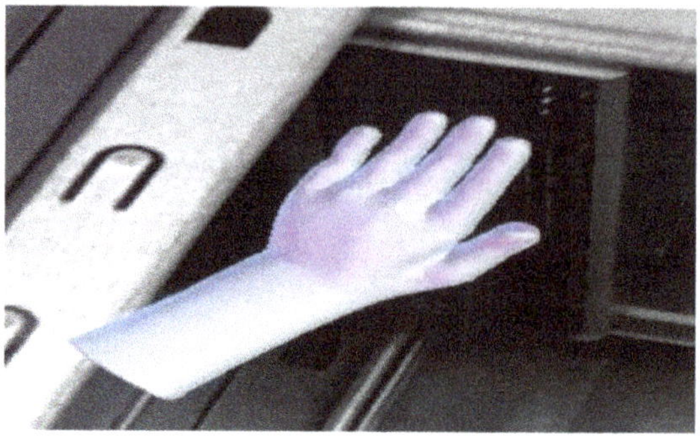

Figure 10. A virtual hand shows that a push on the DVD drive is needed to remove it from computer case.

All the participants could see and read the steps of assembly and disassembly tasks presented by the three types of job instruction without difficulty.

2.5. Procedure

Before the experiment, an experimenter explained the purposes and procedures of the experiment. The components, cables and connectors, and screws were displayed, and the participants were requested to be able to identify them. Each participant attended six trials or treatments comprised of three types of instruction (printed manual, computer display, and AR display) and two tasks (assembly and disassembly). For the assembly task, the components were placed in each assigned location on the table by the experimenter before a trial. For the disassembly task, the location for each component disassembled was marked using colored tapes. For each trial, the participant was instructed that he or she needs to follow the instructions to complete the task. For the printed manual and laptop computer conditions, the printed manual and laptop computer were placed on the left of the computer to be handled (see Figures 11 and 12). For the AR condition, the virtual manual was activated by an experimenter. The participant put on the HoloLens 2 head piece and had an opportunity to practice using the hand gesture to interact with the virtual manual. After the participant felt he or she could use this device without difficulty, he or she then started to perform the task once after hearing a vocal signal (see Figure 13). The trial ended when the participant completed the last step. The time to complete the task was recorded using a stopwatch.

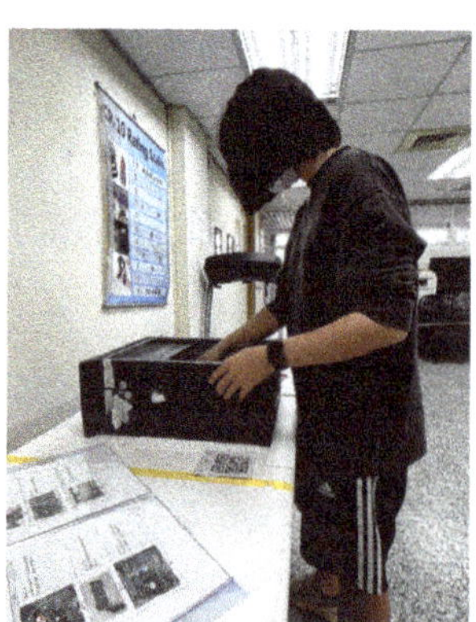

Figure 11. Job instruction on a printed manual.

Figure 12. Job instruction via a video display.

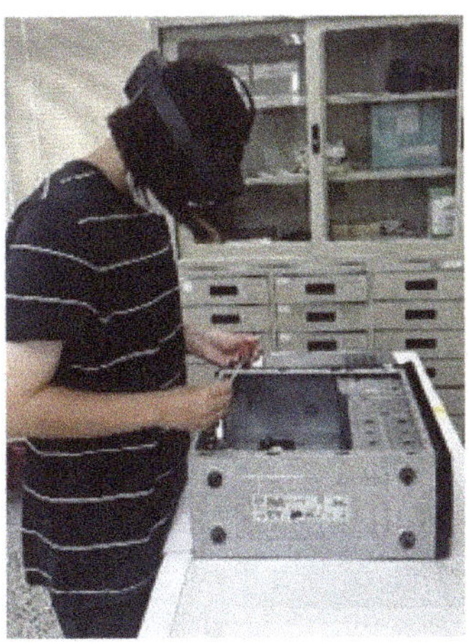

Figure 13. Job instruction via an AR display.

An error in performing the task was recorded when the participant tightened or unscrewed a wrong screw or in a wrong screw hole, plugged or unplugged a wrong cable or in a wrong slot, and picked up or put a screw in a wrong grid in the container. In addition, the participant was required to fix each screw tight enough and press firmly when connecting cables in the assembly task. An error was also recorded if the participant did not put a screw tight enough or did not press the connector on a cable into the slot firmly (this could lead to missed connection) in the assembly task. An experimenter observed the participant performing the trial. She recorded the number of errors without telling the participant.

After the participant completed the trial, he or she filled out a rating scale of system usability (SUS) [42,43] to measure the usability of each job instruction type. The ratings of the SUS include 10 questions (see Appendix A). The participant rated each question on a 5-point Likert scale (1—strongly disagree to 5—strongly agree). After the participant completed one trial, he or she took a break of 4 h or longer before he or she joined the next trial so as to avoid the effects of fatigue on the dependent variables.

2.6. Experimental Design and Data Analyses

A completely randomized block design experiment was performed. As mentioned earlier, the order of the trials was randomized for each participant. A total of 126 trials (21 participants × 3 job instructions × 2 tasks) were performed. The results of the PPT and MDT were recorded to indicate the hand dexterity of the participants. The dependent variables were time to complete the task and rating of the SUS. For the SUS, the score of each question ranged between 1 and 5. The calculations of the total score followed those recommended by Brooke [42]. In addition to the total score, the individual scores of questions 3 (ease of use), 4 (need support from a technical person to use this system), and 10 (need to learn many things before using this system) were also analyzed.

Descriptive statistics were performed. Analyses of variance were performed to determine the significance of the factors affecting the dependent variables. Duncan's multiple range tests were performed if the main effects of the factors reached the $\alpha = 0.05$ level of significance. Pearson's correlation coefficients between the hand dexterity scores and the time to complete the task were calculated to determine the correlation between the variables. Statistical analyses were performed using the SPSS® 20 software (IBM®, Armonk, NY, USA).

3. Results

3.1. Hand Dexterity

Table 2 shows the mean (SD) of the hand dexterity scores of the PPT and MDT for female and male participants. The differences of all the scores between male and female participants was not statistically significant.

Table 2. Mean (SD) scores of the PPT and MDT.

	PPTd	PPTa	MDTd (s)	MDTt (s)
Female	16.70 (1.16)	43.40 (3.08)	62.35 (4.55)	55.80 (8.16)
Male	16.86 (1.75)	41.55 (6.31)	60.55 (6.24)	51.73 (6.86)

Note: PPTd: dominant hand test of the PPT; PPTa: assembly test using both hands of the PPT; MDTd: placing test using dominant hand of the MDT; MDTt: turning test using both hands of the MDT.

3.2. Performance of the Task

The performance of the task was measured using the time to complete the task and the number of errors made in performing the task. These two variables were adopted to indicate the efficiency and quality of the task being performed. Table 3 shows the statistics of the time to complete the task. The ANOVA results of this variable indicated that the effects of type of task ($p < 0.0001$), gender ($p < 0.05$), and type of instruction ($p < 0.05$) were all statistically significant. None of the interaction effects of those three factors were significant at $\alpha = 0.05$ level. The Duncan's multiple range test results showed that the time to complete the assembly task (806.11 ± 234.96 s) was significantly ($p < 0.05$) higher than that of the disassembly task (486.63 ± 161.80 s). The time to complete the assembly and disassembly tasks for females (683.35 ± 272.48 s) was significantly ($p < 0.05$) higher than that of males (612.75 ± 239.35 s). The time to complete the assembly and disassembly tasks for AR instruction (714.55 ± 235.11 s) was significantly ($p < 0.05$) higher than those of the printed manual (627.88 ± 273.94 s) and video instruction (596.69 ± 252.05 s). The time to complete the task between using printed manual and video instruction was not statistically significant.

Table 3. Means and standard deviations (SD) of the time(s) to complete the task.

Task	Gender	Instruction	Mean	SD
Assembly	Female	AR	872.30	253.52
		Printed manual	804.10	182.90
		Video	878.90	257.28
	Male	AR	817.72	173.96
		Printed manual	797.63	333.39
		Video	678.45	159.66
Disassembly	Female	AR	657.90 *	237.49
		Printed manual	482.50	170.80
		Video	404.30	82.99
	Male	AR	519.45	77.96
		Printed manual	430.09	115.07
		Video	433.18	112.94

* $p < 0.05$.

Table 4 shows the number of errors made during performing the task. All those errors occurred in the assembly tasks. The number of errors made by the participants ranged from 0 to 4. The participants did not make any errors in the disassembly tasks. The mean number of errors made by each participant for printed manual, video, and AR instructions were 1.33, 0.62, and 0.33. The participants who followed the printed manual and video instruction made an error four times and almost two times the errors of those following the AR instruction. The ANOVA results showed that the type of instruction significantly ($p < 0.01$) affected the number of errors made by the participants. The difference between male and female participants in making errors was not significant. Duncan's multiple range test results indicated that the AR instruction led to significantly ($p < 0.05$) lower number of errors than that of the printed manual. The difference between that of AR and video instructions and between that of video instruction and printed manual were both insignificant.

Table 4. Number of errors per participant during performing the task.

Gender	AR Instruction	Video Instruction	Printed Manual
Female	0.40	0.70	1.50
Male	0.27	0.55	1.18
total	0.33 *	0.62	1.33

* $p < 0.05$.

3.3. Correlation Analyses

Theoretically, hand dexterity affects the performance of manual assembly and disassembly tasks. Workers with better hand dexterity should have better performance in manual tasks. This was the reason why we measured the hand dexterity scores and calculated the correlation coefficients between these scores and the time to complete the task. Pearson's correlation coefficients between the hand dexterity score and time to complete the task are shown in Table 5. The correlation coefficients between the PPTd and the time of assembly task using video and AR instructions were −0.47 ($p < 0.05$) and −0.52 ($p < 0.05$), respectively. The correlation coefficients between the time to complete the assembly task using video instruction and the MDTd and MDTt were 0.43 and 0.46, respectively, and both were significant at $p < 0.05$.

3.4. Usability Test Results

Table 6 shows the means and standard deviations of the SUS scores. A high SUS score implies good usability. The ANOVA results showed that gender was a significant ($p < 0.0001$) factor affecting the SUS score. The SUS score of male participants (66.1 ± 15.3) was significantly higher than that (55.6 ± 12.8) of female participants. The effects of instruction type were insignificant. Neither were the effects of task significant.

Table 5. Pearson's correlation coefficients between time to complete the task and hand dexterity test score.

	Assembly			Disassembly		
	Printed Manual	Video	AR	Printed Manual	Video	AR
PPTd	0.10	−0.47 *	−0.52 *	−0.37	−0.04	−0.24
PPTa	0.34	−0.25	−0.12	−0.14	0.17	0.11
MDTd	−0.10	0.43 *	0.23	0.25	0.05	0.17
MDTt	−0.02	0.46 *	0.36	0.16	0.09	−0.07

* $p < 0.05$.

Table 6. Means and standard deviations (SD) of the SUS score.

Table	Gender	Instruction	SUS	
			Mean	SD
Assembly	Females	AR	52.00	12.00
		Printed manual	52.75	11.87
		Video	60.75	9.13
	Males	AR	63.63	17.04
		Printed manual	66.59	15.05
		Video	67.72	10.98
Disassembly	Females	AR	52.25	15.91
		Printed manual	54.00	13.29
		Video	62.00	12.57
	Males	AR	73.86	14.67
		Printed manual	65.68	20.55
		Video	71.36	12.61

The participant indicated the ease of use of the job instruction in Question 3 in the SUS. The ANOVA results of this response showed that both gender and interactions of gender and instruction were statistically significant ($p < 0.05$ for both). Duncan's multiple range test results showed that the score of males (3.61 ± 1.09) was significantly higher than that of females (3.18 ± 0.83). Figure 14 shows the interaction effects of gender and job instruction on the score of ease of use.

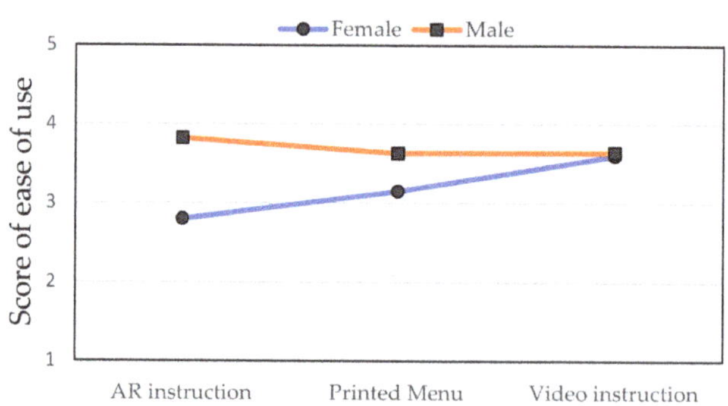

Figure 14. Interaction effects of gender and job instructions on the score of ease of use.

Question 4 in the SUS measures the need of support from a technical person to use a system. The ANOVA results of this response showed that both gender ($p < 0.0001$) and job instruction ($p < 0.001$) affected this score significantly. All other effects were insignificant.

Duncan's multiple range test results indicated that the score of need of support for female participants (2.95 ± 1.13) was significantly higher than their male counterparts (2.09 ± 1.05). For job instruction, Duncan's multiple range test results indicated that the score of need of support for AR instruction (3.05 ± 1.03) was significantly higher than those of the printed manual (2.36 ± 1.18) and video instruction (2.09 ± 1.08). The difference of the score of need support between that of printed manual and video display was not significant.

Question 10 in the SUS measures the need to learn to use a system. The only significant factor was gender ($p < 0.0001$). The score of females (2.90 ± 1.16) was significantly higher than that of males (1.89 ± 1.20). Type of job instruction did not affect this score significantly.

4. Discussion

The order of the assembly/disassembly and types of job instructions were randomly arranged for each participant. It was, therefore, reasonable to assume that learning effects contributed equally to both the task and type of job instruction in this study.

The time to complete the task was adopted as the main variable to indicate the performance of the participants. For inexperienced participants, assembly tasks required significantly ($p < 0.05$) longer time to complete than disassembly. This was not surprising at all. Disassembly is always easier and less time consuming than assembly. Our data showed that disassembling the four components needed only 60% of the time of assembling them. Gender was also a significant ($p < 0.05$) factor affecting the time to complete the assembly/disassembly tasks. Males could complete their tasks faster than their female counterparts. The time to complete the tasks for male participants was only 90% of their female counterparts. This gender difference might be attributed to the motor skills of males and females to tighten and loosen the screws. For job instructions, the participants using AR displays spent significantly ($p < 0.05$) longer time than those of the video display and printed manual conditions. Using AR display, on average, spent 118 and 87 s longer than those video display and printed manual, respectively, in performing the assembly and disassembly tasks. This was consistent with the findings in the literature [32,33] that wearing an AR device might not lead to better efficiency than those who did not wear one and watched a wall-mounted monitor instead. This, however, was inconsistent with the findings of some of the other literature [16,30,31]. All the participants were familiar with clicking a mouse to activate a continue button on a laptop display and to flip a sheet of paper of the manual. On the contrary, none of them were familiar with tapping on the "next page" button on the virtual-displayed instruction on the HoloLens 2. This might result in a longer time accessing the job instructions. The virtual hand(s) to highlight the push/pull movements was one of the innovations of the job instruction we created using the Dynamic 365 app in HoloLens 2 for the tasks. They clearly indicated what the participants should do using their hand(s). However, the participants might be distracted by this innovation and responded slower because of the attraction of the hand(s).

In addition to the time to complete the task, the errors made by the participant was also an important variable to assess the performance of the tasks. There were no errors made in the disassembly tasks. There were, however, 28, 13, and 7 errors in the printed manual, video display, and AR display conditions for the assembly task. The majority of the errors were observed when the participant was assembling the power supply. This component has three cables. Two of them needed to be plugged into two slots on the motherboard and the third needed to be plugged into the DVD drive. Most of the errors were that the participant simply put the connector into the slot without pressing it firmly. An assembly task failed so long as any cable connector was not well connected. The numbers of unsuccessful assembly for the printed manual, video display, and AR display conditions were 14, 8, and 5, respectively, out of 21. Job instruction via AR display apparently led to less failed assembly, which implies better quality of the tasks. This advantage might be attributed to innovation of virtual hand(s) to highlight the hand movements to be performed. Our results in the errors associated with the AR device were consistent with the findings of Tang et al. [29].

The hand dexterity of the participants was measured using both the PPT and MDT. The PPTd was negative correlated with the time to complete the assembly task for both video (-0.47, $p < 0.05$) and AR instructions (-0.52, $p < 0.05$). This was reasonable because high PPT number implies high hand dexterity. High hand dexterity should be associated with less time to complete the task. The correlation coefficients between the PPT scores tested and the time to complete the tasks for other task/job instruction conditions were, however, insignificant. For the MDT results, the MDTd (0.43, $p < 0.05$) and MDTt (0.46, $p < 0.05$) were both positively correlated with the time to complete the mission for video-displayed job instruction tasks. This was also reasonable because low MDT time implies high hand dexterity, which should be associated with less time to complete the task.

Usability is a core item in assessing a device or a system for human use. It has been defined as "the effectiveness, efficiency, and satisfaction with which specified users can achieve goals in particular environments" [44]. We used the SUS scores [42] to assess the usability of the type of job instructions for male and female participants. In addition to the total score, we specifically analyzed the individual score of questions 3, 4, and 10, which assess the ease of use, need support from a technical person, and need to learn, respectively. The score of ease of use of the SUS results showed that both gender and gender × job instruction interactions were significant ($p < 0.05$). Figure 14 shows that the difference of the score of ease of use between males and females was most significant for the AR instruction condition, next for printed manual condition. For the video display condition, there was no difference between the two genders. This implies that AR instruction was easier to use for males than for females. Regarding the question asking the need of support from a technical person to use a system, female participants responded with a significantly higher score than male participants. This implies that females needed support from a technical person more than males. Comparing the need to learn for different job instructions, the score of AR instruction was significantly higher than those of the printed manual and video instruction. This was consistent with what actually happened in that all the participants needed both detailed instructions from the experimenter on how to use the HoloLens 2 to access the job instructions and to practice. For the other two types of job instructions, only simple verbal instruction was required.

There are limitations to the study. There were essential differences among the three types of job instructions. The information displayed on printed manual and AR display were both static and the image displayed on video was dynamic. For printed manual, the participants needed to flip the printed manual three times to read all six pages. For AR display, the participants needed to click 12 and 11 times on the arrow button to go to the next step for the assembly and disassembly tasks, respectively. For video display, the number of clicking the mouse to pause and display was dependent on the participant. The participants might watch multiple steps and make one mouse click to pause. This might reduce their time in performing the tasks. These might be the main reasons that AR display was most time consuming and video display was most efficient. In fact, dynamic information such as video display may also be presented via the HoloLens 2. Future research may be designed to compare the difference between presenting static and dynamic job instruction on an AR device. Another limitation of this study was that we did not consider vocal information in our job instructions. In fact, vocal instruction may be added to both the video and AR displays. Future research may also be designed to study the effects of vocal information in the job instruction on the performance of assembly and disassembly tasks. In addition, we only considered four components in the assembly and disassembly tasks. Assembly and disassembly of all the components in a computer may be investigated in the future using both AR and other types of job instruction presentation. Finally, it should be noted that there are many other AR devices, and we only used the HoloLens 2 in the AR instruction condition. Our results may not be generalized to AR devices other than the HoloLens 2.

5. Conclusions

An experiment was performed to compare the performance and usability of printed manual, video display, and AR display in presenting job instructions for computer assembly and disassembly tasks. The results showed that the time to complete the tasks by reading the job instructions presented via the HoloLens 2 AR glasses was significantly longer than those of the printed manual and video display conditions. Using the AR-based job instructions did not improve the efficiency in assembly and disassembly of computer components under our experimental conditions. On the other hand, the results of the number of errors made by the participants indicated that, among the three types of job instruction, AR-based job instructions led to the best quality of the tasks being performed. The differences of the overall usability score among the three types of job instruction were insignificant. However, more male participants felt that the AR-based job instruction was easy to use than their female counterparts. Support from a technical person was needed more for the AR-based job instruction than the other two traditional instruction types.

Author Contributions: Conceptualization, H.D. and K.-W.L.; methodology, H.D.; software, H.D.; validation, S.K.; formal analysis, H.D. and K.-W.L.; resources, K.-W.L.; data curation, H.D.; writing—original draft preparation, H.D. and K.-W.L.; writing—review and editing, S.K.; visualization, S.K.; project administration, K.-W.L.; funding acquisition, K.-W.L. All authors have read and agreed to the published version of the manuscript.

Funding: This research was funded by National Science and Technology Council of the ROC, grant number MOST 110-2221-E-216-004.

Institutional Review Board Statement: This study was conducted according to the guidelines of the Declaration of Helsinki and was approved by an external ethics committee (Jen-Ai Hospital, 110-55).

Informed Consent Statement: Informed consent was obtained from all subjects involved in the study.

Data Availability Statement: The data presented in this study are available upon request from the corresponding author. The data are not publicly available due to privacy reasons.

Conflicts of Interest: The authors declare no conflict of interests.

Abbreviations

AR	Augmented reality
MDT	Minnesota Dexterity Test
MDTd	Placing test using dominant hand of Minnesota Dexterity Test
MDTt	Turning test using both hands of Minnesota Dexterity Test
PPT	Purdue Pegboard Test
PPTa	Assembly of Purdue Pegboard Test
PPTd	Dominant hand of Purdue Pegboard Test
SD	Standard deviation
SOPs	Standard operating procedures
SUS	System usability scale

Appendix A

Questions of the system usability scale (SUS)

1. I think that I would like to use this system frequently.
2. The system was unnecessarily complex.
3. The system was easy to use.
4. I would need the support of a technical person so as to use this system.
5. Various functions in this system were well integrated.
6. There was too much inconsistency in this system.
7. I believe that most people would learn to use this system very quickly.
8. The system was very cumbersome to use.

9. I felt very confident using this system.
10. I needed to learn many things before using this system.

References

1. Stock, T.; Seliger, G. Opportunities of Sustainable Manufacturing in Industry 4.0. *Procedia CIRP* **2016**, *40*, 536–541. [CrossRef]
2. Rentzos, L.; Papanastasiou, S.; Papakostas, N.; Chryssolouris, G. Augmented Reality for Human-Based Assembly: Using Product and Process Semantics. *IFAC Proc. Vol.* **2013**, *46*, 98–101. [CrossRef]
3. Doil, F.; Schreiber, W.; Alt, T.; Patron, C. Augmented Reality for Manufacturing Planning. In Proceedings of the Workshop on Virtual Environments 2003-EGVE '03, Zurich, Switzerland, 22–23 May 2003; ACM Press: Zurich, Switzerland, 2003; pp. 71–76. [CrossRef]
4. Azuma, R.; Baillot, Y.; Behringer, R.; Feiner, S.; Julier, S.; MacIntyre, B. Recent Advances in Augmented Reality. *IEEE Comput. Graph. Appl.* **2001**, *21*, 34–47. [CrossRef]
5. Boboc, R.G.; Băutu, E.; Gîrbacia, F.; Popovici, N.; Popovici, D.-M. Augmented Reality in Cultural Heritage: An Overview of the Last Decade of Applications. *Appl. Sci.* **2022**, *12*, 9859. [CrossRef]
6. Zhao, C.; Li, K.W.; Peng, L. Movement Time for Pointing Tasks in Real and Augmented Reality Environments. *Appl. Sci.* **2023**, *13*, 788. [CrossRef]
7. Navab, N. Developing Killer Apps for Industrial Augmented Reality. *IEEE Comput. Graph. Appl.* **2004**, *24*, 16–20. [CrossRef]
8. Luh, Y.-P.; Wang, J.-B.; Chang, J.-W.; Chang, S.-Y.; Chu, C.-H. Augmented Reality-Based Design Customization of Footwear for Children. *J. Intell. Manuf.* **2013**, *24*, 905–917. [CrossRef]
9. Nee, A.Y.C.; Ong, S.K.; Chryssolouris, G.; Mourtzis, D. Augmented Reality Applications in Design and Manufacturing. *CIRP Ann.* **2012**, *61*, 657–679. [CrossRef]
10. Plakas, G.; Ponis, S.T.; Agalianos, K.; Aretoulaki, E.; Gayialis, S.P. Augmented Reality in Manufacturing and Logistics: Lessons Learnt from a Real-Life Industrial Application. *Procedia Manuf.* **2020**, *51*, 1629–1635. [CrossRef]
11. Stoltz, M.-H.; Giannikas, V.; McFarlane, D.; Strachan, J.; Um, J.; Srinivasan, R. Augmented Reality in Warehouse Operations: Opportunities and Barriers. *IFAC-Pap.* **2017**, *50*, 12979–12984. [CrossRef]
12. Henderson, S.; Feiner, S. Exploring the Benefits of Augmented Reality Documentation for Maintenance and Repair. *IEEE Trans. Vis. Comput. Graph.* **2011**, *17*, 1355–1368. [CrossRef]
13. Quandt, M.; Knoke, B.; Gorldt, C.; Freitag, M.; Thoben, K.-D. General Requirements for Industrial Augmented Reality Applications. *Procedia CIRP* **2018**, *72*, 1130–1135. [CrossRef]
14. Baird, K.M.; Barfield, W. Evaluating the Effectiveness of Augmented Reality Displays for a Manual Assembly Task. *Virtual Real.* **1999**, *4*, 250–259. [CrossRef]
15. Danielsson, O.; Syberfeldt, A.; Holm, M.; Wang, L. Operators Perspective on Augmented Reality as a Support Tool in Engine Assembly. *Procedia CIRP* **2018**, *72*, 45–50. [CrossRef]
16. Wang, X.; Ong, S.K.; Nee, A.Y.C. A Comprehensive Survey of Augmented Reality Assembly Research. *Adv. Manuf.* **2016**, *4*, 1–22. [CrossRef]
17. Westerfield, G.; Mitrovic, A.; Billinghurst, M. Intelligent Augmented Reality Training for Motherboard Assembly. *Int. J. Artif. Intell. Educ.* **2015**, *25*, 157–172. [CrossRef]
18. Fahim, S.; Maqsood, A.; Das, G.; Ahmed, N.; Saquib, S.; Lal, A.; Khan, A.A.G.; Alam, M.K. Augmented Reality and Virtual Reality in Dentistry: Highlights from the Current Research. *Appl. Sci.* **2022**, *12*, 3719. [CrossRef]
19. Barcali, E.; Iadanza, E.; Manetti, L.; Francia, P.; Nardi, C.; Bocchi, L. Augmented Reality in Surgery: A Scoping Review. *Appl. Sci.* **2022**, *12*, 6890. [CrossRef]
20. Chang, C.-C.; Kuo, C.-H.; Chang, H.-K.; Tu, T.-H.; Fay, L.-Y.; Wu, J.-C.; Cheng, H.; Huang, W.-C. Augmented Reality-Assisted Percutaneous Pedicle Screw Instrumentation: A Cadaveric Feasibility and Accuracy Study. *Appl. Sci.* **2022**, *12*, 5261. [CrossRef]
21. Anastassova, M.; Burkhardt, J.-M. Automotive Technicians' Training as a Community-of-Practice: Implications for the Design of an Augmented Reality Teaching Aid. *Appl. Ergon.* **2009**, *40*, 713–721. [CrossRef]
22. Langley, A.; Lawson, G.; Hermawati, S.; D'Cruz, M.; Apold, J.; Arlt, F.; Mura, K. Establishing the Usability of a Virtual Training System for Assembly Operations within the Automotive Industry. *Hum. Factors Ergon. Manuf. Serv. Ind.* **2016**, *26*, 667–679. [CrossRef]
23. Ikeuchi, K.; Suehiro, T. Toward an Assembly Plan from Observation. I. Task Recognition with Polyhedral Objects. *IEEE Trans. Robot. Autom.* **1994**, *10*, 368–385. [CrossRef]
24. Radkowski, R.; Herrema, J.; Oliver, J. Augmented Reality-Based Manual Assembly Support With Visual Features for Different Degrees of Difficulty. *Int. J. Hum.-Comput. Interact.* **2015**, *31*, 337–349. [CrossRef]
25. Boothroyd, G.; Alting, L. Design for Assembly and Disassembly. *CIRP Ann.* **1992**, *41*, 625–636. [CrossRef]
26. Hahn, J.; Ludwig, B.; Wolff, C. Augmented Reality-Based Training of the PCB Assembly Process. In Proceedings of the 14th International Conference on Mobile and Ubiquitous Multimedia, MUM '15, Linz, Austria, 30 November–2 December 2015; Association for Computing Machinery: New York, NY, USA, 2015; pp. 395–399. [CrossRef]
27. Masood, T.; Egger, J. Adopting Augmented Reality in the Age of Industrial Digitalisation. *Comput. Ind.* **2020**, *115*, 103112. [CrossRef]

28. Werrlich, S.; Nitsche, K.; Notni, G. Demand Analysis for an Augmented Reality Based Assembly Training. In Proceedings of the 10th International Conference on PErvasive Technologies Related to Assistive Environments, Island of Rhodes, Greece, 21–23 June 2017; ACM: Island of Rhodes, Greece, 2017; pp. 416–422. [CrossRef]
29. Tang, A.; Owen, C.; Biocca, F.; Mou, W. Comparative Effectiveness of Augmented Reality in Object Assembly. In Proceedings of the SIGCHI Conference on Human Factors in Computing Systems, CHI '03, Fort Lauderdale, FL, USA, 5–10 April 2003; Association for Computing Machinery: New York, NY, USA, 2003; pp. 73–80. [CrossRef]
30. Fiorentino, M.; Uva, A.E.; Gattullo, M.; Debernardis, S.; Monno, G. Augmented Reality on Large Screen for Interactive Maintenance Instructions. *Comput. Ind.* **2014**, *65*, 270–278. [CrossRef]
31. Masood, T.; Egger, J. Augmented Reality in Support of Industry 4.0—Implementation Challenges and Success Factors. *Robot. Comput.-Integr. Manuf.* **2019**, *58*, 181–195. [CrossRef]
32. Iben, H.; Baumann, H.; Ruthenbeck, C.; Klug, T. Visual Based Picking Supported by Context Awareness: Comparing Picking Performance Using Paper-Based Lists versus Lists Presented on a Head Mounted Display with Contextual Support. In Proceedings of the 2009 International Conference on Multimodal Interfaces, ICMI-MLMI '09, Cambridge, MA, USA, 2–6 November 2009; Association for Computing Machinery: New York, NY, USA, 2009; pp. 281–288. [CrossRef]
33. Theis, S.; Mertens, A.; Wille, M.; Rasche, P.; Alexander, T.; Schlick, C.M. Effects of Data Glasses on Human Workload and Performance during Assembly and Disassembly Tasks. In Proceedings of the 19th Triennial Congress of the IEA, Melbourne, Australia, 9–14 August 2015; Volume 9, pp. 14–21.
34. Wille, M.; Grauel, B.; Adolph, L. Strain Caused by Head Mounted Displays. In Proceedings of the Human Factors and Ergonomics Society Europe, Torino, Italy, 13–15 October 2013; pp. 267–277.
35. Pfendler, C.; Thun, J. *Geografische Orientierung Mit Egozentrischen Und Geozentrischen Karten Auf Einem Head-Mounted Display und Einem Personal Digital Assistant*; Fraunhofer FKIE Bericht: Wachtberg, Germany, 2010.
36. Agnew, J.; Bolla-Wilson, K.; Kawas, C.H.; Bleecker, M.L. Purdue Pegboard Age and Sex Norms for People 40 Years Old and Older. *Dev. Neuropsychol.* **1988**, *4*, 29–35. [CrossRef]
37. Berger, M.A.M.; Krul, A.J.; Daanen, H.A.M. Task Specificity of Finger Dexterity Tests. *Appl. Ergon.* **2009**, *40*, 145–147. [CrossRef] [PubMed]
38. Lawson, I. Purdue Pegboard Test. *Occup. Med.* **2019**, *69*, 376–377. [CrossRef]
39. Rule, K.; Ferro, J.; Hoffman, A.; Williams, J.; Golshiri, S.; Padre, R.; Avila, J.; Coca, C.; Valdes, K. Purdue Manual Dexterity Testing: A Cohort Study of Community-Dwelling Elderly. *J. Hand Ther.* **2021**, *34*, 116–120. [CrossRef] [PubMed]
40. Desrosiers, J.; Rochette, A.; Hébert, R.; Bravo, G. The Minnesota Manual Dexterity Test: Reliability, Validity and Reference Values Studies with Healthy Elderly People. *Can. J. Occup. Ther.* **1997**, *64*, 270–276. [CrossRef]
41. Wang, Y.-C.; Wickstrom, R.; Yen, S.-C.; Kapellusch, J.; Grogan, K.A. Assessing Manual Dexterity: Comparing the WorkAbility Rate of Manipulation Test with the Minnesota Manual Dexterity Test. *J. Hand Ther.* **2018**, *31*, 339–347. [CrossRef]
42. Brooke, J. SUS-A Quick and Dirty Usability Scale. *Usability Eval. Ind.* **1996**, *189*, 4–7.
43. Bangor, A.; Kortum, P.T.; Miller, J.T. An Empirical Evaluation of the System Usability Scale. *Int. J. Hum.-Comput. Interact.* **2008**, *24*, 574–594. [CrossRef]
44. *ISO 9241-11:2018*; Ergonomics of Human-System Interaction—Part 11: Usability: Definitions and Concepts. International Standard Organization (ISO): Geneva, Switzerland, 2018. Available online: https://www.iso.org/obp/ui/#iso:std:iso:9241:-11:ed-2:v1:en (accessed on 18 January 2023).

applied sciences

MDPI

Article

Movement Time for Pointing Tasks in Real and Augmented Reality Environments

Caijun Zhao [1,2], Kai Way Li [2] and Lu Peng [3,*]

[1] School of Safety and Management Engineering, Hunan Institute of Technology, Hengyang 421002, China
[2] Department of Industrial Management, Chung Hua University, Hsin-Chu 30012, Taiwan
[3] College of Information Management, Nanjing Agricultural University, Nanjing 210095, China
* Correspondence: lupeng942@njau.edu.cn; Tel.: +86-25-84395366

Abstract: Human–virtual target interactions are becoming more and more common due to the emergence and application of augmented reality (AR) devices. They are different from interacting with real objects. Quantification of movement time (MT) for human–virtual target interactions is essential for AR-based interface/environment design. This study aims to investigate the motion time when people interact with virtual targets and to compare the differences in motion time between real and AR environments. An experiment was conducted to measure the MT of pointing tasks on the basis of both a physical and a virtual calculator panel. A total of 30 healthy adults, 15 male and 15 female, joined. Each participant performed pointing tasks on both physical and virtual panels with an inclined angle of the panel, hand movement direction, target key, and handedness conditions. The participants wore an AR head piece (Microsoft Hololens 2) when they pointed on the virtual panel. When pointing on the physical panel, the participants pointed on a panel drawn on board. The results showed that the type of panel, inclined angle, gender, and handedness had significant ($p < 0.0001$) effects on the MT. A new finding of this study was that the MT of the pointing task on the virtual panel was significantly ($p < 0.0001$) higher than that of the physical one. Users using a Hololens 2 AR device had inferior performance in pointing tasks than on a physical panel. A revised Fitts's model was proposed to incorporate both the physical–virtual component and inclined angle of the panel in estimating the MT. This model is novel. The index of difficulty and throughput of the pointing tasks between using the physical and virtual panels were compared and discussed. The information in this paper is beneficial to AR designers in promoting the usability of their designs so as to improve the user experience of their products.

Keywords: augmented reality; human–virtual object interaction; index of difficulty; movement time; pointing tasks; revised Fitts's model

1. Introduction

One of the most commonly adopted laws for discussing the behaviors of a human interacting with an object is Fitts's law. This law estimates the time required to touch a target of a certain size over a certain distance. It suggests that the movement time (MT) of a body part from one location to a target depends on the distance of movement and the size of the target. A term, index of difficulty (ID) of the movement, was defined, incorporating the movement distance of the body segment and target size (W), and MT can be estimated using the following equations [1,2].

$$\mathrm{MT} = a + b\,\mathrm{ID} \tag{1}$$

$$\mathrm{ID} = \log_2\left[\frac{2 \times distance}{W}\right] \tag{2}$$

where a and b are constants to be determined.

ID itself is unitless, and a unit of bit is given that may be regarded as the amount of information transferred via the movement. This implies that the difficulty of a target pointing or tapping task may be quantified using the information metric bit.

Fitts's law was originally applied to study the MT of hand and finger movements. These movements may include tapping, pointing, tracking, positioning, and other hand–object contact tasks. They are required in human–machine and human–object interactions. In addition to hand and finger movements, scientists have applied this law to study the MT of other body segments such as the head [3–6], leg [7,8], foot [9–12], and trunk [13,14]. Fitts's law has been applied not only in ordinary environments but also in special environments such as underwater [15–17]. In addition to assessing the MT of body parts for healthy people, Fitts's law has been adopted to assess the body functions of patients for medical and healthcare purposes [18,19].

Since Fitts's law was announced, many variants have been proposed to accurately quantify ID and thus MT. Both Welford [20] and MacKenzie [21] proposed revised equations to predict the MT of moving the right index finger from the starting point to a target. These authors suggested using the terms log2[(A/W) + 0.5] and log2[(A/W) + 1], respectively, to replace the original ID definition. They mentioned that such revision was necessary considering the deviation of the original definition of ID from Theorem 17 of Shannon's information theory [22]. Gan and Hoffman [23] found that MT is linearly related to the square root of the movement amplitude for ballistic movements when the ID is small. This implies that considering the target size is no longer necessary in estimating the MT under such circumstances. Hoffmann [24] investigated the use of an effective target tolerance in an inverted Fitts task. Considering the finite width of the finger probe, he believed that the original Fitts's law could overestimate the ID of the movement. He introduced a finger index pad size into the original ID equation.

Murata and Iwase [25] found that the MT of the 3D pointing task was not satisfactorily explained by the original Fitts's law because the duration of the 3D movements was affected by the direction of the target. They introduced a directional parameter (θ) of the pointing movements into the original Fitts's law. Similar results were reported by Cha and Myung [26]. They added a second directional parameter (θ_2), in addition to the θ in Murata and Iwase [25], to delineate the 3D movement of finger touch. Baird et al. [27] conducted a study to test the effects of probe length on the MT of pointing tasks. They suggested introducing a function of probe length plus a logarithm of the amplitude into the original Fitts's law. They claimed that such additions allow for better prediction of the MT for a worker using a long tool such as a screwdriver or a crowbar.

Virtual reality (VR) devices have been widely used for both entertainment and commercial purposes. VR strives to immerse users completely within a designed scenario so as to provide an intuitive link between a computer and the user [28]. VR users may see the virtual image displayed and interact with virtual objects using one or two handheld remote controllers, which may encompass push buttons, touch pads, joysticks, and so on. They may also manipulate one or two virtual hands and/or raycasting to "contact" virtual targets. Augmented reality (AR), on the other hand, deals with the combination of virtual objects and real environments [29–34]. AR users may see the real world with some virtual, unreal objects in their vision. They may also interact with virtual objects using similar ways as in a VR system. Alternatively, they may "contact" virtual objects using hands directly, such as touching a virtual button using their finger to activate an input command [35,36].

Human interaction with virtual objects is, apparently, quite different from a human interacting with real objects [37]. When studying the MT for VR and AR users, the original Fitts's law and its variants may no longer be valid because human–object interactions are replaced by the more complicated human–virtual object interactions. One of the major issues in interacting with a virtual object is the depth perception deficiencies of stereoscopic displays [38–40]. The literature [41] has shown that the accuracy rate for localizing targets in virtual environments using stereoscopic displays is relatively lower than that in the real world. It was also found that participants tend to overestimate the depth of targets in virtual environments [42,43]. If users cannot locate an object accurately, they need more time to perform a task requiring contact of their hands with objects. This will lead to inferior performance [40,44,45].

Investigations of the MT in virtual environments and human–virtual object interaction tasks have been reported. Ha and Woo [46] performed an MT study of 3D object manipulation techniques based on a virtual hand metaphor in a tangible augmented reality environment. Lubos et al. [38] reported that, when handling a virtual object in an immersive virtual environment, both ID and interaction of target position and movement distance were significant on the MT. They found high selection error along the view axis as compared to those in the other axes. Machuca and Stuerzlinger [45] found that their participants needed significantly more time to select virtual 3D targets than to select identical physical targets. Deng et al. [47] indicated that the MT in operating a handheld controller to perform a target positioning task in a virtual environment may be split into the acceleration, deceleration, and correction phases of hand movement. They included amplitude, object size, and target tolerance in their MT equations and suggested calculating the MT in different phases using different equations. Clark et al. [48] proposed an extended Fitts's law model to consider the effects of the inclined angle of the virtual targets.

Even though there were studies investigating the MT dealing with hand–virtual object interactions, there are still many untouched issues. For example, users have no haptic feedback when "contacting" a virtual object. Lacking haptic feedback could make fine adjustment of the hand/finger before touching the virtual target difficult and thus lead to longer MT than that of contacting a real target. This has not been discussed in the literature and is the main issue considered in the current study. We believe that this could be attributed to a tactile (T) factor. This factor is determined by the presence of tactile feedback. This factor should be incorporated in Fitts's law (Equation (2)) so as to provide good fitness of the MT when the pointing tasks are performed on virtual targets. Equations (3) and (4) show the inclusion of T in the function estimating the MT:

$$\mathrm{MT} = f(distance, \mathrm{W}, T) \tag{3}$$

$$\mathrm{MT} = a + b\ \log_2\left[\frac{2 \times distance}{W}\right] + f(T), \tag{4}$$

where $f(T)$ is a function of the tactile factor T.

The hypothesis of this study was that the MT of pointing tasks on virtual targets in an AR environment should be longer than that on a physical target in a real environment because of the T factor.

The traditional Fitts's law (Equation (1)) used to predict the MT of pointing tasks on virtual targets could underestimate the MT. The literature has recommended incorporating a triangular function of the inclined angle of pointing 3D targets in the original Fitts's equation [25]. We verified whether such an addition was also valid in both physical and virtual pointing tasks. In addition, it was hypothesized that both gender and handedness are significant factors affecting the MT of physical and virtual pointing tasks. The objective of this study was to test these hypotheses. Predictive equations on the MT for the pointing tasks on both virtual and physical targets were established and compared, considering the presence of the T factor, so as to enhance our understanding of the efficiency of giving input commands via a virtual keypad using hand gestures.

This study provides valuable information for AR designers to realize the inefficiency of using a virtual keypad in the AR device tested in this study. Such information is important and may generate new ideas concerning input designs in virtual environments for AR devices.

2. Methods

An experiment was performed in the laboratory. The illuminance on the workbench, measured using a light meter (Lux Meter, Trans Instrument, Singapore), for the physical touch panel was between 600 and 700 lx.

2.1. Participants

A total of 30 adults, 15 men and 15 women, were recruited as human participants. These participants were healthy and had normal corrected vision without color deficiency. Four of them, two male and two female, were lefthanded. The others were all righthanded. The average age and stature of the participants were 21.0 ($\pm$2.5) y and 165.9 ($\pm$7.8) cm, respectively. All the participants signed informed consent forms before joining the experiment.

2.2. AR Device and Pointing Tasks

A Microsoft® Hololens 2 headpiece (Microsoft Inc., Redmond, WA, USA) was used. This device has been reported as the most commonly adopted one in the medical community to display virtual objects [49]. It has a weight of 566 g. The Windows Holographic Operating System was installed. The glasses use see-through holographic lenses. The resolution of the display is 2 k with 3:2 light engines. The holographic density is 2.5 k radians (light points per radian) or higher with eye-based rendering. This device tracks hand location and gestures using built-in cameras and a two-handed fully articulated model with directional manipulation. The world-scale positional tracking and real-time environment mesh are employed in the system for six-degrees-of-freedom tracking and spatial mapping purposes, respectively. The app Graph Calculator (Microsoft Inc., Redmond, WA, USA) was adopted for the pointing tasks.

The pointing tasks were performed in both real (or physical) and AR (or virtual) environments. In the AR environment, the participant wore the Hololens headpiece. The virtual calculator panel appeared in the vision of the participant (see Figure 1). The size, location, ratio of the horizontal to the vertical dimension, and tilt angle of the keyboard were adjustable using hand gestures. The inclined angle of the panel to a flat surface was adjusted to either 90° or 30°. These angles were confirmed using a goniometer. Hand movements of the pointing task included horizontal and vertical movements. There were three sizes (large, medium, and small) of the keys on the panel (see Table 1). Each pointing task was performed by moving the tip of the index finger from the original key to touch a target key. There were three target keys in each of the horizontal and vertical movement trials.

Figure 1. Virtual calculator panel.

Table 1. Size of the keys on the keypad for both real and virtual conditions.

	Horizontal	Movement	Vertical	Movement
	Width	Height	Width	Height
Large	76	14	37	37
Medium	56	14	37	25
Small	37	14	37	14

Note: Size for all the keys except the "0" key; unit: mm.

For horizontal pointing tasks, the key at origin was "C" on the upper left corner of the keypad (see Figure 1). For each of the three key sizes, the third (or "ln"), fifth (or "("), and seventh (or ⌫) keys in the same row were selected as one of the target keys. The sizes (W) of the original and the three target sizes were the widths of the marks on these keys. An averaged W (W_{ave}) was the averaged width of the marks of the origin and each of the target key. For vertical pointing tasks, the original key was the "ln" on the first row on the keypad (see Figure 1). For each of the three key sizes, the second (or "sinh"), fourth (or "tanh"), and sixth (or "0 × 16") keys on the same column were selected as one of the target keys. The sizes (W) of the original and each of the three target sizes were the heights of the marks on these keys. The W_{ave} was also the average height of the marks of the origin and each of the target key and was 4 mm, as the height of the marks on all the original and target keys were this height. The distance between the origin and target key was the distance between the middle of the marks on the two keys. The distance between the origin and a target key, W_{ave}, and ID calculated using Equation (2) of the pointing tasks were summarized in Table 2. The sizes, locations, distances between the origin and end point of the pointing tasks, and distance between each key to the participant were measured using a measuring tape. The angles of the virtual panel were measured using a goniometer.

Table 2. Key size, distance, and index of difficulty (ID) of the pointing tasks.

	Origin-Target Key	Wave (mm)	Distance (mm)	ID
Horizontal				
Large	C-ln	3.5	156	6.5
	C-(	2.5	312	8
	C-⌫	4.5	468	7.7
Medium	C-ln	3.5	116	5.1
	C-(	2.5	232	7.5
	C-⌫	4.5	348	7.3
Small	C-ln	3.5	78	5.5
	C-(	2.5	156	7
	C-⌫	4.5	234	6.7
Vertical				
Large	ln-sinh	4	39	4.3
	ln-tanh	4	117	5.9
	Ln-0 × 16	4	195	6.6
Medium	ln-sinh	4	27	3.8
	ln-tanh	4	81	5.3
	ln-0 × 16	4	135	6.1
Small	ln-sinh	4	16	3
	ln-tanh	4	48	4.6
	ln-0 × 16	4	80	5.3

Note: "ln", "(", and "⌫" were the 3rd, 5th, and 7th keys for horizontal pointing; "sinh", "tanh", and "0 × 16" were the 2nd, 4th, and 6th keys for vertical pointing.

For the pointing tasks in the real environment, the calculator panels mimicked those of the virtual ones and were drawn on plastic boards. These boards were attached on a wood panel that was positioned on a workbench (see Figure 2). The inclined angle of the wood panel was also 30° or 90°. The sizes and distances of the keys and the angles of the panels between the virtual and physical settings were the same and were confirmed using a measuring tape.

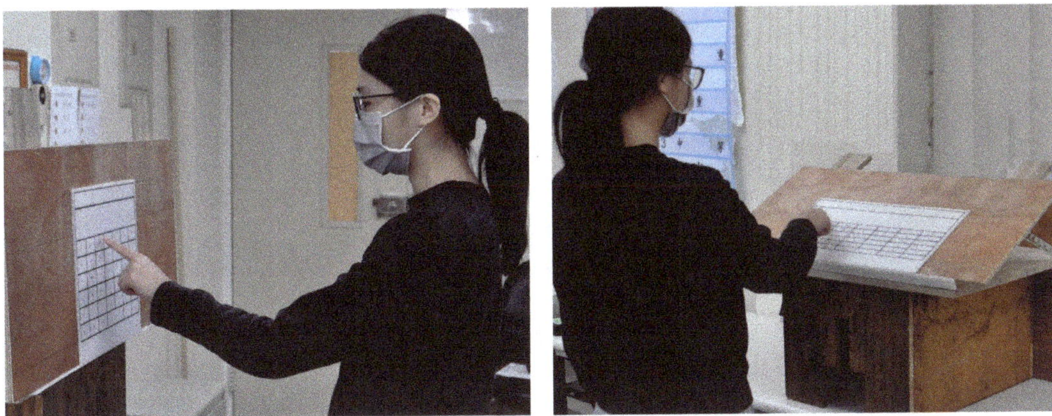

(a) (b)

Figure 2. Calculator panel drawn on a plastic board for the physical pointing tasks: (**a**) 90°, (**b**) 30°.

2.3. Procedure

In the pointing task, the participant was requested to stand approximately 25–35 cm away from the panel. The bottom of the panel was approximately 120 cm above the floor. The heights of the keys were somewhere between the abdomen and chest of the participants.

Before the trial, the participant placed his/her index fingertip on the original key. Upon hearing the verbal start hint, the participant moved his/her fingertip to touch the target key and then returned to touch the original key. The participant was instructed to touch the letter, symbol, or mark on the key so as to ensure the input of that key was accurate. In addition, the participants were requested to complete the task as quickly and accurately as possible. Each pointing task was repeated 10 times and there were 20 fingertip movements between the origin and the target keys. The moment the participant touched the origin key upon completing the 20 movements, he or she gave a verbal hint of the end. The images of these 20 movements were recorded in a laptop computer and the average time of these movements retrieved from the videos in the computer was recorded as the movement time (MT). Both the dominant and nondominant hands of the participant were tested.

Each participant had an opportunity to practice pointing on the virtual key before the first session (see Figure 3). Pointing accurately on a virtual key was found to be relatively difficult as compared to that on a physical key. When the fingertip was approaching a key, the participant could see a virtual ring indicating the location of his or her fingertip and his or her hand became invisible. The ring became smaller when the fingertip drew close to the key (see Figure 4). It became a solid circle when the pointing was done. The participant could hear auditory feedback of the "touch" similar to that on a traditional keyboard at this moment. The virtual image displayed in the AR device was also displayed on a computer monitor so that the research personnel could confirm the success of each key touch.

Because all the participants had no prior experience of using the Hololens device, learning effects would surely affect the MT, especially for the virtual tasks. Randomization of the order of the trials was therefore very important so that the learning effects could be evenly distributed into the treatments. Each participant completed 144 trials. These trials were split into two sessions randomly. Each session included 72 randomly arranged trials and lasted for approximately 2 h. There was a 2-min break for the participant upon completing one trial within a session. Each participant completed one session and returned for the second session at least one day later.

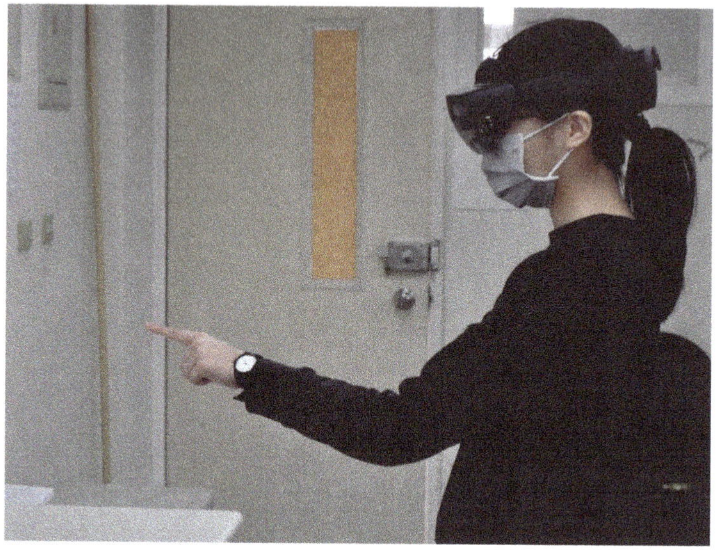

Figure 3. Pointing task on virtual keys wearing the Hololens device.

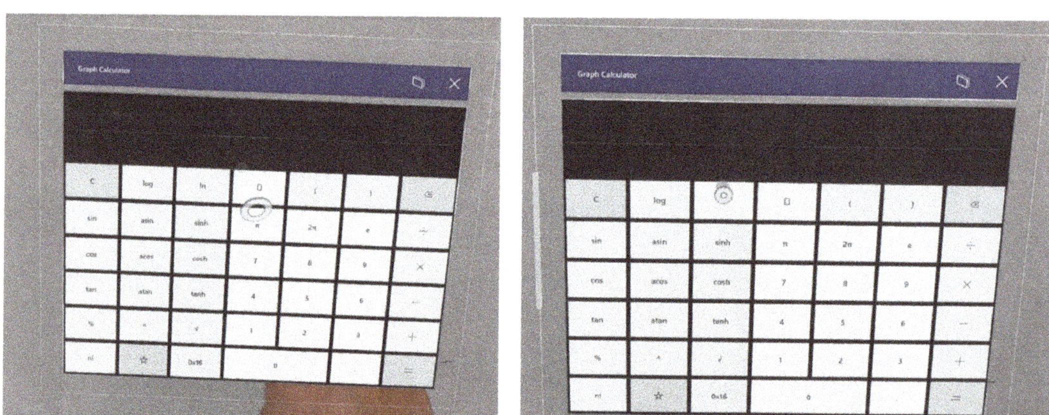

(a) **(b)**

Figure 4. A virtual ring indicating the location of the fingertip in front of a virtual panel (**a**) approaching the key "ln"; (**b**) almost touching.

2.4. Design of Experiment and Data Analyses

A randomized completely block design experiment was performed. Each participant was considered a block. Randomization of the factors was arranged for each participant. These factors included the type of panel (physical and virtual), inclined angle (30° and 90°), hand used (dominant vs. nondominant hand), hand movement direction (horizontal or vertical), key size (large, medium, or small), and target key (3rd, 5th, and 7th for horizontal movement and 2nd, 4th, and 6th for vertical movement). A total of 4320 trials were performed. Descriptive statistics and analysis of variance (ANOVA) were performed to determine the effects of the type of panel, incline angle, gender, and handedness on the MT. The least significant difference (LSD) test was performed for posterior comparisons for the levels for each factor if the factor was significant on MT at the $\alpha = 0.05$ level.

The experiment was conducted for hand movement direction, key size, and target key conditions, but we did not examine the effects of these factors on MT because the distance between the original and target key was confounded within these factors. Discussing whether these factors affected MT was not meaningful because the distance is known to be a significant factor affecting MT.

Regression analyses were performed to fit the Fitts's model of the MT for both hands. In addition to ID, an inclined angle was introduced to incorporate the effects of this parameter. The $f(T)$ in Equation (4) was replaced by a parameter T multiplied by a constant d. We then have Equation (5):

$$MT = a + b\,ID + c\sin\theta + d\,T, \quad (5)$$

where a, b, c, and d are constants to be determined, ID is the index of difficulty in Table 2, and θ is the inclined angle of the panel. T is a dummy variable. This was 1 or 0 depending on whether the participant was pointing on a virtual panel or not:

$$T = 1 \text{ the panel is real,}$$
$$= 0 \text{ otherwise.}$$

The MT equations between the virtual and physical tasks under hand movement direction, gender, and handedness conditions were compared and discussed. Statistical analyses were performed using SPSS 20.0 software (IBM®, Armonk, NY, USA).

3. Results

3.1. ANOVA Results on MT

The ANOVA results showed that the MT of virtual pointing tasks was significantly ($p < 0.0001$) higher than that of physical pointing tasks. The effects of inclined angle were also significant ($p < 0.0001$) on the MT. Both the gender and handedness significantly ($p < 0.0001$) affected the MT of the physical and virtual pointing tasks.

The LSD test results indicated that the MT of the virtual pointing task (1533.5 ms) was significantly ($p < 0.05$) higher than that (628.6 ms) of the physical one. The MT of the 90° inclined angle condition (1118.1 ms) was significantly ($p < 0.05$) higher than that (1049.0 ms) of the 30° condition. For physical pointing tasks, the MT of female participants (639.1 ms) was significantly ($p < 0.05$) higher than that (618.1 ms) of the male participants. The MT of the nondominant hand (642.8 ms) was significantly ($p < 0.05$) higher than that (614.4 ms) of the dominant hand. For virtual pointing tasks, the MT of female participants (1630.2 ms) was significantly ($p < 0.05$) higher than that (1446.7 ms) of the male participants. The MT of the nondominant hand (1581.1 ms) was significantly ($p < 0.05$) higher than that (1495.9 ms) of the dominant hand.

3.2. MT Versus ID

The ID was calculated using Equation (2), while W was substituted using W_{ave} or the average size of the marks on both the origin and target keys. The distance between the origin and the target keys varied depending on both the direction of the hand movement and the specific target key selected. Figures 5 and 6 show the means and standard deviations of the MT versus the ID for the horizontal and vertical hand movement pointing tasks, respectively.

3.3. Regression Models

Regression analyses were performed for each of the gender and handedness conditions for both horizontal and vertical movement tasks using Equation (5). The regression modeling was based on all the MT data using the W_{ave}, distance, and ID in Table 2. The results are shown in Table 3. The R^2 values of the models developed were between 0.73 and 0.79. All the VIF values of b, c, and d were 1, indicating that there was no correlation between any two of the independent variables.

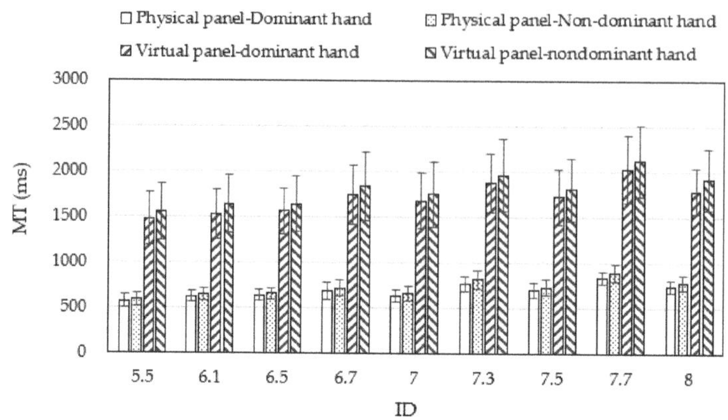

(a)

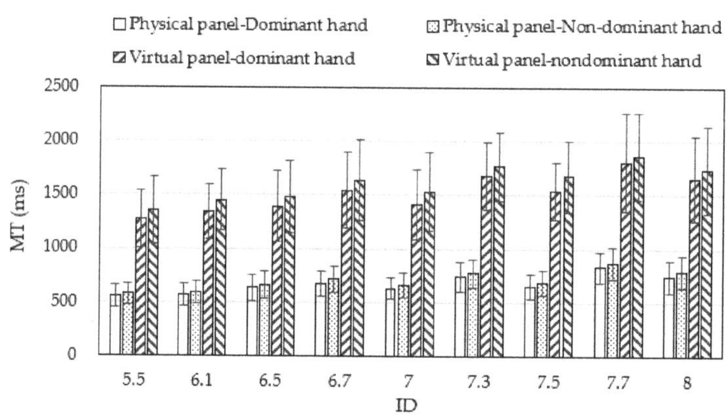

(b)

Figure 5. MT of female (**a**) and male (**b**) participants performing the horizontal pointing task.

Table 3. Results of regression modeling of the MT.

			Regression	Coefficient		
Gender	**Hand**	a	b	c	d	R^2
Horizontal	Movement					
Female	D	633.8	134.4	118.7 *	−970.1	0.76
	ND	660.6	144.1	109.8 *	−1025.9	0.73
Male	D	365.7	141.6	227.3	−846.8	0.75
	ND	454.8	143.4	217.7	−910.1	0.77
Vertical	Movement					
Female	D	1028.6	66.8	68.0 †	−855.6	0.74
	ND	1051.4	74.5	85.7 †	−908.5	0.73
Male	D	878	62.8	133.5	−755.2	0.79
	ND	924.3	67.6	143.8	−803.4	0.78

Note: D: dominant; ND: nondominant. All the regression coefficients, except those marked with * and †, were significant at $p < 0.001$ for a two-tailed t test of $\mu = 0$. * $p < 0.05$; † $p > 0.05$.

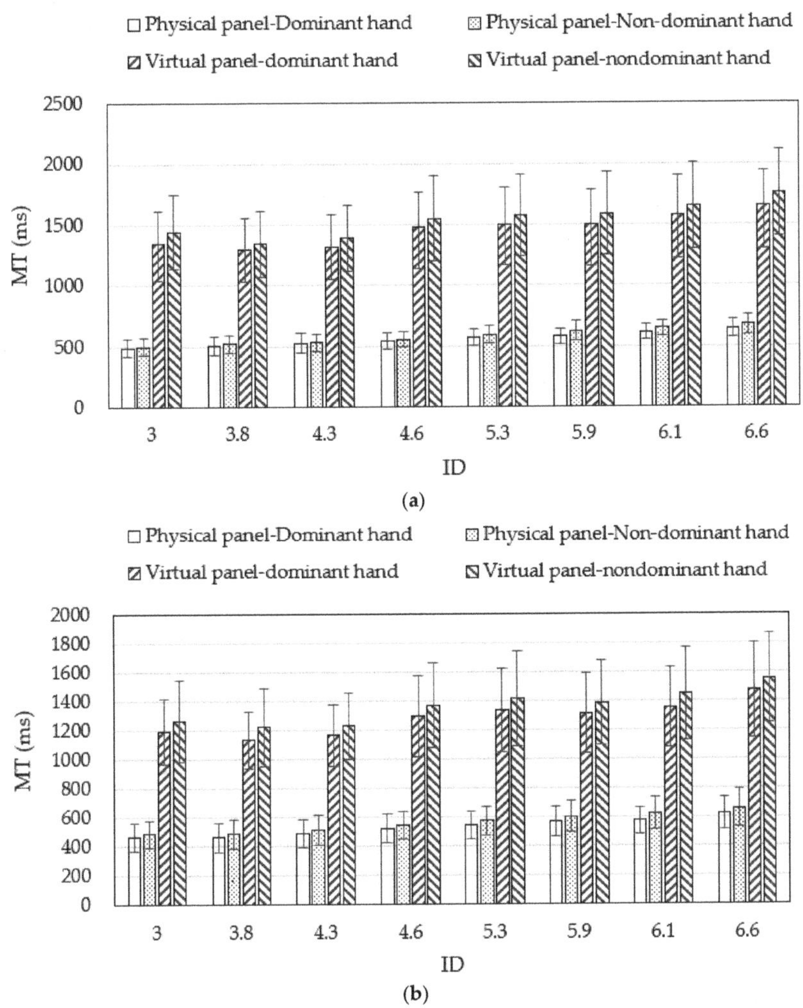

Figure 6. MT of female (**a**) and male (**b**) participants performing the vertical pointing task.

An ID_{adj} was defined using the following equation to incorporate the components of both the inclined angle and T:

$$ID_{adj} = ID + c' \sin\theta + d' T, \quad (6)$$

where c' and d' are constants to be determined.

Both c' and d' may be calculated by dividing c and d, respectively, by the b in Table 3. Table 4 shows the c', d', and $1/b$. Equation (5) may then be rewritten as follows:

$$MT = a + b\, ID_{adj}, \quad (7)$$

where a and b are constants to be determined.

Table 4. Coefficients of ID_{adj} and $1/b$.

Gender	Hand	c'	d'	$1/b$ (bit/s)
Horizontal	Movement			
Female	D	0.88	−7.22	7.4
Female	ND	0.76	−7.12	6.9
Male	D	1.61	−5.98	7.1
Male	ND	1.52	−6.35	7
Vertical	Movement			
Female	D	1.02	−12.81	15
Female	ND	1.15	−12.19	13.4
Male	D	2.13	−12.03	15.9
Male	ND	2.13	−11.88	14.8

Equation (6) shows that ID may be adjusted because of the incline angle and T. By substituting the c' and d' in Table 4 into Equation (6), we obtained the corresponding ID_{adj}. An increase in ID ($ID_{increase}$) using the following equation may be defined:

$$ID_{increase}\ (\%) = \left(\frac{ID_{adj} - ID}{ID}\right) \times 100\%. \tag{8}$$

ANOVA results indicated that both T and movement direction significantly affected the $ID_{increase}$ ($p < 0.0001$). The $ID_{increase}$ of the virtual condition (197.5%) was significantly higher than that (25.3%) of the physical condition. The $ID_{increase}$ of the vertical movement (159.4%) was significantly higher than that (63.4%) of the horizontal movement. These facts imply that T is the dominant factor affecting the increase of ID, and the incline angle was also a factor to be considered. The results indicate that incorporating the T factor increased the difficulty of the pointing tasks more for the vertical movements than for the horizontal movements.

4. Discussion

4.1. Movement Time of Pointing Tasks on Physical and Virtual Targets

Pointing tasks are required when users are using their fingertips to touch a target. Pointing on a physical key and on a virtual one were quite different. For the physical pointing, the control of hand movement was intuitive. When the fingertip touched the target, it was stopped by a board. There was strong tactile feedback. The literature [50] has shown that mechanical interactions of the hand and an object could interfere with the trajectory of the hand and thus affect the MT. For the virtual pointing, the participant needed to move his/her fingertip to the exact location of the target, both in the 2D plane of the virtual panel and in the axis perpendicular to the panel. There was no tactile feedback. Both a visual (the ring became a solid circle) and an auditory signal were presented when a "touch" was successful. However, the feedback, especially the visual one, was not as prominent as that of the tactile one. The fingertip might move over and "penetrate" the virtual panel, as there was no solid to stop it. It was also possible that there may be errors between the visual location of the virtual panel and actual location as recognized by the app in the AR glass [51]. The participant, therefore, needed to make fine adjustments following the cue of visual feedback of the fingertip location in the axis perpendicular to the panel until he or she received visual feedback of a successful input. Such a fine adjustment significantly increased the time for the pointing tasks on the virtual panel.

The regression coefficient d in Table 3 indicates the average extra amount of MT (755–1026 ms) required because the panel was a virtual one. The ratios of MT using the virtual panel versus the one using the physical panel may be calculated using our MT data. It was found that the participants needed to spend 2.23–2.79-fold more time on the MT, on average, to complete the pointing task on the virtual panel than on the physical panel under our experimental conditions. These numbers indicated the effects of the factor T and the inefficiency of using a virtual panel for data input. These results were consistent with the findings in the literature that pointing virtual objects required more time than pointing physical ones [45].

The results of this study support the hypothesis that the MT of pointing tasks on the virtual targets in an AR environment is longer than that on physical targets. This was consistent with the findings in the literature [48]. In addition, the hypotheses that the MT of the pointing tasks was affected by the gender and handedness of the participants were also supported.

4.2. Information Processing Performance

The inverse of the regression coefficient b in Fitts's law equation was adopted to indicate the information processing rate of the hand movement performing the pointing task [52]. The literature has shown that the information processing rate for arm movements was approximately 10 bit/s [25,53]. The inverse of b (or $1/b$) in Table 4 for horizontal and vertical pointing tasks was between 6.9 bit/s and 7.4 bit/s and between 13.4 bit/s and 15.9 bit/s, respectively. The former was approximately half of the latter, indicating that horizontal pointing was less efficient than vertical pointing. It was anticipated that the information processing rate for the virtual pointing tasks should be different from that of the physical ones. However, the $1/b$ values of these two types of tasks in Table 4 were confounded because T was incorporated in $\mathrm{ID}_{\mathrm{adj}}$. Therefore, the inverse of b could not accurately represent the information processing rate when the ID was adjusted with factors in addition to movement distance and target size. Mackenzie [54] also indicated that $1/b$ cannot be used to quantify the information transfer because of the wavering influence of the intercept a and the inconsistency between $1/b$ and Fitts's original definition of index of performance, or throughput. We then calculated the throughput (TP) using the following equation:

$$\mathrm{TP} = \frac{ID_{adj}}{MT}. \quad (9)$$

Table 5 shows the TP for the pointing tasks. The TP values for the physical pointing tasks were between 11.6 and 12.6 bit/s for both the horizontal and vertical movements and were relatively stable. They were between 8.8 and 9.8 bit/s for horizontal or virtual–horizontal pointing tasks and were significantly ($p < 0.001$) lower than those of the physical–horizontal tasks. On the contrary, the TP values of the virtual–horizontal pointing tasks were between 12.2 and 15.0 bit/s and were significantly ($p < 0.001$) higher than those of the physical tasks in the same direction. This implies that the information processing performance between pointing at physical and virtual panels was dependent on the hand movement direction. The participants had inferior information processing performance when pointing at virtual targets than that on physical targets if they were pointing laterally. On the contrary, they had better performance when pointing at virtual targets than on physical targets if they were pointing in the top-down direction. This might be attributed to the fact that the top-down pointing movements involved less movement of the shoulder joint than the lateral pointing movements. It was also likely that the visual depth deficiency problems in the top-down movements were less prominent than in the lateral direction.

Table 5. Throughput (bit/s) for the pointing tasks.

Movement Direction	Gender	Hand	Physical Pointing	Virtual Pointing
Horizontal	Female	D	11.75	8.94
		ND	11.15	8.84
	Male	D	12.54	9.76
		ND	11.9	9.4
Vertical	Female	D	11.6	13.17
		ND	11.29	12.22
	Male	D	12.55	14.98
		ND	11.9	14.09

D: dominant hand; ND: nondominant hand.

4.3. Limitations of the Study

There were limitations to this study. The first was that the participants were requested to touch the letter or the symbol on both the original and target keys to ensure successful input. This resulted in a change in the target size. In fact, even if the participants did not touch the mark on the key, they could still succeed in the pointing task so long as they touched the key without touching the border of the key. Our W_{ave} was, therefore, not the target size in the typical Fitts's law. Instead, it represented the effective target size due to our mark-touching request. This should be considered when interpreting the results of this study. The second limitation was that the virtual panel in this study was generated by the Microsoft Hololens 2 device. The results of this study may, therefore, not be generalizable to the virtual targets generated by other AR/VR devices.

5. Conclusions

This study confirmed that the MT of pointing tasks on virtual targets was significantly ($p < 0.0001$) longer than that on physical ones in real environments. Both gender and handedness significantly affected the MT of the pointing tasks. We have also verified that adding a parameter of the inclined angle of the control panel, for both physical and virtual targets, could increase the fitness of the regression equation in predicting the MT of pointing tasks. A revised equation of Fitts's law, using an adjusted index of difficulty (ID_{adj}), was proposed. The ID_{adj} incorporated both the inclined angle of the panel and the factor T of the tasks. The predictive MT equation proposed in this study may be adopted to estimate the MT of pointing tasks under similar AR environments. AR engineers and designers may consider the pointing behaviors of our participants in touching the keys on the virtual panel in the AR environment so as to promote their designs for better user experience.

Our society is advancing into the world of the metaverse. However, there are many unknowns concerning human interaction with virtual objects. A 2D calculator panel, both physical and virtual, was adopted in the pointing tasks performed in this study. However, there are various 3D objects in the metaverse environments. Future research may be designed to study hand gestures and movement time when in contact with 3D virtual objects for pointing and other tasks. Such research will be helpful to fill the gaps in our knowledge concerning the world of metaverse.

Author Contributions: Conceptualization, C.Z. and K.W.L.; methodology, C.Z. and K.W.L.; validation L.P.; investigation, C.Z.; data curation, C.Z. and K.W.L.; writing—original draft preparation, K.W.L.; writing—review and editing, L.P.; visualization, L.P.; supervision, K.W.L.; funding acquisition, K.W.L. and L.P. All authors have read and agreed to the published version of the manuscript.

Funding: This research was partially supported by research funding from the Ministry of Science and Technology of the ROC (MOST 110-2221-E-216-004) and funding from the 2nd Batch of 2022 MOE of PRC Industry-University Collaborative Education Program (Program No. 202209SZ08, Kingfar-CES "Human Factors and Ergonomics" program).

Institutional Review Board Statement: This study was conducted according to the guidelines of the Declaration of Helsinki and was approved by an external ethics committee (Jen-Ai Hospital, 110-55).

Informed Consent Statement: Informed consent was obtained from all subjects involved in the study.

Data Availability Statement: The data presented in this study are available on request from the corresponding author. The data are not publicly available due to privacy reasons.

Conflicts of Interest: The authors declare no conflict of interests.

Abbreviations

AR	augmented reality
ID	index of difficulty
ID_{adj}	adjusted index of difficulty
$ID_{increase}$	increase in index of difficulty due to pointing on virtual targets
MT	movement time
T	tactile factor
TP	throughput
VR	virtual reality
W_{ave}	average width of the targets

References

1. Fitts, P.M. The information capacity of the human motor system in controlling the amplitude of movement. *J. Exp. Psychol.* **1954**, *47*, 381–391. [CrossRef] [PubMed]
2. Fitts, P.M.; Peterson, J.R. Information capacity of discrete motor responses. *J. Exp. Psychol.* **1964**, *67*, 103–112. [CrossRef] [PubMed]
3. Jagacinski, R.J.; Monk, D.L. Fitts' law in two dimensions with hand and head movements. *J. Motor Behav.* **1985**, *17*, 77–95. [CrossRef] [PubMed]
4. Andres, R.O.; Hartung, K.J. Prediction of head movement time using Fitts' law. *Hum. Factors* **1989**, *31*, 703–713. [CrossRef]
5. Radwin, R.; Vanderheiden, G.C.; Lin, M.L. A method for evaluating head-controlled computer input devices using Fitts' law. *Hum. Factors* **1990**, *32*, 423–438. [CrossRef]
6. Hoffmann, E.R.; Chan, A.H.S.; Heung, P.T. Head rotation movement times. *Hum. Factors* **2017**, *59*, 986–994. [CrossRef]
7. Chan, A.H.S.; Hoffmann, E.R. Effect of movement direction and sitting/standing on leg movement time. *Int. J. Ind. Ergon.* **2015**, *47*, 30–36. [CrossRef]
8. Chan, A.H.S.; Hoffmann, E.R.; Ip, K.M.; Siu, S.C.H. Leg/foot movement times with lateral constraints. *Int. J. Ind. Ergon.* **2018**, *67*, 6–12. [CrossRef]
9. Drury, C.G. Application of Fitts' law to foot-pedal design. *Hum. Factors* **1975**, *17*, 368–373. [CrossRef]
10. Springer, J.; Siebes, C. Position controlled input device for handicapped: Experimental studies with a footmouse. *Int. J. Ind. Ergon.* **1996**, *17*, 135–152. [CrossRef]
11. Chan, A.H.S.; Ng, A.W.Y. Lateral foot-movement times in sitting and standing postures. *Percept. Mot. Ski.* **2008**, *106*, 215–224. [CrossRef] [PubMed]
12. Park, J.; Myung, R. Fitts' law for angular foot movement in the foot tapping task. *J. Ergon. Soc. Korea* **2012**, *31*, 647–655. [CrossRef]
13. Danion, F.; Duarte, M.; Grosjean, M. Fitts' law in human standing: The effect of scaling. *Neurosci. Lett.* **1999**, *277*, 131–133. [CrossRef] [PubMed]
14. Hoffmann, E.R.; Chan, A.H.S. Movement of loads with trunk rotation. *Ergonomics* **2014**, *58*, 1547–1556. [CrossRef] [PubMed]
15. Kerr, R. Diving, adaptation, and Fitts' law. *J. Motor. Behav.* **1978**, *10*, 255–260. [CrossRef]
16. Hancock, P.A.; Milner, E.K. Task performance under water-An evaluation of manual dexterity efficiency in the open ocean underwater environment. *Appl. Ergon.* **1986**, *17*, 143–147. [CrossRef]
17. Hoffmann, E.R.; Chan, A.H.S. Underwater movement times with ongoing visual control. *Ergonomics* **2012**, *55*, 1513–1523. [CrossRef]
18. Drews, F.A.; Zadra, J.R.; Gleed, J. Electronic health record on the go: Device form factor and Fitts' law. *Int. J. Med. Inform.* **2018**, *111*, 37–44. [CrossRef]
19. Melo, F.; Conde, M.; Godinho, C.; Domingos, J.; Sanchez, C. Hand motor slowness in Parkinson disease patients performing Fitts' task. *Annu. Med.* **2019**, *51*, 49. [CrossRef]
20. Welford, A.T. The measurement of sensory-motor performance: Survey and reappraisal of twelve years progress. *Ergonomics* **1960**, *3*, 189–230. [CrossRef]
21. Mackenzie, I.S. A note on the information-theoretic basis for Fitts' law. *J. Motor Behav.* **1989**, *21*, 323–330. [CrossRef] [PubMed]
22. Shannon, C.E. A mathematical theory of communication. *Bell Syst. Tech. J.* **1948**, *27*, 379–423. [CrossRef]
23. Gan, K.C.; Hoffmann, E.R. Geometrical conditions for ballistic and visually controlled movements. *Ergonomics* **1988**, *31*, 829–839. [CrossRef]
24. Hoffmann, E.R. Effective target tolerance in an inverted Fitts' task. *Ergonomics* **1995**, *38*, 828–836. [CrossRef]
25. Murata, A.; Iwase, H. Extending Fitts' law to a three-dimensional pointing task. *Hum. Mov. Sci.* **2001**, *20*, 791–805. [CrossRef] [PubMed]
26. Cha, Y.; Myung, R. Extended Fitts' law for 3D pointing tasks using 3D target arrangements. *Int. J. Ind. Ergon.* **2013**, *43*, 350–355. [CrossRef]
27. Baird, K.M.; Hoffmann, E.; Drury, C.G. The effects of probe length on Fitts' law. *Appl. Ergon.* **2002**, *33*, 9–14. [CrossRef]
28. Guttentag, D.A. Virtual reality: Applications and implications for tourism. *Appl. Ergon.* **2010**, *31*, 637–651. [CrossRef]
29. Carmigniani, J.; Furht, B.; Anisetti, M.; Ceravolo, P.; Damiani, E.; Ivkovic, M. Augmented reality technologies, systems, and applications. *Multimed. Tools. Appl.* **2011**, *51*, 341–377. [CrossRef]
30. Drascic, D.; Milgram, P. Perceptual issues in augmented reality. In Proceedings of the Sterescopic Displays and Virtual Reality Systems III, San Jose, CA, USA, 10 April 1996; Volume 2653, pp. 123–134.

31. Wu, H.K.; Lee, S.; Chang, H.Y.; Liang, J.C. Current status, opportunities and challenges of augmented reality in education. *Comput. Educ.* **2013**, *62*, 41–49. [CrossRef]
32. Ibáñez, M.B.; Kloos, C.D. Augmented reality for STEM learning: A systematic review. *Comput. Educ.* **2018**, *123*, 109–123. [CrossRef]
33. Terhoeven, J.; Schiefelbein, F.P.; Wischniewski, S. User expectations on smart glasses as work assistance in electronics manufacturing. *Procedia CIRP* **2018**, *72*, 1028–1032. [CrossRef]
34. Cardoso, L.F.S.; Mariano, F.C.M.Q.; Zorzal, E.R. A survey of industrial augmented reality. *Comput. Ind. Eng.* **2020**, *139*, 106159. [CrossRef]
35. Milgram, P.; Kishino, F. A Taxonomy of mixed reality visual displays. *IEICE T. Inf. Syst.* **1994**, *77*, 1321–1329.
36. Wang, X.; Ong, S.K.; Nee, A.Y.C. Multi-modal augmented-reality assembly guidance based on bare-hand interface. *Adv. Eng. Inform.* **2016**, *30*, 406–421. [CrossRef]
37. Gattullo, M.; Laviola, E.; Evangelista, A.; Fiorentino, M.; Uva, A.E. Towards the evaluation of augmented reality in the metaverse: Information presentation modes. *Appl. Sci.* **2022**, *12*, 12600. [CrossRef]
38. Lubos, P.; Bruder, G.; Steinicke, F. Analysis of direct selection in head-mounted display environments. In Proceedings of the IEEE Symp 3D User Interfaces, Minneapolis, MN, USA, 29–30 March 2014; pp. 11–18. [CrossRef]
39. Schwind, V.; Leusmann, J.; Henze, N. Understanding visual-haptic integration of avatar hands using a Fitts' law task in virtual reality. In Proceedings of the Mensch und Computer 2019, Hamburg, Germany, 8–11 September 2019; pp. 211–222. [CrossRef]
40. Triantafyllidis, E.; Li, Z. The challenges in modeling human performance in 3d space with Fitts' law. In Proceedings of the Extended Abstracts of the CHI Conference on Human Factors in Computing Systems, Yokohama, Japan, 8–13 May 2021; pp. 1–9. [CrossRef]
41. Lin, C.J.; Woldegiorgis, B.H. Interaction and visual performance in stereoscopic displays: A review. *J. Soc. Inf. Disp.* **2015**, *23*, 319–332. [CrossRef]
42. Swan, J.E.; Singh, G.; Ellis, S.R. Matching and reaching depth judgments with real and augmented reality targets. *IEEE T. Vis. Comput. Gr.* **2015**, *21*, 1289–1298. [CrossRef]
43. Lin, C.J.; Woldegiorgis, B.H. Egocentric distance perception and performance of direct pointing in stereoscopic displays. *Appl. Ergon.* **2017**, *64*, 66–74. [CrossRef]
44. Batmaz, A.U.; Machuca, M.D.B.; Pham, D.M.; Stuerzlinger, W. Do head-mounted display stereo deficiencies affect 3D pointing tasks in AR and VR? In Proceedings of the IEEE Conference Virtual Real 3D User Interfaces, Osaka, Japan, 23–27 March 2019; pp. 585–592. [CrossRef]
45. Machuca, M.D.B.; Stuerzlinger, W. The effect of stereo display deficiencies on virtual hand pointing. In Proceedings of the CHI Conference on Human Factors in Computing Systems, Glasgow, UK, 4–9 May 2019; pp. 1–14. [CrossRef]
46. Ha, T.; Woo, W. An empirical evaluation of virtual hand techniques for 3D object manipulation in a tangible augmented reality environment. In Proceedings of the 2010 IEEE Symposium on 3D User Interface, Waltham, MA, USA, 20–21 March 2010; pp. 91–98. [CrossRef]
47. Deng, C.L.; Geng, P.; Hu, Y.F.; Kuai, S.G. Beyond Fitts' law: A three-phase model predicts movement time to position an object in an immersive 3D virtual environment. *Hum. Factors* **2019**, *61*, 879–894. [CrossRef]
48. Clark, L.D.; Bhagat, A.B.; Riggs, S.L. Extending Fitts' law in three-dimensional virtual environments with current low-cost virtual reality technology. *Int. J. Hum-Comput. St.* **2020**, *139*, 102413. [CrossRef]
49. Barcali, E.; Iadanza, E.; Manetti, L.; Francia, P.; Nardi, C.; Bocchi, L. Augmented reality in surgery: A scoping review. *Appl. Sci.* **2022**, *12*, 6890. [CrossRef]
50. Crossman, E.R.F.W.; Goodeve, P.J. Feedback control of hand movement and Fitts' law. *Q. J. Exp. Psychol.* **1983**, *35A*, 251–278. [CrossRef]
51. El Barhoumi, N.; Hajji, R.; Bouali, Z.; Ben Brahim, Y.; Kharroubi, A. Assessment of 3D models placement methods in augmented reality. *Appl. Sci.* **2022**, *12*, 10620. [CrossRef]
52. Card, S.K.; English, W.K.; Burr, B.J. Evaluation of mouse, rate-controlled isometric joystick, step keys, and text keys for text selection on a CRT. *Ergonomics* **1978**, *21*, 601–613. [CrossRef]
53. Langolf, G.D.; Chaffin, D.B.; Foulke, J.A. An investigation of Fitts' law using a wide range of movement amplitudes. *J. Motor Behav.* **1976**, *8*, 113–128. [CrossRef]
54. Mackenzie, I.S. Fitts' throughput and the remarkable case of touch-based target selection. In *Human-Computer Interaction: Interaction Technologies, Lecture Notes Computer Science, Proceedings of the HCI 2015, Los Angeles, CA, USA, 2–7 August 2015*; Kurosu, M., Ed.; Springer: Berlin/Heidelberg, Germany, 2015; Volume 9170. [CrossRef]

MDPI AG
Grosspeteranlage 5
4052 Basel
Switzerland
Tel.: +41 61 683 77 34

Applied Sciences Editorial Office
E-mail: applsci@mdpi.com
www.mdpi.com/journal/applsci

www.ingramcontent.com/pod-product-compliance
Lightning Source LLC
Chambersburg PA
CBHW040742250926
51850CB00007BB/1385
9783725851539